Neurobiology of Social Communication in Primates

AN EVOLUTIONARY PERSPECTIVE

COMMUNICATION AND BEHAVIOR

AN INTERDISCIPLINARY SERIES

Under the Editorship of **Duane M. Rumbaugh,**
Georgia State University and Yerkes Regional
Primate Research Center of Emory University

DUANE M. RUMBAUGH (ED.), LANGUAGE LEARNING BY A
CHIMPANZEE: THE LANA PROJECT, 1977

ROBERTA L. HALL AND HENRY S. SHARP (EDS.),
WOLF AND MAN: EVOLUTION IN PARALLEL

HORST D. STEKLIS AND MICHAEL J. RALEIGH (EDS.), NEUROBIOLOGY OF SOCIAL
COMMUNICATION IN PRIMATES: AN EVOLUTIONARY PERSPECTIVE

Neurobiology of Social Communication in Primates

AN EVOLUTIONARY PERSPECTIVE

Edited by

HORST D. STEKLIS

Department of Anthropology
Livingston College
Rutgers University
New Brunswick, New Jersey

MICHAEL J. RALEIGH

Neurobiochemistry Laboratory
Brentwood Veterans Administration and Neuropsychiatric Institute
University of California, Los Angeles, School of Medicine
Los Angeles, California

ACADEMIC PRESS

A Subsidiary of Harcourt Brace Jovanovich, Publishers
New York   London   Toronto   Sydney   San Francisco

ACADEMIC PRESS, INC.
111 Fifth Avenue, New York, New York 10003

United Kingdom Edition published by
ACADEMIC PRESS, INC. (LONDON) LTD.
24/28 Oval Road, London NW1 7DX

Library of Congress Cataloging in Publication Data
Main entry under title:

Neurobiology of social communication in primates.

(Communication and behavior : an interdisciplinary
series)
Includes bibliographies and index.
1. Languages––Physiological aspects. 2. Language
and languages––Origin. 3. Neurobiology. 4. Animal
communication. 5. Primates––Behavior. 6. Primates––
Physiology. I. Steklis, Horst D. II. Raleigh,
Michael J. III. Series.
QP399.N48 599'.8'04188 79–21814
ISBN 0–12–665650–9

PRINTED IN THE UNITED STATES OF AMERICA

79 80 81 82 9 8 7 6 5 4 3 2 1

Contents

3

Mechanisms Underlying Vocal Control in Nonhuman Primates   45

DWIGHT SUTTON

4

Central Nervous System Processing of Sounds in Primates   69

JOHN D. NEWMAN

5

Cortical and Subcortical Organization of Human Communication: Evidence from Stimulation Studies   111

GEORGE A. OJEMANN AND CATHERINE MATEER

10

List of Contributors

Preface

During the past two decades there has emerged a renewed interest in the origin and evolution of human social communication. The dramatic growth of modern primatology has added a new dimension to discussions of language evolution by generating comparative behavioral and neurological data from monkeys and apes. This volume presents the most recent evidence on the neural basis of communicative behavior in primates, and explores the implications of these data for the problem of language origins. This book complements *Language Learning by a Chimpanzee: The Lana Project*, edited by Duane M. Rumbaugh in the Communication and Behavior series. That book reevaluated the relationship between human language and animal communication in view of the linguistic abilities of chimpanzees.

The present volume provides a bridge between investigations of the clinical neurology of speech, language, and related gestures, and cognitive abilities and studies of the neural substrates of nonhuman primate communication. Although the book focuses primarily on the neurobiology of vocalization and speech, this orientation is not indicative of any theoretical bias. Rather, it reflects the limits of contemporary experimental knowledge.

The first chapter discusses some of the persistent problems in evolutionary neurobiology of primate communication and summarizes the subsequent chapters. Chapters 2–4 present state-of-the-art reviews of the effects of brain lesions and stimulation on vocalization in New and Old World monkeys, the relation between species differences in peripheral vocal structures and species contrasts in vocal performance, and the anatomy and physiology of the nonhuman primate auditory system. Chapters 5–7 examine the effects of electrical brain stimulation on human verbal communication and facial expression, summarize clinical data pertaining to language pathologies, and discuss the neural mechanisms of manual and oral control. These three chapters also set forth original models of the neural organization of human language. The final three chapters synthesize and place in an evolutionary context the material presented in earlier chapters.

This volume will be of interest to a wide academic audience, including neuroscientists, behavioral biologists, neurologists, and psychiatrists, as well as scholars and educators from the social sciences (especially anthropologists) who are concerned with the problem of the evolutionary heritage of human speech and language.

Acknowledgments

We owe our gratitude to a number of colleagues and staff who have aided the successful completion of this volume. We are especially grateful to the editor of this series, Dr. Duane Rumbaugh, who encouraged us to put together this volume and whose work provided much of the intellectual impetus for this undertaking. We also owe special thanks to Stevan R. Harnad for providing valuable commentary on previous drafts of Chapters 9 and 10 and for sustaining our interest in the topic of language origins. The excellent, professional typing of our manuscripts by Jackie Raffa and Louisa Anthony and the support of the Research Service of the Veterans Administration Medical Center, Brentwood, California, are also gratefully acknowledged. Finally, we thank the editors and editorial staff of Academic Press for their patience and encouragement extended throughout this project.

1

Introductory Comments and Overview

HORST D. STEKLIS   MICHAEL J. RALEIGH

Four types of investigations have rekindled interest in the nature of human verbal and nonverbal communication and their evolutionary relationship to nonhuman primate communication. This interest represents an outgrowth of (*a*) investigations of chimpanzee cognitive and linguistic abilities; (*b*) studies of communicative and cognitive abilities in brain-damaged (e.g., split-brain) and normal humans; (*c*) analyses of social communication among free-ranging and captive nonhuman primates; and (*d*) findings on the neural mechanisms of nonhuman primate communication. Several conferences have explored some of the implications of these data for evolutionary and comparative studies (Rieber, 1976; Harnad, Steklis, and Lancaster, 1976; Dimond and Blizard, 1977). An earlier volume in this series reviewed how the linguistic and communicative abilities of chimpanzees have led to a reevaluation of the relationship between human language and animal communication (Rumbaugh, 1977).

The primary aim of this volume is to draw together data on the neural correlates of human and nonhuman primate vocalization and facial expression. For more than 100 years, clinical observations have provided information about the neural mechanisms underlying human speech. However, systematic controlled studies on the neurobiology of human speech and facial expression have appeared only recently (e.g., Whitaker and Whitaker, 1976a,b; Ekman, 1973). Furthermore, systematic investigations on the neural correlates of communication in nonhuman primates began

1

NEUROBIOLOGY OF
SOCIAL COMMUNICATION IN PRIMATES

less than 20 years ago. As a consequence, this is one of the few volumes containing "state of the art" reviews on the neural mechanisms underlying the production and perception of social communication in humans and other primates.

These presentations make two additional contributions to the study of primate and human communication. First, they aid efforts to specify more precisely the lacunae in research on the behavior and neurobiology of primate communication. Second, they point to the types of evolutionary questions that may profitably be examined.

There are a number of hiatuses in research on primate social communication, at least three of which are underscored in this volume. One is that research has been focused almost entirely on facial expression and vocalization; hence, in this volume many of the behavioral complexities of human and nonhuman communication are not discussed (e.g., see Altman, 1967; Sebeok, 1977). In addition to verbal and facial signals, human social communication utilizes body postures, hand gestures, spatial relations (proxemics), and perhaps signals in other modalities (e.g., olfaction) (see reviews in Weitz, 1974; Harper, Wiens, and Matarazzo, 1978; Siegman and Feldstein, 1978). Many of these may have homologs in nonhuman primate communication, which is similarly a multimodal activity, with vocalization frequently modified by visual and other signals (Marler, 1965). Clearly, as noted by Passingham (Chapter 8) and by Steklis and Raleigh (Chapter 9), this sparsity of data limits the types of comparative analyses that can be made at present.

A second problem is the lack of experimental data on the neurobiology of communication in the great apes. A variety of paleontological, morphological, and biochemical data indicate that the great apes and humans are much more closely related to each other than either is to the Old World monkeys (King and Wilson, 1975; Sarich and Cronin, 1976). In the absence of data on pongids, it is difficult to evaluate some of the apparent contrasts and similarities between humans and other primates. Neurobiological data on apes, for example, could help us determine whether particular neural mechanisms or abilities were evolved independently after the hominid and pongid lineages separated or represent retentions from a time during which humans and great apes shared a common ancestor.

A third difficulty is that behavioral data from different species are rarely directly comparable. Investigators have employed different paradigms in assessing the effects of central nervous system manipulations on communicative behavior. For example, in assessing the effects of brain lesions on human facial expression, some investigators have monitored the alterations in spontaneously occurring expression, whereas others have recorded the changes in expression elicited by commands. In nonhuman primates investigations have focused almost entirely on spontaneous facial or vocal expression, with only a few studies reporting on the effects of stimulation or ablation on (learned) conditioned expression. It may be possible to overcome this methodological problem by utilizing standardized procedures, such as those developed by Miller (1971) and Ekman (1973). Until standardized procedures are employed, it will remain difficult to make conclusive interspecific comparisons.

Despite such limitations, we believe that the comparative and evolutionary approach will lead to important insights into the neurobiology of human and animal social communication. The chapters contained in this volume provide a substantive empirical framework, which we hope will generate further experimentation and theory on the biological basis, behavioral manifestations, and evolutionary history of primate communication. At present these data indicate that the functional and anatomical contrasts between the human and nonhuman primate nervous systems have implications for speculations on the evolution of human speech and language. Several intriguing evolutionary issues can be explored. For example, is there evidence of a dichotomy between the neural mechanisms mediating the production of facial expression and vocalization in humans and those operating in other primates? This and related evolutionary issues are examined in the chapters by Passingham, Kimura, and Steklis and Raleigh.

This volume presents data from both nonhuman and human primates: Chapters 2, 3, and 4 deal primarily with nonhuman primates, while Chapters 5 through 7 focus on humans. Chapters 8, 9, and 10 unify and synthesize some of these data. All contributors were selected on the basis of the relevance of their research to the biological foundations of communication. Each chapter reflects its author's perspective, and thus some are considerably more open-ended and personalized than others. Nonetheless, each chapter represents a substantive contribution to what we believe is an unusual and exciting volume on communication and behavior.

In the second chapter, Jürgens summarizes his investigations into the consequences of stimulation and ablation on vocal behavior of squirrel monkeys. He has mapped the call production system and provided a comparison with the substrate mediating vocalization in other species. Based on these studies, Jürgens concludes that squirrel monkey call production is mediated by four different but interrelated neural systems. The first system includes structures in the pons and the medulla, such as the nucleus ambiguous, involved in the motor coordination of vocalizations. This system is not responsible for the initiation of spontaneous calls. The second system is composed of structures in the caudal periaqueductal gray and adjacent lateral tegmentum. It couples different vocal patterns with appropriate motivational states; spontaneous calls are mediated by this system. The third system consists of anterior limbic cortex, perhaps analogous to the supplemental speech area in humans. This system controls calls independently of specific motivational states and facilitates or inhibits call production. Anterior limbic cortex may be involved in voluntary call initiation. The fourth system is the cortical laryngeal area, concerned with both motor coordination and volitional aspects of call formation. The third and fourth systems have a different set of projections. The third system projects to other limbic vocal areas, whereas the fourth projects to motor and sensory areas related to the oral pharyngeal-laryngeal regions. The third and fourth systems may interact at three loci: *(a)* in the cortex around the anterior cingulate sulcus; *(b)* in the nucleus ventralis anterior of the thalamus; and *(c)* in the parabrachial nuclei within the dorsal–lateral pons. Both systems converge in the nucleus ambiguous. Jürgens' analysis suggests that the fourth system is most likely to be homologous to human speech mechanisms. This system has been the most elaborated of the neural substrates mediating human speech production.

Sutton's presentation in the third chapter focuses on data derived from Old World monkeys, which supplement Jürgens' data. Sutton's results support, at least in part, the importance of paleocortical mechanisms in vocalization in nonhuman primates. Bilateral posterior-parietal and inferior-frontal lesions do not appear to impair spontaneous or learned vocal performance. In contrast, bilateral anterior cingulate removal produces deficits in both learned and spontaneous vocal production. However, with time, spontaneous vocal behavior returns to the preoperative level, whereas more lasting changes are apparent in learned vocalization. The importance of the anterior cingulate region in vocalization is also underscored by Sutton's single-unit data. In addition, he reviews the comparative anatomy of peripheral vocal structures, including innervation ratios, muscle spindle ratios, and a variety of other measures. These anatomical data indicate that differences in peripheral structures at present cannot account for species differences in phonatory performance. Rather, such differences arise mainly because of differences in the central nervous system.

In the fourth chapter, Newman addresses an important question, namely, whether the primate auditory receptor system contains specialized mechanisms similar to those on the effector side. He provides a comprehensive review of the anatomy and physiology of the primate auditory system. Of particular interest is his demonstration that frontal-temporal structures play a role in auditory processing. His review of single-unit data indicates that the superior temporal gyrus contains both specialist and class detector neurons. Found only in the superior temporal gyrus, specialist neurons respond selectively to a particular type of vocal stimulus. Class detectors, in contrast, respond to structurally similar calls. Unlike the superior temporal gyrus, the frontal cortex contains only class detector neurons. Stimulation data indicate that the class detector neurons in the frontal lobe may inhibit the spontaneous and/or driven activity of the superior temporal gyrus neurons involved in auditory processing. It is intriguing that the cingulate gyrus, which has been strongly implicated in effecting primate vocalization, appears to project to the superior temporal gyrus. At this time, however, the nature of the physiological interactions between these areas has yet to be fully specified.

In the fifth chapter, Ojemann and Mateer present data on the effects of electrical stimulation on human communication and utilize these observations to devise a neural model. In combination with Brown's and Kimura's chapters, their presentation underscores the necessity of employing lesion, stimulation, and normative data in constructing evolutionary and functional models. It is noteworthy that Ojemann and Mateer stress the importance of subcortical structures as well as the relationship between language, memory, and attentional mechanisms. Their emphasis on these factors parallels Jürgens' and Sutton's accounts of nonhuman primate neural systems and dovetails with Passingham's and Steklis and Raleigh's comments on the role of cognitive abilities in the evolution of language.

Ojemann and Mateer provide information on three aspects of the neural organization of human communication. First, the general organization of language in the cerebral cortex is assessed by noting the consequences of cortical stimulation on object naming. There is tremendous interindividual variability, as consistent interference was induced only by stimulation of the posterior third of Broca's area. Stimulation of other frontal, temporal, and parietal regions produced variable effects. Second, they attempted to subdivide the large amount of cortex implicated in the mediation of language into different functional systems. The consequences of stimulation on the rate of phoneme production, nonverbal oral-facial movements, phoneme identification, syntax, and verbal memory are discussed. Two functional subdivisions are identified: one associated with motor discrimination, the other with short-term verbal memory. Finally, the effects of ventral-lateral thalamic stimulation were monitored. This region serves as a gating mechanism, influencing access to verbal memory and the likelihood of later retrieval. This functional characteristic is highly lateralized.

Ojemann and Mateer discuss several implications of their results for comparative and evolutionary studies. One is that nonhuman primates may possess analogs to the neural systems underlying human speech and language. Another is that, as in humans, there may be a degree of intra-specific variability in the neural mechanisms underlying nonhuman primate communication.

In Chapter 6, after discussing some difficulties inherent in formulating neural models of language, Brown proposes a model based on clinical data pertaining to language pathologies, as well as related experimental studies and anatomical observations. Brown's major focus is on the effects of brain lesions on language, and thus his presentation complements Ojemann and Mateer's discussion of brain stimulation data.

Brown begins by examining several central issues in the localization-of-function problem. The effects of a brain lesion vary with the patient's emotional state, age, premorbid history, time since the lesion, and other factors. These variables produce changes in the patient's clinical status, and current theories do not adequately relate such changing behavioral states to the (relatively) static brain lesion. After pointing out some of the shortcomings of disconnectionism, Brown maintains that there is need of a comprehensive theory that accounts for recovery of function, right hemisphere compensation, cerebral dominance, and lesion effects.

Subsequently, Brown presents an anatomical model in which the neural structures mediating language are organized hierarchically. The levels correspond to different evolutionary stages through which humans passed, with the progression moving from limbic to paralimbic to generalized neocortex to primary neocortex. Each level is associated with particular behavioral capabilities, and the activities of different areas are coordinated through vertical and horizontal connections. Language is viewed as an emergent process that is bilaterally organized at the limbic and paralimbic levels. In Brown's formulation, cerebral dominance represents an end point of a developmental process. The degree of dominance is reflective of the extent to which particular zones in the left hemisphere have differentiated.

The final two-thirds of Brown's chapter is concerned with clinical material. He considers a variety of disorders resulting from damage to the posterior sector, including confabulation, semantic jargon, neologism, and other "fluent" types of aphasic disturbances. Anterior lesions of the limbic, transitional, or true neocortex produce a different set ("nonfluent" type) of symptoms. Aphasic symptoms result from lesions of particular brain structures, and behavioral consequences of brain damage are indicative of the normal functioning of a particular remaining area. Thus, in Brown's view, pathology unravels the normal emergent process; that is, a brain lesion disrupts a certain level (e.g., limbic, transitional, or neocortex) of processing, which brings to light that level in normal language processing.

In Chapter 7, Kimura presents data on the neural mechanisms of praxia (manual and oral control) and discusses the evolution of communicative behavior in humans. She maintains that the vehicle through which communication is expressed is exceedingly important: Communication is critically dependent on the ability to utilize appropriate movements as communicators. As she notes, this perspective contrasts with the views of Chomsky and others.

On the basis of fossil evidence, reconstructions of peripheral anatomy, and investigations of chimpanzee language learning, she suggests that during human evolution a manual system of communication preceded the vocal one. Similar points of view have been set forth by other investigators (e.g., Hewes, 1973), and an alternative scheme is developed in Chapter 10. Kimura utilizes data on the probable neural mechanisms of human speaking and signing to support her formulation.

In Kimura's view, both speech and manual sign are mediated by left hemisphere systems specialized for motor control. She maintains that apraxia (the deficits in oral and manual movements) associated with left hemisphere damage is best regarded as a movement disorder. The praxis system is particularly important for the selection and execution of new postures, is relatively free from dependence on feedback, and is not critical to fine control of distal muscles. Earlier, widely accepted explanations of the neuroanatomical basis of apraxia emphasized ipsilateral cortico-cortical connections, the corpus callosum, and the pyramidal tracts. Kimura suggests that these conceptions are not fully satisfactory and hints that basal ganglia and thalamic regions are important in motor praxis. From her discussion of the neuroanatomy of oral movement disorders, she concludes that the oral and brachial praxic system is dependent on left posterior Sylvian cortex.

These data are interpreted as showing that the left hemisphere is specialized primarily for certain types of motor functions. It is suggested that this specialization may have evolved in conjunction with the purported asymmetric motor activities associated with human tool fabrication and use.

In Chapter 8, Passingham discusses the relationship between specialized systems and (the possibly unique) human speech and language abilities. He notes that the human brain is approximately three times as large as would be expected on the basis of body size alone. In addition, however, he also points out that a primate brain of this size could be expected to manifest the same proportion of cortical areas and thalamic nuclei as that exhibited in humans. In human evolution, there has not been an unusual alteration in the proportion of any particular association area (although visual cortex appears to be reduced). Similarly, none of the thalamic nuclei is larger or smaller than would normally be expected in a primate with a human-sized brain (except for the lateral geniculate body, which is slightly reduced). While language ability may depend solely on size-consequent reorganization in the proportion of the brain's components, human speech abilities may also be due to the evolution of specialized structures. In Passingham's view, while chimpanzees can be trained to master simple language systems, spontaneous language development does not occur due to lack of sufficient brain size in this organism. At this time, the most likely speech-related anatomical specializations are in Broca's area of the inferior-frontal lobe. Other cortical asymmetries, such as those of the temporal lobe, have a less clear relationship to language ability. After considering language- and speech-related abilities (such as categorical speech perception, auditory–visual associations, voluntary vocal production, and symbol and rule formation), Passingham concludes that present evidence does not persuasively demonstrate the existence of qualitative differences between humans and other primates. He concludes that the type of language chimpanzees can be taught may not rely on specialized left hemisphere mechanisms. Rather, this limited language ability may be similar to language capacities present in the human right hemisphere.

In Chapter 9, we attempt to synthesize material from some of the previous chapters by reviewing a set of issues frequently discussed in considerations of the evolution of human speech and language: (a) whether there are qualitative differences between humans and other primates in the voluntary control of communicative behavior and (b) whether there is evidence of a dichotomy between humans and other primates in terms of the neural correlates of communicative behavior. We begin by reviewing behavioral data from conditioning studies, ethological observations, and developmental investigations. These data suggest that nonhuman primates are quite capable of voluntarily modifying vocalization and facial expression. In contrast to prevailing views, we suggest that the differences between humans and other primates are of a quantitative nature. Subsequently, we compare traditional with more recent conceptions regarding limbic system–neocortex interactions in the mediation of communicative behavior. The concept of encephalization is then examined. Finally, these theoretical observations (on cortical–limbic interactions and encephalization) are applied to data obtained from lesion and stimulation studies. In our view the evidence for a dichotomy in terms of neural mechanisms is not compelling. While there are clearly striking differences between human and nonhuman primate communicative behavior, these probably represent end products of quantitative differences.

In Chapter 10, we attempt to relate some of the information presented in this volume and elsewhere to the problem of language evolution. Putative differences between humans and other primates in language-related cognitive, productive, and perceptual abilities and in their neural correlates are evaluated. Subsequently, several new speculations regarding the origin and evolution of human language and speech are set forth.

Based on our discussion of language-related abilities and neural correlates, we formulated three major conclusions. First, there is no evidence of a qualitative distinction between humans and African apes in language-related cognitive abilities (such as cross-modal perception or propositional symbol use). Second, differences between primates in peripheral vocal anatomy are not likely to represent critical barriers to speech sound production. The critical differences are in the central nervous system: All primates appear to have some neocortical control of the vocal apparatus; however, in humans, additional cortical areas support speech motor mechanisms. In addition, in humans, speech production is hemispherically lateralized and has become linked to mechanisms that govern sequential oral-facial and arm movements. Third, categorical speech perception may be founded in information-processing capacities evolved by all higher primates. In addition, vocal decoding may involve hemispheric specialization in a number of primate species, suggesting that hemispheric specialization evolved prior to the development of human speech.

In our discussion of evolution, we make the necessary assumption that abilities and neural mechanisms shared among living humans and other primates are homologous; that is, these traits were also characteristic of ancestral hominid populations. Going on this assumption, we speculate that human language in the form of "primordial" speech arose early through selection for cognitive elaboration and improved vocal learning abilities. Requisite neural mechanisms for these abilities could have evolved concomitantly with increased encephalization. This may have occurred through intense selection pressures on regulatory DNA and brain growth rates.

References

Altmann, S. A., ed. (1967). "Social Communication Among Primates." Univ. of Chicago Press, Chicago.

Dimond, S. J. and Blizard, D. A., eds. (1977). Evolution and Lateralization of the Brain. *Ann. N.Y. Acad. Sci.*, *299*, 1–501.

Ekman, P. (1973). *In* "Darwin and Facial Expression" (P. Ekman, ed.), pp. 169–222. Academic Press, New York.

Ekman, P., Friesen, W. V., and Tomkins, S. S. (1972). "Emotion in the Human Face. Guidelines for Research and an Integration of Findings." Pergamon Press, New York.

Harnad, S. R., Steklis, H. D., and Lancaster, J. B., eds. (1976). Origins and Evolution of Language and Speech. *Ann. N.Y. Acad. Sci.*, *280*, 1–914.

Harper, R. G., Wiens, A. N., and Matarazzo, J. D. (1978). "Nonverbal Communication." Wiley, New York.

Hewes, G. W. (1973). *Current Anthropology 14*, 5–24.

King, M. C. and Wilson, A. C. (1975). *Science 188*, 107–116.

Marler, P. (1965). *In* "Primate Behavior" (I. DeVore, ed.), pp. 544–584. Holt, Rinehart and Winston, New York.

Miller, R. E. (1971). *In* "Primate Behavior: Developments in Field and Laboratory Research" (L. Rosenblum, ed.), pp. 139–175. Academic Press, New York.

Rumbaugh, D. M., ed. (1977). "Language Learning by a Chimpanzee. The Lana Project." Academic Press, New York.

Rieber, R. W., ed. (1976). "The Neuropsychology of Language." Plenum, New York.
Sarich, V. M. and Cronin, J. (1976). *In* "Molecular Anthropology" (M. Goodman and R. E. Tashian, eds.), pp. 139–167. Plenum Press, New York.
Sebeok, T. A., ed. (1977). "How Animals Communicate." Indiana Univ. Press, Bloomington.
Siegman, A. W. and Feldstein, S., eds. (1978). "Nonverbal Behavior and Communication." Wiley, New York.
Weitz, S., ed. (1974). "Nonverbal Communication." Oxford Univ. Press, New York.
Whitaker, H. and Whitaker, H. A., eds. (1976a). "Studies in Neurolinguistics." Vol 1, Academic Press, New York.
Whitaker, H. and Whitaker, H. A., eds. (1976b). "Studies in Neurolinguistics." Vol. 2, Academic Press, New York.

2

Neural Control of Vocalization in Nonhuman Primates

UWE JÜRGENS

Introduction

Human speech consists of two components: verbal and nonverbal. While it is generally accepted that the verbal component does not have a counterpart among nonhuman primates, the nonverbal component clearly does. The intonation patterns used during scolding, lamenting, caressing, or jubilating as well as nonverbal utterances such as laughing, whining, moaning, and shrieking express emotional states of the utterer and thus must be homologized with animal vocalizations.

The use of monkey calls as a model for studying human speech mechanisms, therefore, although legitimate for the nonverbal, emotional component of speech, becomes questionable for the verbal component. This difficulty is due to three fundamental differences between the two components:

1. The lack of the conceptual aspect in nonverbal utterances.
2. The different role played by articulation: While differentiation of phonemes takes place almost exclusively on the basis of different supralaryngeal movements, nonverbal vocal utterances by man and by monkey are determined predominantly by laryngeal activity, with oral movements playing only a subordinate role.
3. In contrast to verbal speech components, nonverbal emotional components as well as monkey calls seem to be almost completely innate.

11

NEUROBIOLOGY OF
SOCIAL COMMUNICATION IN PRIMATES

For the squirrel monkey this has been shown by Winter *et al.* (1974). These authors compared the vocal development of normally raised squirrel monkey infants with that of infants raised from birth with surgically muted mothers. Despite the fact that the infants placed with mute mothers did not hear adult calls, which could have served as models for vocal learning, all of these infants developed a normal vocal repertoire. In man, psycholinguistic studies have shown that a number of intonation patterns are transcultural, so that recognition of the emotional content of a sentence is possible to some extent without understanding its verbal components (Kaiser, 1962; Kramer, 1964; Ertel and Dorst, 1965). Furthermore, man's capability to enlarge his vocal repertoire by learning new articulatory gestures is highly developed, whereas nonhuman primates seem to lack this ability almost completely (Kellogg, 1968). This is not to say that there is no way of studying certain aspects of verbal speech in animal experiments, but such an attempt requires that each of the different components of speech be studied in a different context: the conceptual component within the framework of a nonvocal, gestural language system (Gardner and Gardner, 1969), the learned articulatory component within a learned nonvocal, oral task (Luschei and Goodwin, 1975), and the phonatory component within a vocal operant conditioning procedure using species-specific calls (Sutton *et al.*, 1973).

The aim of this chapter is to give a comprehensive review of the cerebral stimulation, lesioning, recording, and anatomical studies bearing upon the neural control of vocalization. In most of these studies, monkeys were the subjects. However, from the remarks made above, it should be clear that the results from the studies on monkeys must be viewed not only per se but also as a model for certain aspects of human phonation.

Stimulation Studies

The first question that arises in the study of neural vocalization mechanisms is which brain structures are principally involved in the production of vocalizations. Electrical brain stimulation has proved to be a fruitful method of obtaining a first survey of the areas potentially involved. I will begin this review, therefore, with a discussion of electrically elicited vocalizations and vocal fold movements.

Electrically Elicited Vocalizations and Vocal Fold Movements

Vocalization as a consequence of electrical brain stimulation in primates has been reported in man, chimpanzee, gibbon, rhesus monkey, squirrel monkey, and marmoset. While the rhesus monkey and squirrel monkey have been systematically explored, there are only limited observations for the remaining four species.

Man

In man, two cortical and two subcortical areas have been found to be responsive. In their extensive studies, Penfield and co-workers (Penfield and Bordley, 1937; Penfield and Welch, 1951; Penfield and Rasmussen, 1952; Penfield and Roberts, 1959) showed that stimulating the inferior pericentral cortex and the dorsomedial frontal cortex above the cingulate sulcus—the so-called supplementary motor area—yields vocalization. The type of vocalization is either a long drawn-out vowel sound, sometimes with a consonant component, or a rhythmic sound. In addition, repetitions of words can be observed when the supplementary area is stimulated (see also Brickner, 1940). Both cortical areas are responsive bilaterally. Subcortically, vocalization has been elicited from the posteromedial hypothalamus (Schvarcz *et al.*, 1972) and nucleus ventralis anterior thalami (Hassler *et al.*, 1960; Hassler 1961, 1964; Schaltenbrand, 1965, 1975); nonverbal as well as verbal exclamations were produced from the latter.

Chimpanzee

Several investigators have explored the chimpanzee cortex. While there is general agreement that stimulation of the inferior precentral cortex yields vocal fold movements, the elicitability of vocalization is somewhat disputed. Leyton and Sherrington (1917) concluded that vocalization cannot be elicited from the chimpanzee cortex. In their brain diagrams, however, two stimulation sites are listed that yield "adduction of vocal cords with emission of a sound." Hines (1940) obtained vocalization once from the inferior precentral gyrus; this vocalization was not reproducible. Dusser de Barenne *et al.* (1941), in their brain diagrams, show one vocalization site in the inferior premotor cortex corresponding to Broca's area, which, however, is not further mentioned in the text. Finally, Walker and Green (1938) report that vocalization has been elicited once; the exact stimulation site and the species stimulated (chimpanzee, rhesus, baboon, mangabey, and spider monkey were explored) are not given in their description. All of these studies were carried out under general anesthesia. Subcortically, no systematic exploration has been carried out. The only report on electrically elicitable vocalization I was able to find is that of Brown (1915). Brown stimulated the cut surface of the transected brain stem in the region of the rostral periaqueductal gray and obtained a sound resembling laughter.

Gibbon

Apfelbach (1972) explored three gibbons specifically for vocalization; 194 vocalization-producing electrode positions were found. A total of six different call types was elicited. The effective loci are distributed according to Apfelbach within the following structures: telencephalon—ventral hippocampus, fimbria hippocampi, amygdala, capsula interna; diencephalon —ventral hypothalamus, "thalamus," pulvinar; mesencephalon—area tegmentalis, cerebral peduncle, lemniscus medialis, substantia nigra, colliculus inferior, central gray; pons—nucleus reticularis parvocellularis, nuclei pontis. As there was no histology done in these animals and the stereotaxic coordinates were determined with brain atlases for the chimpanzee and rhesus monkey, the anatomical results of this study should be viewed with reservation. Nevertheless, it is clear from this study that the vocalization-producing electrode positions are not restricted to a few circumscribed brain structures but instead are widely scattered throughout the brain. Furthermore, it appears from Apfelbach's description that there is some correlation between brain structure stimulated and elicited call type.

Rhesus Monkey

The brain of the rhesus monkey has been explored systematically for electrically elicitable vocalizations by Magoun et al. (1937) and Robinson (1967). Magoun and his co-workers restricted their exploration to midbrain, pons, and medulla; Robinson stimulated telencephalon, diencephalon, and rostral midbrain. In addition, casual observations on electrically evoked vocalizations were reported by Smith (1944, 1945), Kaada (1951), Delgado (1955, 1967), Showers and Crosby (1958), and Hughes and Mazurowski (1962). Figure 2.1 (left-hand halves of the brain diagrams) gives a compilation of the data of Magoun et al., Robinson, and Kaada. Only those electrode positions yielding vocalization *during* stimulation are shown; vocalizations appearing after cessation of stimulation ("rebound vocalizations") have been omitted. As Robinson does not distinguish the vocalization loci individually according to elicited call type, no correlation between brain structure and call type is given in the figures. It can be seen from the diagrams that the responsive brain areas extend from the precallosal cingulate cortex through the diencephalon, mesencephalon, and pons down into the medulla, thus confirming the wide distribution already mentioned for the gibbon. The effective loci are not distributed evenly

FIGURE 2.1. *Schematic half sections of the brains of the rhesus monkey (left-hand side) and squirrel monkey (right-hand side) indicating electrode positions producing vocalization during electrical stimulation. [Compiled from Magoun et. al., 1937; Kaada, 1951; Robinson, 1967; Jürgens and Ploog, 1970.]*

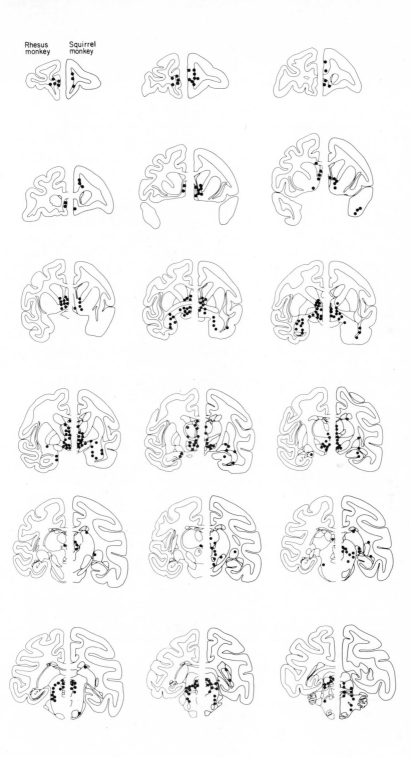

Rhesus
monkey

Squirrel
monkey

throughout the brain, however. In the telencephalon, they are limited to the anterior limbic cortex, septum, preoptic region, diagonal band of Broca, amygdala, stria terminalis, rostral hippocampus, fornix, and the ventromedial edge of the internal capsule. All of the neocortex, the basal ganglia (except around the ventromedial capsula interna), the posterior limbic cortex, and posterior hippocampus are free. In the diencephalon, the vocalization loci are dispersed throughout the hypothalamus—from there they can be followed into the ventral tegmental area of Tsai. Dorsally, in the thalamus and subthalamus, vocalization can be elicited only from nucleus anterior, nucleus ventralis anterior, and midline thalamus. In the brainstem, finally, the responsive electrode positions group around the aqueduct in the central gray, from where they descend further caudalward into the lateral pontine and medullary reticular formation.

While stimulation of the rhesus neocortex has never produced vocalization, it has been reported by several authors to yield vocal fold movements (Walker and Green, 1938; Sugar *et al.*, 1948; Hast *et al.*, 1974). The responsive area lies in the inferior motor cortex between sulcus centralis caudally and sulcus arcuatus rostrally just above the anterior end of the Sylvian fissure. According to Hast and co-workers (1974), different laryngeal muscles are represented separately within this area (Figure 2.2).

Squirrel Monkey

The right-hand halves of the brain diagrams in Figure 2.1 show the vocalization-eliciting electrode positions in the squirrel monkey (Jürgens *et al.*, 1967; Jürgens and Ploog, 1970). In these diagrams, only those positions

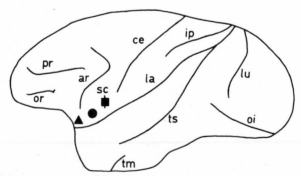

FIGURE 2.2. *Lateral view of the rhesus monkey cortex. Electrical stimulation of the areas marked with a triangle, circle, and square produces movements of the cricothyroid muscle, thyroarytenoid muscle, and extrinsic laryngeal muscles, respectively. (Adapted from Hast et al., 1974.] ar, sulcus arcuatus; ce, sulcus centralis; ip, sulcus intraparietalis; la, sulcus lateralis; lu, sulcus lunatus; oi, sulcus occipitalis inferior; or, sulcus orbitalis; pr, sulcus principalis; sc, sulcus subcentralis anterior; tm, sulcus temporalis medius; and ts, sulcus temporalis superior.*

that (a) yielded vocalization at least three times consecutively and (b) could be reproduced in a second animal from a comparable area have been mapped. Again, the loci producing only rebound vocalizations have been omitted. (The only area from which rebound vocalizations were obtained exclusively was the di- and mesencephalic ventral tegmentum.)

A comparison between the left-hand (rhesus) and right-hand (squirrel monkey) halves of the brain diagrams of Figure 2.1 reveals an almost identical distribution. The only differences concern a few loci connecting the amygdala with the temporal pole along the uncinate fasciculus and some others near the tr. opticus/lateral geniculate, which were not found in the rhesus monkey. On the other hand, some positions in the fornix and nucleus anterior thalami were found in the rhesus but not in the squirrel monkey. Furthermore, the effective positions in the region of the nucleus ventralis anterior thalami are limited to the ventromedial margin of this nucleus in the squirrel monkey, whereas they are scattered throughout this structure in the rhesus monkey.

In the squirrel monkey study, all of the vocalizations elicited were subjected to spectrographic analysis. It was found that, in general, these vocalizations appeared to be within the range of variation of natural, spontaneous calls. Only in the case of the caudal midbrain, pons, and medulla were some of them of a more or less artificial character. Figure 2.3 summarizes schematically the relations between elicited call type and brain area stimulated. For this purpose, all of the vocalizations elicited have been classified into four call classes (cackling calls; growling calls; trilling/chirping calls; shrieking/quacking/groaning calls) and their cerebral distributions projected onto a sagittal plane. It can be seen that there are areas yielding only one call type, such as the midline thalamus, which exclusively produces chirping, or the cortex around the anterior sulcus cinguli, which produces only growling. Two call types can be obtained from other brain structures, for instance, growling and shrieking from the stria terminalis and growling and cackling from the inferior thalamic peduncle. Finally, there are areas from which representatives of all electrically elicitable call classes can be evoked, such as the midbrain central gray and the ventromedial capsula interna at the level of the anterior commissure.

Figure 2.3 also shows that the brain areas producing the same call type do not necessarily form a continuous system. Chirping calls, for instance, can be evoked from five separate regions, namely, the gyrus subcallosus–nucleus accumbens area, midline thalamus, rostral hippocampus, central gray, and along the lower spinothalamic tract. Despite the fact that there are a number of direct neuroanatomical connections between these five regions, no vocalizations can be obtained along them. In contrast to chirping, the substrate yielding cackling does form a continuous system.

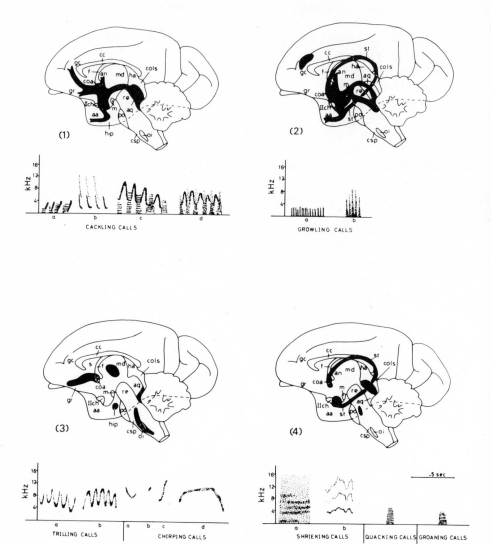

FIGURE 2.3. *Sagittal sections of the squirrel monkey brain showing the "cerebral representation" of four different groups of electrically elicitable calls. Below each brain diagram, the calls elicitable from the black areas are represented as frequency-time spectrograms. aa, area anterior amygdalae; an, nucleus anterior thalami; aq = substantia grisea centralis; cc = corpus callosum; coa = commissura anterior; cols = colliculus superior; csp, tractus cortico-spinalis; f, fornix; gc, gyrus cinguli; gr, gyrus rectus; ha, habenula; hip, hippocampus; m, corpus mamillare; md, nucleus medialis dorsalis thalami; oi = oliva inferior; po, griseum pontis; re, formatio reticularis mesencephali; s, septum; st, stria terminalis; and llch, chiasma nervorum opticorum.*

Beginning in the anterior limbic cortex it follows the internal capsule into the diencephalon, where it branches into three components: two following the inferior thalamic peduncle ventrolaterad and dorsomediad into the amygdala and rostromedial thalamus, respectively; the third continuing caudalward in a periventricular position through the dorsomedial hypothalamus and central gray down to the midbrain–pons transitional zone.

Measurements of the latencies between stimulus onset and beginning of vocalization show extreme variability. This variability depends not only on the stimulus parameters (intensity, frequency, and duration of electrical pulses) but also on the relative position of the electrode within a given brain structure and probably also on the individual animal stimulated. (The latter point is difficult to prove because it is practically impossible to place an electrode in precisely the same locus in two animals; it is found, however, that the average latencies of different animals with numerous comparable electrode positions differ significantly.) The shortest latencies ever encountered (50–60 msec) were obtained from the central gray of the midbrain. Latencies between 60 and 100 msec are found, apart from the central gray, only in the dorsolateral midbrain and pons tegmentum and the medulla. Latencies between 100 and 200 msec are obtained, in addition to the above structures, from the ventrolateral midbrain tegmentum, diencephalic periventricular gray, midline thalamus, and precallosal cingulate cortex. All of the other vocalization-eliciting brain structures have "minimal latencies" above 300 msec, sometimes even up to several seconds. For comparison, it should be mentioned that the vocal reaction after application of a strong painful stimulus (external electric shock) occurs with a latency of 100 to 200 msec.

Like the rhesus monkey, the squirrel monkey possesses a neocortical area whose stimulation yields vocal fold movements but not vocalization (Hast and Milojevic, 1966; Jürgens, 1974). This area occupies a position homologous to that of the rhesus monkey, namely, just above the anterior Sylvian fissure within Brodmann's area 6bα.

Marmoset

Lipp and Hunsperger (1978) explored the hypothalamus of the marmoset (*Callithrix jacchus*) and found that stimulation of the ventromedial hypothalamic nucleus produces a "geckering" sound typical of threatening behavior.

Nonprimate Mammals

Electrically elicited vocalization has been obtained not only from primates but also from other mammals, such as the dog (Clark *et al.*, 1949;

Skultety, 1962), cat (e.g., Ingram et al., 1932; Kabat, 1936; Gibbs and
Gibbs, 1936; Magoun et al., 1937; Hess and Brügger, 1943; Kaada, 1951;
MacLean and Delgado, 1953; Spiegel et al., 1954; Hunsperger, 1956; Hilton
and Zbrozyna, 1963; Romaniuk, 1965; and many others), rat (Woodworth, 1971; Waldbillig, 1975; Yajima et al., 1976), guinea pig (Martin,
1976), bat (Suga et al., 1973), and opossum (Roberts et al., 1967). Not all
of these species were explored thoroughly. Nevertheless, it is clear that
there is an extensive overlap between the vocalization-producing areas
described for primates and those of other mammalian orders. The periaqueductal gray and laterally adjacent tegmentum, for instance, yielded
vocalization in all species tested in that region (dog, cat, rat, guinea pig,
and bat). The same holds for the hypothalamus and preoptic region (cat,
rat, guinea pig, and opossum). In general, it can be said that all areas
reported to yield vocalization in nonprimate mammals have also been
found to be responsive in primates. The reverse also seems to be true with
one exception: the anterior limbic cortex. Despite extensive testing of this
area in the cat, rat, and guinea pig, no vocalization has been elicited from
it. In the dog, the results are somewhat controversial. While Kremer (1947)
and Kaada (1951) deny the elicitability of vocalization from this region,
Clark et al. (1949) report a single case of stimulation-induced vocalization
without giving, however, any further information on reproducibility,
latency, or type of elicited call. A number of extensive early stimulation
studies on the dog's cortex were also negative with respect to limbic cortex-
elicited vocalization (Fritsch and Hitzig, 1870; Krause, 1884; Katzenstein,
1908).

The neocortical larynx area described above for primates also exists in
cat, dog, and rabbit (Katzenstein, 1905; Kirikae et al., 1962; Krause, 1884;
Milojevic and Hast, 1964; Munk, 1884). It lies in the lateral precrucial
gyrus, thus corresponding to the motor cortex between forelimb and face
representation. In contrast to the monkey, however, the representation of
the laryngeal muscles is less differentiated in other mammals. While in the
squirrel monkey it is possible to elicit isolated movements of the vocal
folds, and in the rhesus monkey even of single laryngeal muscles, the effect
obtained in cat, dog, and rabbit is a global contraction involving intrinsic
and extrinsic laryngeal muscles as well as soft palate, tongue, and neck
muscles.

Nonmammalian Vertebrates

An appreciation of the phylogenetic age of the vocalization-eliciting
system in primates can be obtained from an examination of data from other
vertebrate classes.

In birds, vocalization has been elicited in the chicken (von Holst and

Saint Paul, 1960; Putkonen, 1966, 1967; Cannon and Salzen, 1971; Peek and Phillips, 1971; Phillips *et al.*, 1972; Andrew, 1973), dove (Åkerman, 1965; Delius, 1971), gull (Delius, 1971), quail (Potash, 1970), red-winged blackbird (Brown, 1972), and house finch (Epro, 1977). Despite some differences in the overall distribution, there is general agreement that the vast majority of vocalization-eliciting electrode positions form a continuous system. It runs from the posteromedial archistriatum along the tractus occipitomesencephalicus into the preoptic region and hypothalamus, where it is joined by a second component from the posteroventral septum. Both components then continue along the occipitomesencephalic tract (which runs along the dorsolateral border of the periventricular fiber system) into the midbrain, where the effective loci are clustered at the medial and ventromedial borders of the nucleus mesencephalicus lateralis dorsalis. From here the system descends into the ventrolateral pons and medulla. As the posteromedial archistriatum is considered to be homologous to the mammalian amygdala, and the nucleus mesencephalicus lateralis dorsalis homologous to the inferior colliculus, it becomes evident that there is an extensive overlap between the mammalian and avian vocalization systems. This overlap is particularly surprising because the peripheral vocal apparatus and its innervation are different in mammals and birds. While mammals use the nervus vagus- and nervus accessorius-innervated larynx, birds produce their calls with the syrinx, a special avian organ situated below the larynx at the junction of the two bronchi, which is innervated by the hypoglossal nerve.

In reptiles, vocalization has been elicited by Kennedy (1975) in the gecko. The effective electrode positions group around the torus semicircularis, the reptilian counterpart of the mammalian colliculus inferior.

In amphibians, the extensive studies of Schmidt (1966, 1968, 1971, 1974, 1976) have revealed two areas yielding species-specific calls. One is the preoptic region, which produces mating calls; the other is the dorsolateral pontine tegmentum, just ventrocaudal to the inferior colliculus, which produces releasing calls.

Finally in fish, despite a totally different sound-producing mechanism, Demski and Gerald (1974) were able to elicit whistling and grunting calls by stimulation of the medial border of the inferior colliculus and ventrocaudally adjacent lateral pons of the toadfish (*Opsanus*).

Taking these results together, we can conclude that some of the vocalization-eliciting brain areas in primates have a very great phylogenetic age. The dorsal midbrain–pons transitional zone seems to be related to call production from the very beginning of this function. Other vocalization areas, such as the preopticohypothalamic region, also have an age certainly much greater than that of the mammalian class. Furthermore, from the fact that a number of vocalization areas are the same in different vertebrate classes despite different call production mechanisms, it may be concluded that the motor coordination of species-specific calls takes place outside these areas. This seems to be corroborated by the observation that, apart from some loci in the lower brainstem, the electrically elicited vocalizations always sound like natural (nonartificial) calls.

Motivational Concomitants of Electrically Elicited Vocalizations

If the assumption is correct that the motor coordination of vocalization takes place outside the forebrain and midbrain vocalization areas, then how should the vocalizations elicitable from these areas be interpreted? Two interpretations seem possible. One is that the vocalization is a secondary reaction due to a stimulus-induced motivational change. The stimulation, for instance, induces pain, and the animal reacts to this pain with a shrieking call. The other possibility is that the stimulation facilitates or triggers directly the activity of a "vocalization center" neuroanatomically connected with the stimulation site.

To test these hypotheses it is necessary to analyze the motivational effects accompanying the electrically elicited vocalizations. In nonhuman primates only one study of this type has been conducted, and this study is of an indirect nature (Jürgens, 1976a). Instead of testing the *specific* motivational changes evoked by the stimulation, a more general emotional indicator was chosen: the reinforcement effect, that is, the pleasurable or aversive quality of the vocalization-eliciting stimulation. The hypothesis underlying this study assumes that there is a high probability that electrically elicited motivational changes strong enough to induce vocalization as a secondary reaction do have positive or negative reinforcing qualities. To state this more specifically: An increase in motivation, no matter which type of motivation, without the possibility of performing the adequate consummatory act (because of lack of the goal object) is assumed to be negatively reinforcing; a decrease in motivation is assumed to be positively reinforcing. This procedure has the advantage that the vocalization sites do not have to be tested for each type of motivation separately, an impossible approach in an animal with a behavioral repertoire as complex as that of the squirrel monkey. In this experiment the animals had the opportunity of switching the vocalization-eliciting stimulation on and off themselves. The study revealed that there are two groups of vocalization-producing brain areas: one group in which the electrically elicited vocalization is independent of the accompanying reinforcement effect (if any), and a second group in which vocalization and reinforcement are strictly correlated. The first group consists of the anterior limbic cortex, ventromedial capsula interna, caudalmost periaqueductal gray, and adjacent parabrachial region. The second group includes all of the rest of the vocalization-eliciting areas. The results are schematically represented in Figure 2.4. From these results it can be concluded that the areas of the first group probably represent primary vocalization areas in the previously mentioned sense, namely, that the stimulation triggers directly the activity of a neuroanatomically related "vocalization center." There are three possible interpretations of the vocalizations elicitable from areas of the second group:

1. They represent secondary reactions due to stimulus-induced motivational changes.
2. They represent primary reactions as well, which are accompanied, however, by the typical motivational states (e.g., the stimulation elicits an aggressive motivation together with its vocal expression).
3. The vocalization as well as the reinforcement effect are triggered directly, but their correlation is merely a coincidence.

The possibility that electrically elicited vocalizations can be accompanied by the corresponding motivational states has been reported by a number of authors for different nonprimate species. The growling and hissing

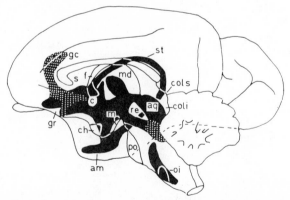

FIGURE 2.4 *Sagittal section of the squirrel monkey brain. Black areas indicate brain structures in which electrically elicited vocalization and motivation (as measured by the self-stimulation technique) are correlated. White-stippled areas indicate brain structures producing vocalization and motivation as independent effects. [From Jürgens and Ploog, 1976.] aa, area anterior amygdalae; an, nucleus anterior thalami; aq, substantia grisea centralis; c, commissura anterior; cc, corpus callosum; coa, commissura anterior; coli, colliculus inferior; cols, colliculus superior; csp, tractus cortico-spinalis; f, fornix; gc, gyrus cinguli; gr, gyrus rectus; ha, habenula; hip, hippocampus; m, corpus mamillare; md, nucleus medialis dorsalis thalami; oi, oliva inferior; po, griseum pontis; re, formatio reticularis mesencephali; s, septum; st, stria terminalis; and Ilch, chiasma nervorum opticorum.*

behavior of cats is perhaps the best documented example. If elicited from the hypothalamus, it is almost always accompanied by all of the other signs typical of defense behavior, such as lowering the head, laying the ears back, hunching the back, pupillary dilatation, and piloerection (Hess and Brügger, 1943). Attack occurs in the presence of a stuffed cat (Hunsperger and Bucher, 1967). If the opportunity is given to kill a rat at the end of a Y-maze, a cat under stimulation learns and carries out this task (Robert and Kiess, 1964). Electrically elicited attack from growling and hissing sites has also been reported from the periaqueductal gray (Brown et al., 1969). Conversely, single-unit recording from the growling/hissing-eliciting area in the periaqueductal gray and adjacent tegmentum has shown that this area contains cells that are exclusively active during spontaneous threat behavior. This activity, however, is not correlated specifically with growling or hissing, but with the threat reaction in general (Adams, 1968).

Another example is the preoptic region, an area known to play a crucial role in male sexual behavior. It contains a large number of testosterone-accumulating cells (Kelley et al., 1975). Injection of minimal amounts of testosterone into this area induces male sexual behavior in male and female rats (Miller, 1965); its destruction eliminates male sexual behavior (Heimer and Larsson, 1966). In the opossum, electrical stimulation of this area yields clicking vocalizations, which under normal conditions are uttered during courting (Robers et al., 1967). If a stuffed animal is introduced during stimulation, copulatory attempts occur.

Stimulation of the septum has been reported to yield purring in the cat (Meyer and Hess, 1957). The vocalization was accompanied by nestling behavior against the experimenters.

In man, stimulation of a number of areas yielding vocalization in monkeys produces motivational changes. From the amygdala and hypothalamus, rage as well as fright can be obtained (Chapman et al., 1954; Heath and Mickle, 1960; Feindel, 1961; King, 1961; Stevens et al., 1969; Sano et al., 1975). The septum has been reported to produce a feeling of well-being (King, 1961). Stimulation of the periaqueductal region may evoke euphoric laughing and joy as well as depression, fright, and horror (Sem-Jacobsen and Torkildsen, 1960; Nashold et al., 1969). Stimulation of the lateral midbrain tegmentum in the vicinity of the spinothalamic tract produces localized sharp pain (Spiegel and Wycis, 1961). No specific emotional effects were obtained from the areas yielding vocalization in man, that is, inferior pericentral cortex, supplementary motor area, nucleus ventralis anterior thalami, and posteromedial hypothalamus. At the latter two sites, however, a strong general arousal effect accompanied the elicited vocalizations (Hassler et al., 1960; Schvarcz et al., 1972).

It is clear from these observations that electrically elicitable vocalizations do not represent a uniform phenomenon. Some of them are elicited in a very indirect way; others are triggered fairly directly. Of the latter, some have to be considered only as components of a more complex reaction; others, in fact, represent independent stimulus responses. Areas that seem to be related rather directly to vocalization are the cortical larynx area, supplementary motor area, anterior limbic cortex, ventromedial capsula interna, ventromedial part of the nucleus ventralis anterior thalami, caudal periaqueductal gray, and laterally adjacent tegmentum.

Neuroanatomical Studies

Because of the wide distribution of the vocalization-eliciting loci, it is not possible to describe here all of the projections of each single vocalization area. I will concentrate, therefore, on a few especially relevant ones.

An area of special importance according to the motivational studies described above, and at the same time the most rostral vocalization area common to the rhesus and squirrel monkeys, is the anterior limbic cortex. Anatomical studies of this region have shown that its projections are very widespread. From a single vocalization locus within that area more than 50 projection fields can be found (Müller-Preuss and Jürgens, 1976); from the whole area more than 80 can be found (Jürgens and Müller-Preuss, 1977). Figure 2.5 shows the projection system of a vocalization point in the dorsal precallosal cingulate gyrus and its overlap with the whole vocalization-eliciting system in the squirrel monkey. It can be seen from the diagrams that there is only partial overlap between the two systems: Not all vocalization loci lie within the cingular projection system, and, conversely, not all of the projection system yields vocalization when electrically stimulated. Areas of overlap are the anterior limbic cortex, preoptic region, dorsal hypothalamus, midline thalamus, inferior thalamic penduncle, central amygdaloid nucleus, periventricular and periaqueductal gray, and the lateral midbrain tegmentum. Whether all of these structures form a functionally coherent system in which some vocalization areas are dependent upon the intactness of others cannot be decided from the anatomical data alone. The question is further complicated by the fact that some vocalization areas receive their afferents from the cingulate gyrus via different routes simultaneously. For example, the central amygdaloid nucleus receives fibers from the external and internal capsules, and the parabrachial nucleus receives periventricular fibers and fibers ascending from the pyramidal tract through the lateral pons. In regard to the call types elicitable from these areas, they all belong to the cackling type, except those from the midline thalamus, which belong to the chirping group.

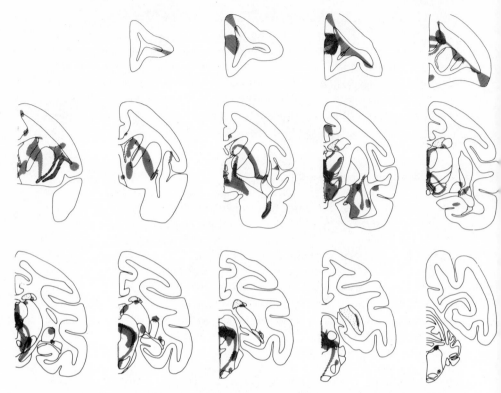

FIGURE 2.5.   *Frontal half sections of the squirrel monkey brain. Hatching from the upper right to the lower left corner indicates projections of the dorsal cingular vocalization area; hatching from the upper left to the lower right corner indicates vocalization-eliciting areas. Cross-hatching shows the overlap of both systems.*

If one compares the projections from the dorsal cingulate gyrus with those of other vocalization loci within the anterior limbic cortex, namely, from the ventral cingulate gyrus, subcallosal gyrus, and gyrus rectus, a number of additional vocalization areas are found that receive direct limbic cortex afferents. These are the head of the caudate nucleus around the ventromedial capsula interna, nucleus accumbens, septum, stria terminalis, basal amygdaloid nucleus, fasciculus uncinatus, and large parts of the hypothalamus. Nevertheless, the general picture given above remains the same. All of the limbic-cortex projections taken together do not cover the entire vocalization-producing substrate, and by no means all projection areas produce vocalization when electrically stimulated. The areas of overlap described for the dorsal cingulate projection system, however, receive direct connections from *all* of the vocalization loci within the anterior limbic cortex. (Only in the midline thalamus is there some variation in the exact position of the terminal fields of the different projection systems.)

The fact that not all vocalization-eliciting areas are mutually connected with each other suggests that these areas do not form a unitary system. The question then arises whether there is *any* vocalization area receiving direct projections from all of the other vocalization areas. At the moment this question cannot be answered definitively because the projections of all of the vocalization areas have not yet been thoroughly investigated. Nevertheless, it can be stated that in the squirrel monkey all vocalization areas studied to date do have a direct connection with the periaqueductal gray–parabrachial vocalization area at the dorsal midbrain–pons transition (Jürgens and Pratt, 1979a). The following vocalization areas have been studied in this respect: anterior limbic cortex, ventromedial rostral caudatum, nucleus accumbens, septum, diagonal band of Broca, substantia innominata, central amygdaloid nucleus, basal amygdaloid nucleus, nucleus striae terminalis, preoptic region, dorsomedial and lateral hypothalamus, midline thalamus, periventricular gray, and dorsolateral and ventrolateral midbrain tegmentum.

The act of vocalization involves movements of the vocal folds, of the respiratory musculature, and of the oral region. This means that the vocalization-eliciting areas must in some way be connected with the motoneurons controlling these movements. The motoneurons for vocal fold movements lie in the nucleus ambiguus, those for respiratory control in the anterior horn of the cervical, thoracic, and lumbar spinal cord, and those for articulatory movements in the trigeminal motor nucleus, facial nucleus, rostral ambiguus, and hypoglossal nucleus. Because the involvement of articulation in monkey calls is quite variable, and respiratory motoneurons are widely scattered, only the connections to the nucleus ambiguus will be discussed here.

In the literature, direct connections to the nucleus ambiguus have been reported from the nucleus solitarius (Morest, 1967; Cottle and Calaresu, 1975), nucleus spinalis nervi trigemini (Stewart and King, 1963), lateral medullary reticular formation (Holstege *et al.*, 1977), midbrain tegmentum immediately ventrolateral to the periaqueductal gray (Bebin, 1956), dorsal hypothalamus (Saper *et al.*, 1976), and cortical larynx area (Kuypers, 1958a,b). The last connection should be received with some reservation, however, as it definitely does not exist in the squirrel monkey (Jürgens, 1976b) or cat Szentágothai and Rajkovits, 1958). Kuypers considered it equivocal in the rhesus monkey and sparse in the chimpanzee (terminals were only found in the most rostral part of the nucleus ambiguus, which innervates *extra* laryngeal muscles), and in man the lesion examined in his anatomical study reached deep into the basal ganglia and was thus by no means a pure lesion of the cortical larynx area. Unpublished results of ours showed that in the squirrel monkey the most rostral vocalization area projecting directly to the nucleus ambiguus is the dorsal hypothalamus, thus confirming the results of Saper *et al.* (1976). The connecting fibers follow the course of the periventricular fiber system down to the caudal periaqueductal gray, where they sweep lateralward into the parabrachial nuclei of the dorsolateral pons. From here they descend in the immediate vicinity of the lateral lemniscus to the dorsomedial border of the superior olive, where they sweep again caudalward, passing the dorsomedial border of the facial nucleus and finally reach the nucleus ambiguus. This fiber bundle, however, neither originates nor ends exclusively in the dorsal hypothalamus and nucleus ambiguus, respectively. All along its course new fibers join the bundle, and others terminate before reaching, or after having passed, the nucleus ambiguus. For example, direct projections to the nucleus ambiguus also come from the periaqueductal gray, parabrachial nuclei, and lateral pontine and medullary reticular formation. Vocalization can be elicited along the whole course of this fiber system. The type of call, however, is not always the same. Furthermore, in the caudal part of its course, the calls often sound artificial, some of them even showing a frequency modulation in the rhythm of the stimulus frequency.

It has been mentioned here that in the squirrel monkey there is no direct connection from the cortical larynx area to the nucleus ambiguus. Therefore, there must be at least one relay within the pathway. Of the structures mentioned above that have been reported to have direct connections with this nucleus, two receive direct afferents from the cortical larynx area. One is the parabrachial area in the dorsolateral pons. This projection, however, is extremely weak. The other is the nucleus solitarius, the sensory relay nucleus for (among others) laryngeal afferents. Whether either of these regions is involved in the transmission of the cortically induced vocal fold activation is still unclear.

In man, Furstenberg (1937) has tried to trace the pathway from the cortical larynx area to nucleus ambiguus by analyzing brain lesions that produce glottal paralysis. According to him, the cortical larynx efferents enter the internal capsule and follow this fiber system down into the pons. Here, some fibers leave the main bundle and continue caudalward in a position between pyramidal tract ventrally and medial lemniscus dorsally until they reach the nucleus ambiguus. Other fibers remain within the corticospinal fiber system and follow its course caudalward. At the level of the nucleus ambiguus they also leave the pyramidal tract to course dorsolateralward until reaching their destination.

In the cat, Kirikae *et al.* (1962) have recorded the electromyographic (EMG) activity of the vocal muscle (musculus thyroarytenoideus) while stimulating the cortical larynx area and its presumed output. According to their study, the cortical larynx efferents run somewhat ventromedial to the internal capsule in the forebrain. (How this position is reached remains unclear from their description.) The fibers then enter the cerebral peduncle and follow it down to the upper pons level. Here, they leave the pyramidal tract dorsally and continue through the pontine and medullary reticular formation to the nucleus ambiguus.

The mean latencies of the stimulus-induced EMG activity are 8.6 msec from the subcortical forebrain and midbrain, 4.3 msec from the lower pons and medulla, and 1.1 msec from nervus recurrens. There is, therefore, at least one synapse within the cortex–ambiguus pathway located at pons level. Unfortunately, Kirikae *et al.* do not give the latency for the cortex itself, so that it remains unclear whether there is a first synapse in the striatum. The finding that in the cat the cortical larynx efferents do not run within the pyramidal tract is corroborated by Rudomín (1965). Rudomín obtained a short-latency evoked potential (7–9 msec) from nervus recurrens when he stimulated 2–5 mm dorsal to the medullary pyramidal tract. Cutting of the latter was without effect on the evoked potential.

Lesion Studies

Lesion Effects on Electrically Evoked Vocalizations

The only study in primates dealing with the effects of lesions on electrically elicited vocalizations stems from Kaada (1951). Kaada found that large lesions within the motor cortex, including the cortical larynx area, do not extinguish the vocalizations elicitable from the anterior cingulate gyrus and rostral hippocampus in the rhesus monkey. Two conclusions can be drawn from these results. First, the cortical larynx area does not form a relay station in the vocalization pathways from the cingulate and hippocampal areas to the lower brainstem. Second, the cortical larynx area does not play an essential role in the motor coordination of the evoked calls.

Jürgens and Pratt (1979b) attempted to determine the cingular vocalization pathway in the squirrel monkey. In this study the elicitability of vocalization from the precallosal cingulate gyrus was tested after placing lesions all along the known projections of this region. The results (Figure 2.6) show that the fibers responsible for vocalization, on leaving the cingulate gyrus, traverse the prefrontal white matter and enter the capsula interna dorsolaterally at about the level of the genu of the corpus callosum. Within the capsula interna, they run in a mediocaudal direction, so that they reach its ventromedial edge at about the level of the anterior com-

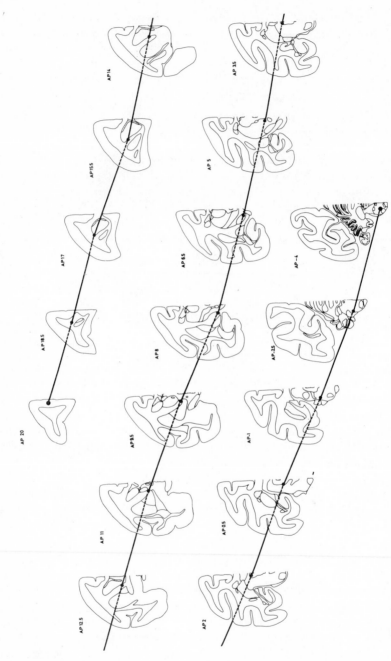

FIGURE 2.6. Course of the fibers responsible for call production evoked from the precallosal cingulate gyrus in the squirrel monkey. (The sterotaxic frontal planes are indicated for each brain diagram in the upper left corner.)

missure. The effective fibers further follow the internal capsule down to the caudal diencephalon, where they suddenly sweep dorsalward and cross the substantia nigra and diencephalic/mesencephalic tegmentum to reach the periventricular fiber system at about the level of the posterior commissure. They continue within the periaqueductal gray to the caudal midbrain, where they sweep lateralward into the parabrachial area and further follow the course of the periaqueductal gray–nucleus ambiguus connection described above.

In the same study, it was also decided to test the effects of periaqueductal lesions on a number of other vocalization areas in addition to the cingular one. The following areas have been studied: amygdala, stria terminalis, preoptic region, substantia innominata, inferior thalamic peduncle, midline thalamus, dorsal hypothalamus, and periventricular gray. After destruction of the periaqueductal gray, all of these vocalization areas became mute.

In nonprimate mammals, several studies on the effects of lesions on stimulation-evoked vocalizations exist in the cat. De Molina and Hunsperger (1962) showed that hypothalamic lesions abolish amygdala-evoked vocalization, but are without effect on vocalizations evoked from the periaqueductal gray. Lesions within the periaqueductal gray abolish amygdala- as well as hypothalamus-evoked vocalization. Lesions within the amygdala are without effect on hypothalamic and periaqueductal gray vocalizations. Amygdala, hypothalamus, and periaqueductal gray, therefore, represent a hierarchical system in which the amygdala-initiated vocalization depends on the intactness of the hypothalamus and periaqueductal gray and elicitability of vocalizations by hypothalamic stimulation depends on the intactness of the periaqueductal gray. As there are direct connections between all three structures, their close functional relationship becomes neuroanatomically understandable. A point of controversy, however, is the question of which of the two pathways interconnecting amydala and hypothalamus is responsible for the elicitation of the amygdalar vocalization. De Molina and Hunsperger (1962) claim that it is the stria terminalis, Hilton and Zbrozyna (1963) that it is the ventral amygdalofugal fiber system. The first authors destroyed the nucleus striae terminalis and obtained an extinction of the amygdala-evoked vocalization, but no effect on the hypothalamically induced vocalization. Hilton and Zbrozyna placed three electrodes in succession into the stria terminalis, coagulated the middle electrode, and tested the other two for elicitability of vocalization. They found that the electrode nearer to the amygdala still elicited vocalization, whereas the electrode farther from the amygdala became mute after the lesion. Lesions within the ventral amygdalofugal fiber system abolished amygdalar vocalization; interruption of the stria terminalis had no effect.

Another study made in the cat should be cited here as a helpful supplement to the monkey data. Kanai and Wang (1962) attempted to trace the vocalization pathway in the lower brainstem of the cat. In their extensive exploration work they found that the sites yielding vocalization when electrically stimulated extend from the periaqueductal gray into the lateral tegmentum below the inferior colliculus. From here they descend into ventrolateral and then ventromedial pons, where they occupy a position between pyramidal tract ventrally and medial lemniscus dorsally. In this same relative position the effective loci can be traced caudalward into the posterior medulla. A small lesion within this area, that is, immediately ventral to the caudal medullar lemniscus medialis, extinguishes all vocalizations elicitable along this pathway as well as those elicitable from the hypothalamus. Kanai and Wang interpret this vocalization system as the descending pathway for vocalization. Such an interpretation seems to be doubtful, however, as most motoneurons involved in vocalization lie rostral to the caudal medulla. The only vocalization component controlled by motoneurons caudal to the medulla is respiration. It is more probable, therefore, that the effective lesion in the medulla interferes with the respiratory component of vocalization instead of with vocalization as an integral reaction. In which way the vocalizations elicitable from this area are to be interpreted remains an open question. An important observation in this context, however, is the finding that transection of the brainstem between midbrain and pons does not abolish these vocalizations. Kanai and Wang did not study the effects of changes of stimulus frequency on the evoked call (a soft mewing cry), which would have been of great interest in the light of a study by McGlone *et al.* (1966). The latter authors elicited the same call type from the lateral midbrain tegmentum ventral to the inferior colliculus after transection through the rostral midbrain. Systematic variation of the stimulus frequency revealed a clear correlation between stimulus frequency and the fundamental frequency of the elicited call (increases of up to three times the lowest fundamental frequency were obtained).

Finally, it may be added that in the squirrel monkey a lesion of the medullary medial lemniscus–pyramidal tract area does not interrupt more rostrally evoked vocalizations. The effective area in this animal lies dorsolateral to the medial lemniscus in the region of the nucleus ambiguus.

Lesion Effects on Conditioned Vocalizations

Sutton *et al.* (1974) trained rhesus monkeys (*Macaca mulatta*) and stump-tailed macaques (*Macaca speciosa*) to emit calls of a specified intensity and duration to obtain a food reward. Then, lesions of different cortical areas were made to study their role in voluntary call production. The conditioning procedure required the animal to vocalize only in the presence of a white signal light. After a correct vocalization, the animal was rewarded with applesauce and the signal light was switched off for 2.8 or 4 sec. Vocalization during this time-out period prolonged the time-out for an additional period of 2.8 or 4 sec, respectively. A session was completed when 50 rewards were obtained. After bilateral removal of the cortical larynx area and rostrally adjacent Broca's area, the conditioned vocalizations were still normal. Bilateral ablation of the temporo–parieto–preoccipital cortex, corresponding to Wernicke's sensory speech center, also did not impair the conditioned vocalizations. In contrast to these neocortical lesions, bilateral ablation of the anterior limbic cortex, including anterior cingulate and subcallosal gyrus, severely interfered with the conditioned vocal task. Both animals tested were unable to fulfill the preoperative criterion. They uttered only a very few calls, which, in addition, were much shorter and weaker than those uttered preoperatively. One of these animals had been trained preoperatively to press a lever to switch on the white signal after the 4-sec time-out period. This conditioned lever-pressing, in contrast to the conditioned vocalization, was not impaired. These results indicate an essential and specific role played by the anterior limbic cortex in the voluntary control of phonation.

Lesion Effects on Spontaneous Vocalizations

Bilateral ablations of the cortical larynx area have been reported by Green and Walker (1938), Myers (1976), and Sutton *et al.* (1974) to have no effect on spontaneous vocalization in the rhesus monkey. Green and Walker, however, also found that ablation of the whole facial motor cortex, including the cortical larynx area, produces an impairment in phonation together with a paralysis of the lower facial muscles and distal half of the tongue. The impairment is characterized by an absence of the typical noisy chatter and variation of intonation; the remaining calls are feeble chirps or sound low-pitched and husky. In contrast to "pure" lesions, therefore, lesions extending much beyond the cortical larynx area do have an effect on vocalization in the rhesus monkey (Figure 2.7).

In man, it is well known that lesions within the inferior frontal cortex posterior to Broca's area, and thus corresponding to the cortical larynx and face representation, can cause speech disturbances (Conrad, 1948, 1954; Bay, 1957; Brain, 1961; Konorski *et al.*, 1961; Hécaen and Angelergues, 1964; Luria, 1964; Benson, 1967; Lecours and Lhermitte, 1976). These disturbances are characterized mainly by articulatory difficulties (dysarthria). They can be seen after left- and right-sided lesions (Kohlmeyer, 1969) and are usually accompanied by a more general orofacial apraxia.

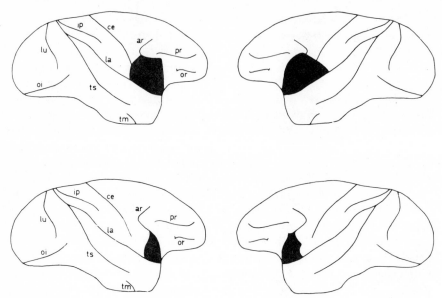

FIGURE 2.7.   *Lower diagrams: lesions of the cortical larynx area in the rhesus monkey that are without effect on vocalization. Upper diagrams: lesions of the cortical larynx and face area in the rhesus monkey affecting vocalization. [Adapted from Green and Walker, 1938.] ar=sulcus arcuatus; ce=sulcus centralis; ip=sulcus intraparietalis; la=sulcus lateralis; lu=sulcus lunatus; oi=sulcus occipitalis inferior; or=sulcus orbitalis; pr=sulcus principalis; sc=sulcus subcentralis anterior; tm=sulcus temporalis medius; and ts=sulcus temporalis superior.*

The anterior limbic cortex, despite its importance in voluntary phonation control, seems to play only a minor role in spontaneous vocalization. Sutton *et al.* (1974) report that their limbic monkeys produced a normal vocal reaction to fear-inducing stimuli. Furthermore, the overall vocal activity of these animals showed only a slight reduction. Myers (1976) also found a slight reduction of vocal utterances after ablation of the entire limbic cortex of rhesus monkeys. Showers and Crosby (1958), in contrast, report an increase in vocal activity after ablation of the anterior cingulate gyrus in the same species. In the squirrel monkey, MacLean (1978) has studied the effects of different lesions on the so-called mirror display. This display behavior represents a species-specific dominance gesture, consisting of lateral thigh spreading, penile erection, and vocalization, performed against a mirror. It was found that bilateral ablation of the anterior cingulate cortex did not affect the display vocalization.

In man, the question of which area functionally corresponds to the cingular vocalization area in monkeys is still open. Stimulation of the anterior limbic cortex in man neither yields vocalization nor has any effect on speech production (Talairach *et al.*, 1973). Electrically elicitable vocalizations and speech disturbances can be obtained, however, from the supplementary speech area. This area lies somewhat dorsocaudal to the anterior limbic cortex, that is, just above the cingulate sulcus and about 4 cm in front of the central sulcus (Penfield and Welch, 1951). Although, in the squirrel monkey, vocalizations can also be elicited from above the cingulate sulcus, the effective sites lie at about the same level as the genu of the corpus callosum and thus are localized clearly rostral to the supplementary speech area in man. The problem of comparability seems to resolve itself, however, by the fact that anterior cingular as well as supplementary motor lesions affect speech in a practically identical manner. In both cases, the readiness to speak is reduced (Barris and Schuman, 1953; Botez and Barbeau, 1971; Buge *et al.*, 1975; Rubens, 1975). This reduction manifests itself in extreme cases as akinetic mutism, in less severe cases as a lack of spontaneous speech with retained ability to read aloud and answer questions with short replies, and in still milder cases as a more or less manifest inertia to speak. The more severe cases are also characterized by a lack of intonation.

Apart from the areas known from stimulation studies to play some role in monkey call production, there are others whose influence can be detected only by lesion studies. Larson *et al.* (1978) report that lesions within different parts of the cerebellum in the rhesus monkey cause slight changes in the quality of "coo" calls. The changes are characterized either by a somewhat reduced intensity and duration of the call or by an alteration of the relation between intensity and fundamental frequency in calls of different intensity. (Usually an increase in intensity is paralleled by a specific increase in fundamental frequency.) The authors interpret these changes on the basis of their cricothyroid EMG recordings as a disorganization in motor coordination of respiratory and glottal component of phonation. For a comparison of these results with the well-known dysarthric disturbances after cerebellar lesions in human patients, the reader is referred to Chapter 6.

A finding that is possibly related to the cerebellar effects of Larson *et al.* is reported by Newman *et al.* (1977). These authors observed a quite severe deterioration of the "isolation peep" in squirrel monkeys after lesions within the diencephalic tegmentum. The isolation peep is a call uttered by animals who are separated from other group members and who are trying to regain contact. It has a fundamental frequency of about 10 kHz and a duration of several hundred milliseconds. After tegmental lesions this call is very breathy, or its frequency course becomes irregular, or noisy components appear superimposed on the normally harmonic frequency structure. One of the main fiber bundles traversing the diencephalic tegmentum is the brachium conjunctivum, a pathway connecting the cerebellum with the ventrolateral thalamus. As Newman *et al.* do not give exact data on the site of the effective lesion, and other fiber bundles also traverse the diencephalic tegmentum, it is not possible to give a conclusive interpretation of their findings.

In the rhesus monkey one additional area has been reported to affect vocalization: the cortex of the frontal pole (including the rostral orbital cortex). According to Franzen and Myers (1973), ablation of this area diminishes vocal activity drastically. Inspection of the data, however, reveals that, of the three vocalization types listed, only two are regularly reduced; the third can even be enhanced after ablation. Furthermore, the spontaneous vocalization rate in two of the three animals studied is very low (four calls per hour before operation versus two calls per hour after operation). Therefore, the importance of the frontal pole in call production should not be overestimated.

There are hardly any studies on the effects of lesions below the diencephalon in primates. One exception is a short notice by MacLean (1978) that states that lesions within the rostral periaqueductal gray do not affect the mirror display and concomitant vocalization in the squirrel monkey. Some observations from other mammals will be cited to supplement the monkey data.

If the brainstem is cut at the level of the rostral midbrain in cats or dogs, no effects on spontaneous vocalization are observed (Onodi, 1902; Bazett and Penfield, 1922; Keller, 1932; Bard and Macht, 1958). Transection between superior and inferior colliculi dorsally and the exit of nervus oculomotorius ventrally abolishes mewing in the cat, but leaves hissing, growling, and purring intact. Transection caudal to the inferior colliculi abolishes all spontaneous vocalizations in cats and dogs. In addition to these transection studies, there are others investigating the effects of more restricted lesions. Skultety (1958, 1965, 1968) has carried out the most exhaustive study of this kind in cats. By varying the site of the lesions systematically, and using a large number of animals to test reproducibility, he explored the whole midbrain. The only area from which he obtained mutism was the periaqueductal gray. Only a very small percentage of animals, however, showed total, permanent mutism (2 cats out of 46). In these animals the lesion encroached slightly on the laterally adjacent midbrain tegmentum. Some of the animals did not vocalize spontaneously but did to a painful stimulus. The majority showed only transient mutism. Among these animals there were subjects with lesions that included all of the tissue destroyed in the totally mute animals. The muting effect of periaqueductal lesions is confirmed by other authors for the cat (Kelly *et al.*, 1946; Adametz and O'Leary, 1959; Randall, 1964) and rat (Chaurand *et al.*, 1972) and by Skultety himself (1962) for the dog. Sprague *et al.* (1961) and Randall (1964), in addition, hold that lateral midbrain lesions also may have a muting effect. The only animal showing total mutism in the Sprague *et al.* study, however, had a lesion that severely invaded the caudolateral periaqueductal gray. Randall's lateral midbrain lesions spared the periaqueductal gray, but destroyed large parts of the ventrolateral pons in addition to the lateral midbrain.

Recording Studies

There are two studies dealing with electrical brain activity during monkey vocalization. One is by Sutton *et al.* (1978), who recorded the single-unit activity in rhesus monkeys during conditioned vocalization. These authors found a number of cells in the anterior cingulate gyrus that always changed their spontaneous activity 200–800 msec *before* vocalization. No cell within the cingulate gyrus showed a strict correlation with the vocalization itself. The other study stems from Szirtes *et al.* (1977). These authors recorded the electroencephalographic activity from different cortical areas during vocalization in the rhesus monkey. A negative potential was seen in correlation with vocalization from the entire precentral gyrus, but not from the parietal and occipital cortex. Simultaneously with this potential, however, an increase in the electromyographic activity of the eye and neck muscles occurred.

Conclusions

The neural control of vocalization in nonhuman primates seems to be organized in a hierarchical manner. The lowest level is represented by the cranial nerve nuclei involved in phonation, the respiratory motoneurons, and the interneurons connecting these areas with each other. This level, therefore, consists of a fairly diffuse system reaching from the pons down into the lumbar spinal cord. It includes the trigeminal motor nucleus for opening the mouth, the facial, rostral ambiguus, and hypoglossal nuclei for articulatory movements of the lips, soft palate, and tongue, the caudal nucleus ambiguus for vocal fold movements, the nucleus solitarius and the nucleus retroambigualis/respiratory anterior horn cells for afferent and efferent respiratory control, and the lateral pontine and medullary reticular formation for integration of the activity of all of these structures. No essential role seems to be played by the afferents of the larynx ending within the nucleus solitarius (Testerman, 1970; Remmers and Gautier, 1972; Tanabe *et al.*, 1975). From the transection studies mentioned under Lesion Effects on Electrically Evoked Vocalizations (p. 29) and under Lesion Effects on Spontaneous Vocalizations (p. 33), it may be concluded that the integration of the different components involved in phonation takes place at this lowest level. It is still unclear, however, if the specific neural activity patterns characterizing different call types are also generated here. Furthermore, it follows that this integration system is not able by itself to produce the complete vocalization; this requires the facilitatory influence of the next integration level, or that of electrical stimulation.

The next hierarchical level within the vocalization system is the caudal periaqueductal gray and laterally adjacent tegmentum between inferior colliculus and brachium conjunctivum. Regarding its possible role in call production, the following conclusions may be drawn.

1. From the observation that the shortest latencies for electrically evoked vocalizations are found here, and from the fact that these vocalizations last as long as the stimulation continues, it follows that the periaqueductal–parabrachial area exerts a very direct influence on call production mechanisms. This is further corroborated by the neuroanatomical fact of a direct connection between this area and the nucleus ambiguus.

2. The finding that transection of the brainstem immediately anterior to this area in cats leaves vocalization intact indicates that this area can function independently of higher levels of the vocalization system.

3. Because of the nonartificial character of the evoked calls and their elicitability in species with different types of vocal apparatuses, it must be assumed that this area does not play an essential role in motor coordination.

4. The fact that practically all vocalization-eliciting areas, most of which simultaneously represent motivation-eliciting areas, project directly to the periaqueductal–parabrachial area points to a relationship of this area with motivation.

5. This relationship, however, seems to be of a more general and indirect nature, because this area yields very different call types from the same electrode position; furthermore, the correlation between vocalization and simultaneously elicited motivation, while present in the rostral periaqueductal gray, is lost in its caudal part.

Taking these considerations together, it may be concluded that the caudal periaqueductal gray and laterally adjacent tegmentum probably serve to couple motivations to the adequate vocal expressions. The formation of a pain call, for instance, necessitates (*a*) that the painful stimulus generates a specific neural activity pattern that subjectively is perceived as pain and objectively represents a behavioral tendency to escape the stimulus and (b) that this characteristic neural state be transformed into a specific vocal pattern. The function of the periaqueductal–parabrachial area in this process probably is to determine (according to the momentary motivational state) and trigger the different vocal patterns whose motor integration then takes place within the pons and medulla.

The third integration level of the vocalization system seems to be the anterior limbic cortex in monkey and supplementary speech area in man. In lower mammals, the corresponding area has not yet been identified. The self-stimulation studies in the squirrel monkey (see under Motivational Concomitants of Electrically Elicited Vocalization, p. 22) have shown that the vocalizations elicitable from this area cannot be interpreted as secondary reactions due to stimulus-induced motivational changes. The relationship of this area to call production, therefore, seems to be quite direct. On the other hand, the negligible effect of limbic cortex lesions on the structure of calls and the fact that the neural activity in this brain structure precedes vocalization suggest that this area is not involved in motor coordination of vocalization. The severe effects of anterior limbic cortex lesions on the initiation of conditioned vocalization, finally, point to an important role in voluntary call production. While the periaqueductal–lateral tegmental area seems to control call production in situations with a rigid stimulus–response characteristic, the cingular vocalization area controls vocal activity in a more general sense, that is, independently of specific motivational states. The cingular vocalization area thus represents a higher order integration level for vocalization that by its direct connection with the periaqueductal–parabrachial area can exert a facilitatory or inhibitory influence on the latter.

An area of a somewhat equivocal significance in the monkey is the cortical larynx area. If it is ablated alone, no vocal disturbances can be observed; if it is ablated together with the surrounding face area, clear changes in vocalization occur. Ablation of the total cortical face and larynx area in lower mammals seems to be without effect on vocalization. Lesions of the larynx area alone can cause severe dysarthric disturbances in man. From these observations, it can be concluded that the cortical larynx area gains increasing importance from lower mammals to man. Furthermore, the fact that, in the rhesus monkey, this area when electrically stimulated yields isolated movements of different laryngeal muscles with very short latencies points to an active role of this area in motor coordination. The type of lesion effect occurring in man gives a further indication about its function. This function seems to be the voluntary control of call formation. Such an interpretation would explain why the cortical larynx area is dispensable in animals with a completely innate vocal repertoire and becomes more and more important with increasing vocal learning capacity. The cat, for instance, which does not have an isolated cortical larynx representation, also does not have voluntary control of call formation. Its innately preprogrammed vocalizations are always the same; they can be initiated voluntarily (Molliver, 1963), but not modified in form. In man, the severe lesion deficits after infarction of the cortical larynx area are paralleled by an enormous vocal learning capacity in normal individuals. Emotional nonverbal vocal utterances as well as highly stereotyped and automatized verbal utterances, like curses and salutation formulas, in which voluntary control of call formation plays only a minor role, often survive destruction of this area. The rhesus monkey, in this respect, takes an intermediate position between cat and man. Behavioral studies have shown that macaques undergo vocal development during ontogeny (Takeda, 1965, 1966; Chevalier-Skolnikoff, 1974). The formation of new calls, however, seems to be mainly due to maturational factors and only in minor degree to learning, that is, voluntary factors (Newman and Symmes, 1974).

The projections of the voluntary call formation area (pericentral larynx and face cortex) and voluntary call initiation area (anterior limbic cortex and possibly supplementary motor cortex) are very different (Jürgens, 1976b). While the latter projects mainly to other limbic vocalization areas, the former is connected predominantly with motor and sensory areas related to the oro–pharyngo–laryngeal region, namely, nucleus ventralis posterior medialis thalami, nucleus solitarius, medial pontine relay nuclei (and thus indirectly with the cerebellum), and ventromedial part of nucleus ventralis lateralis thalami. Nevertheless, there are three levels at which direct interactions between the two systems are possible: (1) within the cortex (the cortical larynx area projects to the cortex around the anterior cingulate sulcus); (2) in the ventromedial part of nucleus ventralis anterior thalami; and (3) in the parabrachial nuclei within the dorsolateral pons. Both systems finally converge after at least one further synapse in the nucleus ambiguus.

References

Adametz, J:, and O'Leary, J. L. (1959). *Neurology, Minneap. 9,* 636–642.
Adams, D. B. (1968). *Arch. Ital. Biol. 106,* 243–269.
Åkerman, B. (1965). *Behaviour 26,* 323–350.
Andrew, R. J. (1973). *Brain Behav. Evol. 7,* 424–446.
Apfelbach, R. (1972). *Z. Tierpsychol. 30,* 420–430.
Bard, P. and Macht, M. B. (1958). *In* "Neurological Basis of Behaviour" (G. E. W. Wolstenholme and C. M. O'Connor, eds.) pp. 55–71. Churchill, London.

Barris, R. W. and Schuman, H. R. (1953). *Neurology, Minneap. 3*, 44–52.

Bay, E. (1957). *Dtsch. Z. Nervenheilk. 176*, 553–594.

Bazett, H. C. and Penfield, W. G. (1922). *Brain 45*, 185–265.

Bebin, J. (1956). *J. comp. Neurol. 105*, 287–332.

Benson, D. F. (1967). *Cortex 3*, 373–394.

Botez, M. J. and Barbeau, A. (1971). *Int. J. Neurol. 8*, 300–320.

Brain, R. (1961). *Brain 84*, 145–166.

Brickner, R. M. (1940). *J. Neurophysiol. 3*, 128–130.

Brown, J. L. (1972). *Behaviour 39*, 91–127.

Brown, J. L., Hunsperger, R. W., and Rosvold, H. E. (1969). *Exp. Brain Res. 8*, 113–129.

Brown, T. G. (1915). *J. Physiol., Lond. 49*, 195–207.

Buge, A., Escourolle, R., Rancurel, G., and Poisson, M. (1975). *Rev. Neurol., Paris 131*, 121–137.

Cannon, R. E. and Salzen, E. A. (1971). *Anim. Behav. 19*, 375–385.

Chapman, W. P., Schroeder, H. R., Geyer, G., Brazier, M. A. B., Fager, D., Poppen, J. L., Solomon, H. C., and Yakovlev, P. J. (1954). *Science 120*, 949–950.

Chaurand, J. P., Vergnes, N., and Karli, P. (1972). *Physiol. Behav. 9*, 475–481.

Chevalier-Skolnikoff, S. (1974). *Contrib. Primatol. 2*, 1–174.

Clark, G., Chow, K. L., Gillaspy, C. C., and Klotz, D. A. (1949). *J. Neurophysiol. 12*, 459–463.

Conrad, K. (1948). *Arch. Psychiat. Z. Neurol. 180*, 54–104.

Conrad, K. (1954). *Brain 77*, 491–509.

Cottle, M. K. W. and Calaresu, F. R. (1975). *J. comp. Neurol. 161*, 143–158.

Delgado, J. M. R. (1955). *J. Neurophysiol. 18*, 261–275.

Delgado, J. M. R. (1967). *J. nerv. ment. Dis. 144*, 383–390.

Delius, J. D. (1971). *Exp. Brain Res. 12*, 64–80.

De Molina, A. F. and Hunsperger, R. W. (1962). *J. Physiol., Lond. 160*, 200–213.

Demski, L. S. and Gerald, J. W. (1974). *Brain Behav. Evol. 9*, 41–59.

Dusser de Barenne, J. G., Garol, H. W., and McCulloch, W. S. (1941). *J. Neurophysiol. 4*, 287–303.

Epro, R. A. (1977) *Behaviour 60*, 75–97.

Ertel, S. and Dorst, R. (1965). *Z. exp. angew. Psychol. 12*, 557–569.

Feindel, W. (1961). *In* "Electrical Stimulation of the Brain" (D. E. Sheer, ed.), pp. 519–532. Univ. of Texas Press, Austin, Texas.

Franzen, E. A. and Myers, R. E. (1973). *Neuropsychol. 11*, 141–157.

Fritsch, G. and Hitzig, E. (1870). *Arch. Anat. Physiol. wiss. Med. 37*, 300–332.

Furstenberg, A. C. (1937). *Ann. Otol. Rhinol. Laryng., St. Louis 46*, 39–54.

Gardner, R. A. and Gardner, B. T. (1969). *Science 165*, 664–672.

Gibbs, E. L. and Gibbs, F. A. (1936). *J. comp. Neurol. 64*, 209–211.

Green, H. D. and Walker, A. E. (1938). *J. Neurophysiol. 1*, 262–280.

Hassler, R. (1961). *Dtsch. Z. Nervenheilk. 183*, 148–171.

Hassler, R. (1964). *In* "Progress in Brain Res." Vol. 5 (W. Bargmann and J. P. Schade, eds.), pp. 1–32. Elsevier, Amsterdam.

Hassler, R., Riechert, T., Mundinger, F., Umbach, W., and Ganglberger, J. A. (1960). *Brain 83*, 337–350.

Hast, M. H., Fisher, J. M., Wetzel, A. B., and Thompson, V. E. (1974). *Brain Res. 73*, 229–240.

Hast, M. H., and Milojevic, B. (1966). *Acta oto-laryng., Stockh. 61*, 196–204.

Heath, R. G. and Mickle, W. A. (1960). *In* "Electrical Studies on the Unanesthetized Brain" (E. R. Ramey and D. S. O'Doherty, eds.), pp. 214–242. P. B. Harper (Hoeber), New York.

Hécaen, H. and Angelergues, R. (1964). *In* "Disorders of Language" (A. V. S. de Reuck and M. O'Connor, eds.) pp. 223–256. Churchill, London.

Heimer, L. and Larsson, L. (1966). *Brain Res. 3*, 258–263.

Hess, W. R. and Brügger, M. (1943). *Helv. physiol. pharmacol. Acta 1*, 33–52.

Hilton, S. M. and Zbrozyna, A. W. (1963). *J. Physiol., Lond. 165*, 160–173.

Hines, M. (1940). *J. Neurophysiol. 3*, 442–466.

Holstege, G., Kuypers, H. G. J. M., and Dekker, J. J. (1977). *Brain 100*, 265–286.

Hughes, J. R. and Mazurowski, J. A. (1962). *Electroenceph. clin. Neurophysiol. 14*, 477–485.

Hunsperger, R. W. (1956). *Helv. physiol. pharmacol. Acta 14*, 70–92.

Hunsperger, R. W. and Bucher, V. M. (1967). *Progr. Brain Res. 27*, 103–127.

Ingram, W. R., Ranson, S. W., Hannett, F. I., Zeiss, F. R., and Terwilliger, E. H. (1932). *Arch. Neurol. Psychiat. 28*, 513–541.

Jürgens, U. (1974). *Brain Res. 81*, 564–566.

Jürgens, U. (1976a). *Exp. Brain Res. 26*, 203–214.

Jürgens, U. (1976b). *Exp. Brain Res. 25*, 401–411.

Jürgens, U., Maurus, M., Ploog, D., and Winter, P. (1967). *Exp. Brain Res. 4*, 114–117.

Jürgens, U. and Müller-Preuss, P. (1977). *Exp. Brain Res. 29*, 75–83.

Jürgens, U. and Ploog, D. (1970). *Exp. Brain Res. 10*, 532–554.

Jürgens, U. and Ploog, D. (1976). *Arch. Psychiat. Nervenkr. 222*, 117–137.

Jürgens, U. and Pratt, R. (1979a). *Brain Res. 167*. 367–378.

Jürgens, U. and Pratt, R. (1979b). *Exp. Brain Res. 34*. 499–510.

Kaada, B. R. (1951). *Acta physiol. scand. Suppl. 83*, 1–285.

Kabat, H. (1936). *J. comp. Neurol. 64*, 187–208.

Kaiser, L. (1962). *Synthese 14*, 300–319.

Kanai, T. and Wang, S. C. (1962). *Exp. Neurol. 6*, 426–434.

Katzenstein, J. (1905). *Arch. f. Physiol. 29*, 396–400.

Katzenstein, J. (1908). *Arch. f. Laryng. 20*, 501–524.

Keller, A. D. (1932). *Amer. J. Physiol. 100*, 576–586.

Kelley, D. B., Morrell, J. I., and Pfaff, D. W. (1975). *J. comp. Neurol. 164*, 47–78.

Kellogg, W. N. (1968). *Science 162*, 423–427.

Kelly, A. H., Beaton, L. E., and Magoun, H. W. (1946). *J. Neurophysiol. 9*, 181–183.

Kennedy, M. C. (1975). *Brain Res. 91*, 321–325.

King, H. E. (1961). *In* "Electrical Stimulation of the Brain" (D. E. Sheer, ed.), pp. 477–486. Univ. of Texas Press, Austin, Texas.

Kirikae, J., Hirose, H., Kawamura, S., Sawashima, M., and Kobayashi, T. (1962). *Ann. Otol. Rhinol. Laryng., St. Louis 71*, 222–241.

Kohlmeyer, K. (1969). *Medizin. Klinik 64*, 2079–2086.

Konorski, J., Koźniewska, H., Stepień, L., and Subczyński, J. (1961). *In* "Pathophysiological Mechanism of Disorders of Higher Nervous Activity after Brain Lesions in Man" (J. Konorski, H. Koźniewska, L. Stepień, and Subczyński, eds.), Polish Acad. of Sci. (Warsaw).

Kramer, E. (1964). *J. abnorm. soc. Psychol. 68*, 390–396.

Krause, H. (1884). *Arch. f. Physiol. 8*, 203–210.

Kremer, W. F. (1947). *J. Neurophysiol. 10*, 371–379.

Kuypers, H. G. J. M. (1958a). *Brain 81*, 364–388.

Kuypers, H. G. J. M. (1958b). *J. comp. Neurol. 110*, 221–255.

Larson, C. R., Sutton, D., and Lindeman, R. C. (1978). *Exp. Brain Res. 33*, 1–18.

Lecours, A. R. and Lhermitte, F. (1976). *Brain Lang. 3*, 88–113.

Leyton, A. S. F. and Sherrington, C. S. (1917). *Quart. J. Exper. Physiol. 11*, 135–222.

Lipp, H. P. and Hunsperger, R. W. (1978). *Brain Behav. Evol. 15*, 260–293.

Luria, A. R. (1964). *In* "Disorders of Language" (A. V. S. de Reuck and M. O'Connor, eds.), pp. 143-167. Churchill, London.

Luschei, E. D. and Goodwin, G. M. (1975). *J. Neurophysiol. 38*, 146-157.

MacLean, P. D. (1978). *Brain Res. 149*, 175-196.

MacLean, P. D. and Delgado, J. M. R. (1953). *Electroenceph. clin. Neurophysiol. 5*, 91-100.

Magoun, H. W., Atlas, D., Ingersoll, E. H., and Ranson, S. W. (1937). *J. Neurol. Psychopath. 17*, 241-255.

Martin, J. R. (1976). *J. comp. physiol. Psychol. 50*, 1011-1034.

McGlone, R. E., Richmond, W. H., and Bosma, J. F. (1966). *Folia phoniat. 18*, 109-116.

Meyer, A. E. and Hess, W. R. (1957). *Helv. physiol. pharmacol. Acta 15*, 401-407.

Miller, N. E. (1965). *Science 148*, 328-338.

Milojevic, B. and Hast, M. H. (1964). *Ann. Otol. Rhinol. Laryng., St. Louis 73*, 979-988.

Molliver, M. E. (1963). *J. exp. Anal. Behav. 6*, 197-202.

Morest, D. K. (1967). *J. comp. Neurol. 130*, 277-300.

Müller-Preuss, P. and Jürgens, U. (1976). *Brain Res. 103*, 29-43.

Munk, H. (1884). *Arch₁ f. Physiol. 8*, 470-480.

Myers, R. E. (1976). *In* "Origins and Evolution of Language and Speech" (S. R. Harnad, H. D. Steklis, and J. Lancaster eds.), pp. 745-757. Ann. N.Y. Acad. Sci.

Nashold, B. S., Wilson, W. P., and Slaughter, D. G. (1969). *J. Neurosurg. 30*, 14-24.

Newman, J. D. and Symmes (1974). *Developm. Psychobiol. 7*, 351-358.

Newman, J. D., MacLean, P. D., and Gelhard, R. E. (1977). *Proceed. 7th Ann. Meeting Neurosciences Society.*

Onodi, A. (1902). "Die Anatomie und Physiologie der Kehlkopfnerven." Coblentz, Berlin.

Peek, F. W. and Phillips, R. E. (1971). *Brain Behav. Evol. 4*, 417-438.

Penfield, W. and Bordley, E. (1937). *Brain 60*, 389-443.

Penfield, W. and Rasmussen, T. (1952). "The Cerebral Cortex of Man." Macmillan, New York.

Penfield, W. and Roberts, L. (1959). "Speech and Brain Mechanisms." Princeton Univ. Press, Princeton, N.J.

Penfield, W. and Welch, K. (1951). *Arch. Neurol. Psychiat., Chicago 66*, 289-317.

Phillips, R. E., Youngren, O. M., and Peek, F. W. (1972). *Anim. Behav. 20*, 689-705.

Potash, L. M. (1970). *Behaviour 36*, 149-167.

Putkonen, P. T. S. (1966). *Experientia 22*, 405-407.

Putkonen, P. T. S. (1967). *Ann. Acad. Sci. fenn., A.V. 130.*

Randall, W. L. (1964). *Behavior 23*, 107-139.

Remmers, J. E. and Gautier, H. (1972). *Resp. Physiol. 16*, 351-361.

Roberts, W. W. and Kiess, H. O. (1964). *J. comp. physiol. Psychol. 58*, 187-193.

Roberts, W. W., Steinberg, M. L., and Means, L. W. (1967). *J. comp. physiol. Psychol. 64*, 1-15.

Robinson, B. W. (1967). *Physiol. Behav. 2*, 345-354.

Romaniuk, A. (1965). *Acta Biol. Exper. 25*, 177-186.

Rubens, A. B. (1975). *Cortex 11*, 239-250.

Rudomín, P. (1965). (1965). *Acta physiol. lat.-amer. 15*, 180-190.

Sano, K., Sekino, M., Hashinoto, J., Amano, K., and Sugiyama, H. (1975). *Confin. neurol. 37*, 285-290.

Saper, C. B., Loewy, A. D., Swanson, L. W., and Cowan, W. M. (1976). *Brain Res. 117*, 305-312.

Schaltenbrand, G. (1965). *Brain 88*, 835-840.

Schaltenbrand, G. (1975). *Brain Lang. 2*, 70-77.

Schmidt, R. S. (1966). *Behaviour 26*, 251-285.

Schmidt, R. S. (1968). *Behaviour 30*, 239-257.

Schmidt, R. S. (1971). *Behaviour 39*, 288-317.

Schmidt, R. S. (1974). *J. comp. Physiol. 92*, 229-254.

Schmidt, R. S. (1976). *J. comp. Physiol. 108*, 99-113.

Schvarcz, J. R., Driollet, R., Rios, E., and Bethi, O. (1972). *J. Neurol. Neurosurg. Psychiat. 35*, 356-359.

Sem-Jacobsen, C. and Torkildsen, A. (1960). In "Electrical Studies on the Unanesthetized Brain." (E. R. Ramey and D. S. O'Doherty, eds.), Harper (Hoeber), New York.

Showers, M. J. C. and Crosby, E. C. (1958). *Neurology, Minneap. 8*, 561-565.

Skultety, F. M. (1958). *J. comp. Neurol. 110*, 337-365.

Skultety, F. M. (1962). *Arch. Neurol., Chicago 6*, 235-241.

Skultety, F. M. (1965). *Arch. Neurol., Chicago 12*, 211-225.

Skultety, F. M. (1968). *Arch. Neurol., Chicago 19*, 1-14.

Smith, W. K. (1944). *Fed. Proc. 3*, 43.

Smith, W. K. (1945). *J. Neurophysiol. 8*, 241-255.

Spiegel, E. A., Kletzkin, M., and Szekeley, E. G. (1954). *J. Neuropath. exp. Neurol. 13*, 212-220.

Spiegel, E. A. and Wycis, H. T. (1961). In "Electrical Stimulation of the Brain." (D. E. Sheer, ed.), pp. 37-44. Univ. of Texas Press, Austin, Texas.

Sprague, J. M., Chambers, W. W., and Stellar, E. (1961). *Science 133*, 165-173.

Stevens, J. R., Mark, V. H., Erwin, F., Pacheco, P., and Suematsu, K. (1969). *Arch. Neurol., Chicago 21*, 157-169.

Stewart, W. A. and King, R. B. (1963). *J. comp. Neurol. 121*, 271-286.

Suga, N., Schlegel, P., Shimozawa, T., and Simmons, J. (1973). *J. Acoust. Soc. Am. 54*, 793-797.

Sugar, O., Chusid, J. G., and French, J. D. (1948). *J. Neuropath. exp. Neurol. 7*, 182-189.

Sutton, D., Larson, C., and Lindeman, R. C. (1974). *Brain Res. 71*, 61-75.

Sutton, D., Samson, H. H., and Larson, C. R. (1978). "Brain mechanisms in learned phonation of Macaca mulatta." Proc. 6th Intern. Congr. Primatol. Soc. in Cambridge, England 1976, Adademic Press, London.

Sutton, D., Larson, C., Taylor, E. M., and Lindeman, R. C. (1973). *Brain Res. 52*, 225-231.

Szentâgothai, J. and Rajkovits, K. (1958). *Arch. Psychiat. Nervenkr. 197*, 335-354.

Szirtes, J., Marton, M., and Urbán, J. (1977). *Electroenceph. clin. Neurophysiol. 42*, 852.

Takeda, R. (1965). *Primates 6*, 337-380.

Takeda, R. (1966). *Primates 7*, 73-116.

Talairach, J., Bancaud, J., Geier, S., Bordas-Ferrer, M., Bonis, A., Szikla, G., and Rusu, M. (1973). *Electroenceph. clin. Neurophysiol. 34*, 45-52.

Tanabe, M., Kitajima, K., and Gould, W. J. (1975). *Ann. Otol. Rhinol. Laryng., St. Louis 84*, 206-212.

Testerman, R. L. (1970). *Exp. Neurol. 29*, 281-297.

von Holst, E. and v. Saint Paul, U. (1960). *Naturwissenschaften 47*, 409-422.

Waldbillig, R. J. (1975). *J. comp. physiol. Psychol. 89*, 200-212.

Walker, A. E. and Green, H. D. (1938). *J. Neurophysiol. 1*, 152-165.

Winter, P., Handley, P., Ploog, D., and Schott, D. (1974). *Behaviour 47*, 230-239.

Woodworth, C. H. (1971). *Physiol. Behav. 6*, 345-353.

Yajima, I., Hada, J., and Yoshii, N. (1976). *Med. Osaka Univers. 27*, 25-32.

3

Mechanisms Underlying Vocal Control in Nonhuman Primates[1]

DWIGHT SUTTON

Introduction

The evolution of speech has intrigued scientists for virtually as long as the concept of evolution itself. As a process, speech (or communicative phonation) does not evolve (Klopfer, 1976). Rather, the evolution of structures on which the process rests must provide the basis for identifying how vocal functions developed. A major problem in assessing the generality of mechanism in vocal communication involves determining the limits of activity of structures (both central and peripheral) that have been derived from common origins. This problem is not one that can be resolved immediately.

In this chapter some of the anatomical–physiological characteristics of the phonatory system of human and other primates will be considered, with an effort toward further understanding of the mechanisms involved in phonation. The chapter examines suggestions that contemporary primates offer sources of information relating to the origins of speech, and it considers evolutionary ties that may relate members of the primate order with respect to the development of communicative phonation in progressively more complex patterns.

[1] Preparation of this chapter was supported by NIH Grants NS 11870 and NS 12165, and by the William G. Reed Fund.

45

NEUROBIOLOGY OF
SOCIAL COMMUNICATION IN PRIMATES

The following discussion leads through a survey of data indicating that most primates have similar phonatory structures at the periphery, as well as sharing a number of similarities in central nervous system organization. The chapter further considers evidence exploring communalities in vocal signaling behavior evident in various primate species, comparing nervous system support of this activity among them.

The Peripheral Vocal Apparatus

Phylogeny

Evolutionary modifications of the vocal tract and upper airway are presumed to reflect important influences upon the emergence of speech in hominids. These changes in the vocal tract are seen by some authors as comprising the most significant determinant of speech capability (Keleman, 1948; Spuhler, 1959; Zenker and Anzenbacher, 1962; Hill, 1972). From this perspective, the major reason for failure of nonhuman primates to achieve human speech lies in the limits of peripheral vocal apparatus.

DuBrul (1958) traced the evolution of the speech apparatus through the primate order, emphasizing the gradual change in osteology of the skull, with concomitant modification of soft tissues surrounding and controlling the upper airway. His work does not deal with specific phyletic stages that might be related to different levels of complexity in phonatory function.

In Wind's view (1970), there is a clear recognition of the difficulty in specifying steps in evolution of the organizational framework of the laryngeal elements most significant for the elaboration of vocal function. Wind points out that a study comparing current species with fossil specimens provides only a general indication of that species' phylogenetic status. The peripheral vocal system of many of the higher primates is well developed. Wind proposes that there is sufficient structural complexity in many species to provide the support necessary for speech activity, leaving open the problem of how this could be validated. Until more detailed comparative information is available, such a position is difficult to dispute.

Negus (1962) emphasized the comparative studies demonstrating that virtually all nonhuman primates have continuity between larynx and oral cavity, with little or no intervening pharyngeal component. The human, in contrast, exhibits separation of oral from laryngeal chamber by the intervening mobile pharynx. This anatomical divergence is considered to be highly important in limiting the extent of phonational activity in the nonhumans.

The functional limits implied by upper airway geometry have guided attempts to specify the range of speech communication in earlier hominids.

Of course, we cannot study directly either the upper airway anatomy or the physiology of fossil forms, but a cross-species examination of the peripheral vocal apparatus in extant primates may provide useful information relevant to questions addressing the evolution of phonatory mechanisms. Contemporary species with structural resemblance to fossil hominid osteology might be used to estimate the functional limitations imposed on the extinct forms. Such an examination could perhaps identify characteristics revealing the uniqueness of the human system. This is the basis of the approach employed by Lieberman and associates (1972). They reasoned that, if models of the upper airway of extinct hominids could be constructed, vocal characteristics could be inferred.

Measures from endocasts of the upper airway of the rhesus monkey were treated by a computer algorithm to determine the theoretical range of movement of that animal's vocal tract (Lieberman *et al.*, 1972). This yielded an estimate of the vocal sounds available to the animal. The results indicated that the monkey has limited capacity for production of human speech sounds (e.g., vowels) due principally to the restriction in size of the pharyngeal area. It was presumed that most other primates would exhibit similar limitations. By extension of this view, hominids (including extinct forms) whose vocal tract structure could be inferred from known osteology could then be indirectly analyzed for range of vocal sounds.

Serious difficulties attend efforts to use reconstruction of soft-tissue relations in extinct hominids for the purpose of inferring vocal characteristics. Attempts to deduce the phonatory activity of extinct species assume not only that we have valid analyses of the functional range of the phonation in contemporary animals used as models, but also that the upper airway geometry of fossil forms has been correctly established from the reconstruction of the cranial fragments. For example, DuBrul (1977) has called attention to defects in the reconstruction of portions of the basicranium of La Chapelle aux Saints (an early hominid specimen). This questionable restoration of the basal bone geometry makes inferences regarding the configuration of oropharyngeal soft tissue quite uncertain. In turn, the error interjects confusion in deducing the correlations in vocal capabilities that might have characterized the specimen.

Jordan's (1971) studies of chimpanzee vocal sounds weigh against a strong distinction between human and nonhuman vocal tract capabilities. He used an analysis technique that produced a sound spectrum of the animal's vocalizations. Jordan observed that the chimpanzee makes several vowel-like sounds, together with vowel–consonant combinations. These sounds were identified as phonetically similar to the sounds of human speech. If confirmed, Jordan's observations would suggest that anthropoids (and perhaps other primates) have evolved the capacity to produce a diverse array of vocal sounds not predictable from anatomical models.

On the basis of the preceding considerations, we might conclude that there are a number of pitfalls in seeking details of phonation characteristics of extinct hominids through study of contemporary primates. With care, a general picture can be formed of the progression in structural modifications leading to the current form of hominid peripheral phonatory apparatus. There is as yet insufficient scope in analysis of various primate species to generate reasonable inferences about the essential steps occurring in the process of hominid phonatory evolution.

Comparative Allometry of Larynx and Upper Airway

The morphology of the peripheral vocal apparatus of various species can be compared through allometric analyses. Comparative measures may show the extent of common relationships in the dimensions of the primate larynx, and they may reveal directions of differentiation among species.

A beginning along this line has been made, with studies available on human infants and on newborn monkeys. Klock and Beckwith (1970) obtained multiple measures of dimensions of the upper airway of human infants in the early postnatal period. We followed their protocol closely in determining similar measures from the neonatal monkey (*Macaca mulatta, M. nemestrina*) (Taylor *et al.*, 1976). Table 3.1 presents a comparison of data from the two studies.

Ten separate measures are included from the larynx of the macaque, together with corresponding measures from the human infant. The ratio of these values is presented to indicate the extent of consistency in scale for

TABLE 3.1 *Allometric Analyses of the Morphology of the Upper Airway of Human Infants and Neonatal Monkeys*

	MACAQUE, 350–800 G (TAYLOR ET AL., 1976)	HUMAN, 60–70 CM FROM CROWN TO HEEL (KLOCK & BECKWITH, 1970)	H/M RATIO
Supraglottic width, maximum	4.0 ± .9	8.2 ± 1.3	2.05
Supraglottic depth	4.6 ± .4	5.9 ± .9	1.28
Subglottic depth	5.4 ± .6	7.0 ± .8	1.30
Subglottic width, maximum	2.8 ± .4	4.6 ± 1.7	1.64
Subglottic width, minimum	.7 ± .2	2.6 ± .8	3.71
Epiglottic height	5.0 ± .7	15.8 ± 1.0	3.16
Vestibular fold depth	2.1 ± .4	6.8 ± 1.0	3.24
Ventricle depth	2.6 ± .4	4.5 ± .5	1.73
Trachea width	3.6 ± .3	5.3 ± .9	1.47
Larynx height, superior	10.4 ± 1.3	17.5 ± 1.0	1.68

each measure. It can be seen that there is a large range in ratios of the measures. Supraglottic depth is only slightly greater in the human infant, whereas subglottic width is greater by a factor of nearly 4. Clearly, the infant macaque is not simply a scaled-down version of the human with respect to measures of the larynx.

There are no data to indicate whether further specific dimensional changes might occur selectively during the maturation period. A more general analysis including other species and ages would be interesting from the standpoint of determining divergence from a possibly more homogeneous "module" in early fetal stages.

Comparative Anatomy: Microscopic

Human speech involves many modulations and rapid transitions indicating that the laryngeal musculature is capable of relatively swift and precise dynamic adjustments. Comparative studies of histological features of the larynx are limited. The available data do not identify crucial structural elements as uniquely suited for speech, thus segregating humans from other primates. This may be revised as additional information is obtained.

MUSCLE SPINDLE MORPHOLOGY

The capacity for fine motor control has been attributed in part to the presence of muscle spindles. These spindles provide the basis for detection of static or dynamic changes in muscle (Granit, 1970). Spindles have been found in postural muscles of many species (Cooper, 1960). Cooper and Daniel (1963) point out that there is not a unique spindle structure in the human, aside from a somewhat greater length and slightly greater numbers of intrafusal fibers in the spindle. Neither of these characteristics has been associated with improved sensitivity to dynamic change or to muscular control in physiologically normal situations.

Several studies have identified spindles in intrinsic laryngeal muscles of the human (Paulsen, 1958; Bowden et al., 1960; Keene, 1961; Rossi and Cortesina, 1965; Grim, 1967; Baken and Noback, 1971). Grim (1967) reported that the human posterior cricoarytenoid (PCA) contained spindles with both nuclear bag and nuclear chain fibers. Their configuration closely resembles that found in postural muscles.

The density of spindles per gram weight of muscle is considerably higher in laryngeal than in postural muscles of the human. Spindles in the latissmus dorsi (1.4 spindles/g) and trapezius (2.2 spindles/g) (Cooper, 1960) are sparse in comparison to PCA, which has 4.8 to 11.2 spindles/g (Grim, 1967). However, human lumbricals may have an even higher density (12.2 to 19.7 spindles/g), as do muscles of mastication. For the latter, Cooper (1960) noted a density in masseter of 11.2 spindles/g; in medial pterygoid the value reached 20.3.

High variability in spindle density from muscle samples obtained from different individuals might eliminate apparently large species differences. We examined laryngeal muscles of baboon and macaque, finding spindles in each species (Larson *et al.*, 1974). The density of muscle spindles varied from 0 in some specimens to fairly high values. There were 40.0 spindles/g in the PCA of one *M. mulatta*; in another there were 6.6. With generally limited numbers of samples in all species, it is difficult to infer whether there is an orderly progression of spindle density in the different intrinsic muscles of the larynx and whether such a pattern may characterize the different species.

Our study also found both bag and chain fibers in the spindles of these animals. There appear to be fewer fibers per spindle in baboon and macaque than were reported for the human (Cooper and Daniel, 1963), but, again, variability may render this observation insignificant. At present it is unclear if such measures could reflect important differences in control over phonatory processes.

INNERVATION RATIO

Another facet of the concept of fine motor control concerns the ratio of extrafusal muscle fibers to motor nerve fibers (innervation ratio or motor unit size). Postural muscles have rather large innervation ratios, with values in the hindlimb ranging from almost 2000 (cat medial gastrocnemius, McPhedran *et al.*, 1965) to 127 (rat tibialis anterior, Edstrom and Kugelberg, 1968). Other muscles that serve special senses or highly differentiated motor functions have smaller values. Thus, innervation ratios of middle ear muscles range from 4 to 27 in the human, whereas extraocular muscles have ratios of 2 to 10 (Blevins, 1968).

English and Blevins (1969) compared innervation ratios of human with feline intrinsic laryngeal muscles. They found that the innervation ratio of the human cricothyroid muscle ranged from 23 to 30, compared to a range of 41 to 50 for this muscle in the cat. The lateral cricothyroid of the human has innervation ratios of 48 to 67, as compared to values of 68 to 90 in the cat. These data might suggest that primates have smaller innervation ratios than do nonprimates. There is not adequate information to reveal whether differential values may exist among different primate species in association with complexity of phonation. We recently began a study of *M. mulatta* laryngeal muscle innervation ratio. The value for PCA muscle is approximately 19, a value that is consistent with the low ratios found among humans. An extension of this investigation among other primate forms will be useful in offering further insights into comparative phonation mechanisms, but we should not construe small innervation ratio as a sufficient criterion of differential, complex function in muscles. For example, Krnjevic and Miledi (1958) reported that rat diaphragm has innervation ratios of from 10 to 20. Fine motor control does not appear to be a characteristic of the diaphragm. Whether fine control of laryngeal musculature is directly related to the size of the motor unit has not been established.

MUSCLE FIBER DIMENSION

The size of individual extrafusal muscle fibers in laryngeal muscles varies. We have found that, for the human, the vocalis muscle fiber diameter averages $29 \pm 8\mu$m. In M. *nemestrina*, vocalis muscle fibers are of a similar size ($29 \pm 6\mu$m). The average PCA muscle fiber diameter of the human is smaller than the vocalis ($16 \pm 4\mu$m), whereas the human cricothyroid has a somewhat larger fiber diameter ($35 \pm 10 \ \mu$m). Histochemical studies of these muscles may reveal additional functional properties.

OTHER HISTOLOGICAL FEATURES

Ontogeny of the laryngeal epithelium in the macaque exhibits a pattern that is somewhat dissimilar to that of the human. Both species appear to develop thickened cuboidal epithelium in the glottal regions during the early postnatal period, but the human does not exhibit a prominent transformation to ciliated columnar cells at the posterior commissure (Sutton *et al.*, 1977). In the adult monkey this region is composed of ciliated pseudo-stratified columnar cells. Perhaps this reflects specifically the more frequent use of the voice by the human, as well as a less complete closure of the posterior glottal chink in the nonhuman primate. In either case, in the monkey there would be less frequent mechanical contact of the opposing regions of the vocal cord.

Laryngeal Physiology

MUSCLE TWITCH PROPERTIES

The intrinsic laryngeal muscles are called upon to make rapid adjustments during the production of complex vocal signals. Speed of response by these muscles is a factor that might help define the functional limitations of phonation. Experimental data from several species suggest potential differences in laryngeal function, based on muscle physiology.

Twitch time ranges from slow to moderately fast for the various intrinsic laryngeal muscles. Hast (1967) reported that the cricothyroid of the cat is a relatively slow muscle (twitch time, 53 msec). He found a less sluggish response in the dog cricothyroid (39 msec). However, Martensson and Skoglund (1964) reported no difference between cat and dog cricothyroid twitch time. This discrepancy emphasizes the requirement for careful control of experimental variables. Specious differences may obscure species similarity.

Hast (1969) observed that twitch time of cricothyroid muscle of the Hynan gibbon was 39 msec, which is close to the value he found for *M. mulatta* (36.4 msec). In contrast, he reported that the squirrel monkey had a considerably more rapid cricothyroid twitch time (18.8 msec).

It appears that the twitch time of thyroarytenoid muscle is more rapid than cricothyroid. In the cat, Hast (1969) found a twitch time of 22 msec, whereas the dog and three primate species all exhibited twitch times of 13 to 14 msec.

The observations indicate that primates may have generally speedier twitch properties of laryngeal muscles than might be found in other mammals. Obviously it would be helpful to have data regarding this action from a wide range of species, including all of the anthropoids. This information could be used for correlations with microscopic structure on the one hand and with phonatory characteristics on the other.

In summary, the morphological and physiological measures of the larynx and balance of the upper airway indicate both intra- and interspecies variability. It is difficult to relate one or more of these measures uniquely to vocal functions that pertain to speech. We need further studies establishing the parameters of function of the elements of the upper airway during phonation and speech if we wish to rely on peripheral factors for help in classifying species into orderly relations with respect to the evolution of speech. Even if such efforts yield positive results, further questions remain concerning the role of the central nervous system (CNS) in vocal functions. Let us consider next the CNS factors that have been proposed to be significant in speech evolution.

Central Nervous System

Phylogeny

Correlations between cortical microstructure and phylogenetic position have been sought by many investigators. Ramón y Cajal (1911) proposed that cortical phylogenesis includes progressive elaboration of the basilar

dendritic tree of pyramidal cells. His views (reflected by Lorente de No, 1949) also suggested that phylogenesis was expressed in an increase in the number of cells with short axons. There are relatively few applications of this concept in studies of cytoarchitectonics of cortex of primates. A number of other possibilities could also distinguish subtle evolutionary steps. Some of these have been pointed out by Holloway (1968): (*a*) decrease in neuron density; (*b*) increased average size of neurons; (*c*) increased dendritic branching; (*d*) possible increase in glia/neuron ratio. Our increasing technological capabilities will likely add yet other considerations that may be applicable to exploration of neuronal structures that evolved to provide important mechanisms for phonation (Brazier and Petsche, 1978).

It is well known that restricted regions of the human cerebral cortex are important to normal speech functions. The comparative cytoarchitectural studies have devoted insufficient effort toward establishing or characterizing the distinguishing features of homologous regions, particularly within the primate order. A useful step in this direction was made by Walker (1940), who examined the cytoarchitecture of the prefrontal cortex of the macaque. He characterized the inferior frontal convolution as homologous to Area 45 of the human, but failed to mention a homolog of the human Area 44 (Broca's area) in the monkey. Walker noted large pyramidal cells in Layers 3 and 5 of the inferior frontal convolution of the monkey, an observation also applicable to the human. This observation is also consistent with material derived from the squirrel monkey. Rosabal (1967) identified Area 44 in this species, observing large pyramidal cells in Layers 3 and 5 along with small granule cells in Layers 4 and 6.

Sanides (1970, 1975) studied the cytoarchitecture of squirrel monkey brain to develop some general concepts regarding the neurology of phonation. The squirrel monkey exhibits cytoarchitectural developments that Sanides indicates are closely identifiable with human cortical structure. He suggested that there are common cytoarchitectonic characteristics of integration cortex of many taxa, proposing that the elaboration of *primary* sensory and motor cortex occurred relatively recently in evolution. From this standpoint it might be inferred that his suggestion of the early phyletic emergence of association cortex (i.e., isocortex) implies that the capacity for complex behavioral functions is widespread. In such a framework, it is possible that behaviors including communicative vocal patterns might be supported by the brain of relatively primitive specimens.

Sanides (1975) also points out in particular the elaboration of integration cortex of the temporal lobe of primates. He indicates that there is not sufficient development of the integrative cortex in nonhuman primates to support *language* capacity, although this seems a difficult judgment to sustain on purely morphological evidence. Indeed, Rosabal (1967) concluded from his anatomical study of the frontal lobe cytoarchitecture of the squirrel monkey that there is "striking similarity in the cytological characteristics of each cortical area throughout the primate order."

Other cortical regions homologous to speech-related areas of the human are even less well characterized in primates. However, Beheim-Schwarzbach (1975) reported that the dorsal surface of the first temporal convolution of chimpanzee and orangutan could not be partitioned into the several subdivisions present in this region of a human, thereby establishing at least one possible distinctive feature segregating the latter species from others. We have no independent criteria to establish the minimum structural complexity needed for support of speech functions. The process might well depend on the connectivity among various brain regions. Furthermore, speech may rely on mechanisms that are somewhat different from those employed in phonation.

Clearly, there is a need for closer attention to the subtle details of CNS structure, based on systematic comparative studies of the type represented by Passingham (1973).

Let us now explore more specifically the behavioral aspects of phonation in nonhuman primates, seeking for evidence that may reveal details of the neural mechanisms governing vocal function.

A diverse literature encompassing both field and laboratory studies covers many aspects of vocal behavior of the nonhuman primate. The following sections survey the principal factors that help determine the phonatory behavior. A major objective is to establish whether there is flexibility or plasticity in such behaviors sufficient to indicate a possible link with human phonation.

Vocal Behavior in Nonhuman Primates

Repertoire Size

The bulk of information regarding the types and variability of calls of primates has been derived from field studies. Conclusions regarding the categories and the principal features of vocal signals depend on the adequacy of sampling and technical limitations. These considerations must be kept in mind when interpreting the data.

Table 3.2 lists a variety of studies that have sought to describe the variety of calls in the primate repertoire. The apparent variation among investigations on a particular species serves to emphasize the difficulty associated with securing an appropriate test framework. Not all studies reflect a spectrum of social situations important in eliciting the widest possible spectrum of calls. Furthermore, few studies have approached the

TABLE 3.2 *Vocal Repertoire Size in Primates*

SPECIES	REFERENCE	NUMBER OF CALLS
Cebuella pigmaea	Pola and Snowden, 1975	10
Cercopithecus		
C. aethiops	Struhsaker, 1967	36
	Andrew, 1963a,b	5–8
C. ascanius	Marler, 1973	5
C. cephus	Andrew, 1963a,b	5–8
C. diana	Andrew, 1963a,b	5–8
C. neglectus	Andrew, 1963a,b	5–8
C. nigroviridis	Andrew, 1963a,b	5–8
C. mitis	Andrew, 1963a,b	5–8
	Marler, 1973	7
C. nictitans	Andrew, 1963a,b	5–8
Macaca		
M. mulatta	Rowell and Hinde, 1962	20–30
	Altmann, 1962	7–17
	Andrew, 1963a,b	5–8
M. nemestrina	Grim, 1967	27
M. fuscata	Green, 1975a	37
	Nishimura, 1973	11
	Itani, 1963	37
Saimiri sciurius	Winter *et al.*, 1966	26
Erythrocebus patas	Hall *et al.*, 1965	11
Aotus trivirgatus	Moynihan, 1964	9–11
	Andrew, 1963a,b	4
Colobus guerza	Marler, 1972	5
Presbytis entellus	Jay, 1962	10
	Vogel, 1973	6
Miopithecus talapoin	Gautier, 1974	31
Alouatta palliata	Carpenter, 1934	20
	Altmann, 1959	20
	Baldwin and Baldwin, 1976	26
Theropithecus gelada	Richman, 1976	11+
Hylobates lar	Carpenter, 1940	9
	Marshall and Marshall, 1976	10
Pan troglodytes	Goodall, 1965	23
	Reynolds and Reynolds, 1965	12
	Andrew, 1962, 1963a,b	6–7
	Marler, 1969	7
	Yerkes and Learned, 1925	32
Papio sp.	Hall and DeVore, 1965	14
	Andrew, 1963a,b	4–5
Gorilla gorilla	Schaller, 1963	22

exhaustive situational correlations of the type presented by Green (1975a).
His careful observations document not only the diversity of calls of M.
fuscata, but also the social contexts in which the calls usually are emitted.

Note in the table that several species have more than 20 different calls.
This confers on them the potential for highly complex vocal signaling.
Species with the greatest variety of vocal calls do not fall into a readily
discernible hierarchy relating to other indexes of relative phylogenetic
status. For example, ceboid and cercopithecine radiations emerged at different points in geological history, yet they do not exhibit obvious differences in size of the vocal repertoire. Both genera contrast with an apparently limited range in vocal output of marmosets.

According to Struhsaker (1970), calls may provide a good index of
phylogenetic affinity. At this point, however, we have no yardstick permitting classification on such a basis.

The studies listed in Table 3.2 offer little to indicate the nature of factors
that enter into the control over vocal signals. An extensive repertoire could
simply indicate that a given species has a large array of preprogrammed
vocal signals. Such a view suggests that specific environmental cues elicit
essentially fixed-action patterns. Variability in acoustic properties of calls
by a given individual, and variation in calls given by different members of
a species within a single context, would provide better evidence that there
is a potential for flexibility in use of the vocal signals. If an animal is
capable of producing vocal calls with several variants in a general class,
there is an opportunity for a broad range of information transfer.

Marler (1975) suggests that species might be distinguished on the basis of
discrete or graded vocal calls. Certain species have a more extensive repertoire of graded calls than do others. Marler proposed that M. *fuscata, Colobus badius*, and *Pan troglodytes* are examples that provide many parallels
to the human vocal functions inasmuch as these animals utilize numerous
graded calls. Some of the calls of chimpanzee, as noted earlier (Jordan,
1971), have vocal elements resembling human vowels and vowel–consonant blends.

Systematic manipulation of environmental factors results in modification
of the phonatory repertoire. This effect further suggests that the nonhuman
primate is not restricted to fixed-action pattern responding in all situations.

Isolation and Deafness

Human vocal communicative patterns are learned slowly during infancy
and childhood. Congenital deafness imposes great handicaps on the
development of normal speech. Without feedback of self-generated vocal
sounds, the individual suffers defective speech patterns.

The ontogeny of vocalization in nonhuman primates under conditions of minimal stimulation or feedback has received limited attention. Winter *et al.* (1973) reported that deafening a squirrel monkey early in the postnatal period produced little impairment in subsequent phonatory development. Normal calls were present at their appropriate time in the maturational period. The report by Winter and associates is relatively sketchy, and evidence for total deafness was not provided. However, the work clearly suggests that self-generated vocalization or stimulation produced by conspecifics may not be prerequisite for normal phonation in this species.

Our own observations (unpublished) of vocalization in deafened *M. nemestrina* indicate an apparent divergence between species. We deafened three newborns within 24 hr. of birth by performing a bilateral total labyrinthectomy. A follow-up period of 3 months revealed abnormalities in vocal output. The variety of calls given and the frequency with which they were emitted were greatly reduced compared to a normal infant *M. nemestrina* reared in the same environment.

Acoustic isolation of individual *M. mulatta* may also result in abnormal vocal behavior. Newman and Symmes (1974) found that early isolation of the infant rhesus resulted in subsequent failure in appearance of certain calls. They also noted that there were unusual dynamic changes during emission of the clear calls.

The evidence suggests that species may vary in the requirement for feedback vis à vis the ontogenesis of a normal vocal repertoire and implies that the acoustic environment can influence the characteristics of vocal signals, at least in some species. Possibly species that are susceptible to the effect may exhibit a correlated capacity for voluntarily modifying vocal output. Perhaps there are also species differences in mechanisms for monitoring acoustic signals.

Imitation of Vocal Signals

The human acquires most of the vocal repertoire through the process of imitation. The extent to which the nonhuman may be capable of a corresponding process is unclear. Early investigations were directed toward evaluating the capability for imitation of human speech by nonhuman primates. These studies have provided no strong evidence for successful speech imitation (Hayes and Hayes, 1952; Kellogg, 1968). Virtually no words could be produced by the chimpanzee, even with prolonged training [and despite the suggestion that vowels and vowel–consonant blends may be indigenous components of the vocal repertoire of this species (Jordan, 1971)].

In a broader perspective, Andrew (1962, 1976) proposed that the ability to mimic the vocal sounds of conspecifics is an essential aspect of human language evolution. Andrew (1962) pointed out that *Lemur catta* will answer loud, short calls with like calls. *Theropithecus gelada* produces a segmented grunt in response to grunts uttered by a human, and a similar pattern is reported for the gibbon (Andrew, 1962).

Green (1975a) found that a playback of intertroop roars elicited the roaring call in *Colobus guerza*, and *M. fuscata* responded to a playback of coos by emitting a similar coo call. Winter *et al.* (1966) noted that the isolation peep of the squirrel monkey was often produced by presenting a recorded specimen of this call to isolated animals. Other calls also elicited a counterpart call resembling that which was presented. In any case, none of these studies demonstrates imitation.

The field observations made by Green (1975b) identify a situation that more clearly suggests imitative phonation in *M. fuscata*. He reported that unique vocal dialects were specific to each of several geographically isolated troops. Such calls were primarily related to the feeding situation. Since the various groups that were the subjects of Green's observations originated from a common stock only a few generations removed, it is plausible that immediate factors (e.g., the reward contingencies associated with feeding) helped to shape the development of the regional dialects.

In all of the preceding observations, the vocal responses are not specifically indicative of imitation. Some of them might be interpreted as antiphonal calls in response to specific signals. Well-controlled laboratory studies using unique combinations of components synthesized from elements in the normal vocal repertoire could help establish whether imitation of vocal sounds is within the capability of nonhuman primates. Such evidence would be significant with respect to identifying precursor conditions for language in nonhumans.

Conditionability of Vocal Signals

Learning by the Intact Animal

Mechanisms involving learning and selective production of vocal calls in new situations, or their modification to form new signals, must be considered as a significant feature in primate evolution. These processes are important to the formation of primitive vocal communication systems with open-ended features, a property thought to be related to the emergence of language (Peters, 1972). The experimental evidence suggests that systematic changes in vocal behavior of nonhuman primates can be achieved by suitable minipulation of reward.

The issue of learning of vocal behavior has been approached in several ways to demonstrate that stimulus control of specific features of vocalization can be obtained. Conditions under which vocal activity occurs can be modified by making food or other reinforcement contingent on this activity.

Myers *et al.* (1965) obtained increases in frequency of vocalization of cebus monkeys by imposing the requirements of a simple ratio schedule of food reinforcement in the presence of discriminative stimuli. Response rates associated with the positive stimulus were considerably greater than during intervals in which a negative stimulus was present. Cue reversal was followed by appropriate changes in vocal responding.

Leander *et al.* (1972) also obtained good schedule control over vocalizations (barks, chirps, or whistles) from cebus monkeys. Under fixed-ratio reinforcement, response rates increased 1.5- to 25-fold over the baseline rate. Vocalization under variable-interval reinforcement conditions was lower than a nonvocal behavior (bar press) studied under this same schedule, but the vocalization response exhibited a typical variable interval pattern under this program of reinforcement.

Wilson (1975) showed that the frequency of specific vocal calls (bark or yip) could be increased by differential reward in the presence of visual cues. The response rate of reinforced calls increased from 2- to 60-fold during training, and the discrimination ratio (S^D response rate/S^D + $S^\triangle$ response rate) was high. Other calls were rarely emitted in the context of the test situation. Cue reversal produced appropriate shifts in vocal behavior.

Randolph and Brooks (1967) employed social reinforcement (play) to modify vocal behavior of a young chimpanzee. They demonstrated that response latency and frequency were directly related to acquisition and extinction trials. Discriminative responding was readily obtained, extinguished, and reacquired.

In the preceding studies, vocalizations from the normal repertoire were brought under stimulus control. Such results indicate that learning extends to include control over vocal activity. A complementary approach entails the question of modifiability of physical parameters of a call. We tested the rhesus monkey's ability to manipulate the duration and amplitude of a clear call (coo), achieving a more prolonged and somewhat louder coo than is normal for the animal. We found that these animals can make modifications in their normal vocalization, producing a revised response in the presence of an arbitrary visual stimulus (Sutton *et al.*, 1973; Larson *et al.*, 1974).

We have also conducted a study requiring *different* phonatory signals as responses to visual cues (Sutton *et al.*, 1978). Rhesus monkeys were trained to emit a coo to a red light and a bark to a green light. The example shown in Figure 3.1 indicates that the rhesus monkey is capable of producing a specified vocal response to a specific visual cue. (Other tasks were included in this test to evaluate whether phonation could be associated exclusively with a given cue.) Each of the responses was usually issued in the context of the appropriate signal, revealing good discrimination.

Errors in vocal responding were relatively uncommon. The short bars distributed above each of the top two lines indicate each coo or bark trial. Note that the animal consistently produced the appropriate call. If an incorrect vocal response occurred, it was not usually repeated within the same trial. We rarely observed vocal responses in the presence of signals for nonvocal behaviors.

A Discrimination Index for coo calls was computed in order to identify the relative amount of responding associated with the discriminative stimulus. The Discrimination Index improved during training, reaching values of .6 for one animal and .9 for another. These levels of performance resemble those obtained from cebus monkeys in a simple discrimination test (Wilson, 1975).

The data from cebus and rhesus monkeys and the chimpanzee reveal that vocal performance is subject to learning. The studies have demonstrated that reward contingencies regulate the rate and type of vocalization. Of course, such performances are far more primitive than human speech, but they offer evidence that the processes regulating phonation in the nonhuman primate may be homologous to the basic processes underlying human phonation.

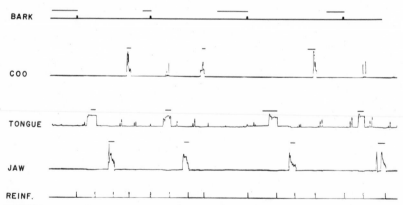

FIGURE 3.1. *Polygraph recording showing four separate behaviors, each cued by a distinctive signal. Cue lights, represented by horizontal bars above the respective lines, elicited reinforced behavior with minimal error.*

Cortical Mechanisms in Spontaneous and Learned Phonation

An important consideration with respect to the possibility of similar neural processes across various species relates to their mediation by homologous structures. Analysis of the role of cortex in vocal activity of monkeys may lay the groundwork for understanding mechanisms of phonation relevant to all primates, including the basic organization for humans. There is little detailed information concerning neural patterns of cortex or other regions; indeed, the identification of relevant regions involved in phonation is probably incomplete (see Chapter 2).

Human speech mechanisms are known to involve several neocortical as well as subcortical regions: Destruction or electrical stimulation of specific areas disrupts speech (Penfield and Roberts, 1959). Interruption or repetition of words is produced by stimulation over a wide area of inferior frontal and parietotemporal cortex of either hemisphere. Vocalization (isolated vowel sounds) is elicited from stimulation of somewhat more restricted sites higher on the frontal convexity and on the medial surface (Brickner, 1940; Penfield and Welch, 1951).

Electrical stimulation of midline periallocortex (anterior cingulate gyrus) has yielded little phonatory responding (Lewin and Whitty, 1960; Talairach *et al.*, 1973). This result may reflect the inaccessibility of the most anterior segment of the anterior cingulate gyrus (Brodmann's Areas 24 and 25) as much as it defines a lack of responsiveness.

Electrical stimulation of neocortex in nonhuman primates has produced variable evidence for elicited vocalization. In the chimpanzee, Leyton and Sherrington (1917) and Dusser de Barenne *et al.* (1941) found that stimulation of inferior frontal or opercular cortex resulted in vocalization. Walker and Green (1938) obtained vocalization from neocortical stimulation in an unidentified species.

Subsequent studies using awake squirrel monkeys (Jürgens *et al.*, 1967, 1970) and rhesus monkeys (Robinson, 1967) reported that neocortical stimulation elicits no vocalization. These studies do not make clear whether the entire area of inferior frontal cortex and all of the parietotemporal cortex were explored. However, they suggest that the neocortex of the monkey is less directly involved in vocalization than is the case for the human. This is not to imply that the vocal apparatus is unresponsive to cortical stimulation in these animals. Direct pathways run from the inferior frontal region to the oral and laryngeal musculature (Vogt and Vogt, 1919). Hast *et al.* (1974) demonstrated that individual laryngeal muscles can be activated by stimulating selected points within the inferior frontal cortex of *M. mulatta*.

Phylogenetically more primitive cortex of the monkey may be associated with several "systems" that are anatomically distinct, each relating to a particular class of vocal calls (see Chapter 2). It has been shown that stimulation of the anterior cingulate cortex or of structures connected to the anterior cingulate cortex in squirrel monkeys results in growling calls (Jürgens and Ploog, 1970); in rhesus monkeys stimulation in cingulate-related structures yields coo calls (Robinson, 1967). Stimulation within another set of structures with close interconnections might produce only cackles or geckers.

Other information concerning cortical mechanisms in primate phonation is derived from the effects of lesions. Green and Walker (1938) noted that removal of cortex from the "lateral convexity" of the rhesus monkey resulted in a weak voice. We also used this species in an investigation of the effects of lesion damage to neocortex (Sutton et al., 1974). Learned phonation was not disrupted by bilateral damage that was confined mainly to inferior frontal or parietotemporal cortex. There were no significant changes in duration or amplitude of calls emitted in the test situation, and the animals appeared to vocalize normally within the colony environment.

Phonatory control mechanisms involve areas outside isocortex, and these may, in part, be distinguishable from speech mechanisms. Clinical studies reveal that patients with traumatic or neoplastic lesions of the cingulate gyrus (particularly the anterior cingulate sites) may develop mutism (Fulton, 1951; Nielsen and Jacobs, 1951; Barris and Schuman, 1953; Guidetti, 1957). Even painful stimuli had little influence in eliciting vocal exclamations in these patients. Other clinical studies, however, have not mentioned mutism in patients undergoing extirpation of the anterior cingulate cortex for treatment of behavioral disorders (LeBeau, 1954) or relief of intractable pain (Wilson and Chang, 1974). Without histological data, correlation of phonation disturbance with the location and extent of damage cannot be established in these patients undergoing cingulectomy.

We tested anterior cingulate control over phonation in rhesus monkeys by creating bilateral lesions in animals that had learned a simple discriminative phonatory task (Sutton et al., 1974). In postoperative testing, previously learned coo calls could not be elicited in response to a visual stimulus. This result is markedly different from the continued high performance following neocortex lesions.

Additional work completed recently indicates that anterior cingulate cortex destruction in M. mulatta has a rather general influence on learned phonation. Animals trained to emit two different calls (i.e., bark or coo) to appropriate visual stimuli were unable to produce either call after bilateral destruction of the area.

Figure 3.2 compares preoperative with postoperative performance of vocal and nonvocal response measures obtained from two rhesus monkeys in which bilateral anterior cingulectomy was carried out. In both cases, the

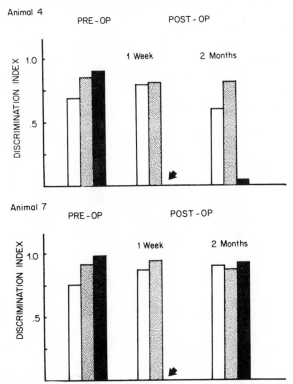

FIGURE 3.2.   *Effects of anterior cingulectomy on performance of a discriminative vocal response (coo, hatched bar) and two oromotor behaviors (jaw bite, open bar; tongue thrust, stippled bar). The preoperative Discrimination Index for all behaviors was high in both cases (Animals 4 and 7). Surgery produced an immediate loss of vocal responding (arrows), leaving the nonvocal behaviors unimpaired. The animals recovered ability to vocalize in approximately 2 months. In one case (Animal 7), the discrimination returned to preoperative levels, but in the other (Animal 4) there was continued poor discrimination.*

Discrimination Index was high preoperatively for the coo (no similar measure obtained for the bark), indicating that the animals rarely made errors in responding (see also Figure 3.1). When errors in vocal responding were made, they tended to be unbiased. That is, coos were substituted for barks at about the same frequency as the reverse. Nonvocal behaviors were similarly well discriminated.

In the 2 or 3 weeks immediately following the lesion, the animals failed to vocalize in test sessions lasting up to an hour in duration. Neither barks nor coos appeared during these early postoperative sessions, although in this interval the animals continued to perform discriminated nonvocal tasks at the preoperative level.

The first indication of return of vocal function was noted in the colony. Here, calls that resembled chirps or hoarse screeches were emitted. At this point there were still no vocalizations obtained in the daily test sessions.

After the lapse of an additional 1 to 3 weeks accompanied by intermittent testing, there were attempts to vocalize during the test. The few calls given were brief and of low intensity. Later tests produced more calls, but a normal dynamic range did not appear during the subsequent 2-month testing period.

A distinct bias in call type emerged on the part of both animals used in the experiment. Normally, each of the calls occurred about equally often such that the bark/coo ratio was approximately unity. After recovery from the totally mute stage, both animals had a preference for production of barks to *both* signals requiring a vocal response. Perseveration of this response was maintained. The preoperative ratio of approximately 1:1 for each animal was altered markedly with ratios of 5:1 (Animal 7) and 20:1 (Animal 4).

These data suggest that the processes governing the production of barks are modified less dramatically by anterior cingulate damage than are the mechanisms governing coos.

We have recorded single-unit discharges from anterior cingulate cortex of rhesus monkeys (Sutton *et al.*, 1978). Discharge patterns undergo changes as much as 900 msec prior to sound production. Frequently, these altered firing rates persist for the duration of the vocal call given by the animal. Evidence of this nature further implicates the anterior cingulate region as a participant in phonation mechanisms.

Summary

This chapter has examined mechanisms relating to phonation in the primate order. Peripheral vocal structures of most primates are similar, although the data are limited concerning fine details. Further work must examine subtle organizational features that may distinguish the human from other primates.

The primate larynx appears to have the potential for action relevant to complex phonation. There is progressive modification of the supraglottic geometry in contemporary anthropoids. Possibly fossil hominids also fit the progression, but it is very difficult to establish a pattern in extinct forms.

The evidence concerning versatility of phonation suggests that many primate species have a variety of graded calls, signifying broad capability in control over the vocal system. Furthermore, several species have been shown to be capable of discriminative use of the vocal repertoire, including the capacity to learn vocal responses to cues presented in the environment.

Phonatory control in nonhuman primates is apparently under limited neocortical influence, but is strongly regulated by periallocortex. The human may have a neocortically organized speech system overlaid on a more primitive periallocortex vocal regulatory system. Whatever the specific mechanisms, essential structural features establishing the specific characteristics of vocal function at each step of primate evolution have not yet been identified.

References

Altmann, S. A. (1959). *J. Mammalogy 40*, 317–330.
Altmann, S. A. (1962). *Ann. N.Y. Acad. Sci. 102*, 338–435.
Andrew, R. J. (1962). *Ann. N.Y. Acad. Sci. 102*, 296–315.
Andrew, R. J. (1963a). *Behaviour 20*, 1–105.
Andrew, R. J. (1963b). *Symp. Zool. Soc. Lond. 10*, 89–101.
Andrew, R. J. (1976). *Ann. N.Y. Acad. Sci. 280*, 678–293.
Baken, R. J. and Noback, C. R. (1971). *J. Speech Hear. Res. 14*, 513–518.
Baldwin, J. D. and Baldwin, J. I. (1976). *Folia Primatol. 26*, 81–108.
Barris, R. W. and Schuman, H. R. (1953). *Neurol. 3*, 44–52.
Beheim-Schwarzbach, D. (1975). *Z. Mikrosk.-Anat. Forsch. Leipzig 89*, 759–776.
Blevins, C. E. (1968). *Arch. Otolaryng. 86*, 249–254.
Bowden, R. E. M., Keene, M. F. L., Gooding, M., Mahran, Z. Y., and Withington, J. L. (1960). *Anat. Rec. 132*, 168.
Brazier, M. A. B. and Petsche, H. (1978). "Architectonics of the Cerebral Cortex." Raven, New York.
Brickner, R. M. (1940). *J. Neurophys. 3*, 128–130.
Carpenter, C. R. (1934). *Comp. Psychol. Monogr. 10*, 1–168.
Carpenter, C. R. (1940). *Comp. Psychol. Monogr. 16*, 1–212.
Cooper, S. (1960). *In* "Structure and Function of Muscle" (G. H. Bourne, ed.), pp. 381–420. Academic Press, New York.
Cooper, S. and Daniel, P. M. (1963). *Brain 86*, 563–586.
DuBrul, E. L. (1958). "Evolution of the Speech Apparatus." Charles C. Thomas, Springfield, Ill.
DuBrul, E. L. (1977). *Brain and Language 4*, 365–381.
Dusser de Barenne, J. G., Garol, H. W. and McCulloch, W. S. (1941). *J. Neurophysiol. 4*, 287–303.
Edstrom, L. and Kugelberg, K. E. (1968). *J. Neurol. Neurosurg. Psychiat. 31*, 424–433.
English, D. T. and Blevins, C. E. (1969). *Arch. Otolaryng. 89*, 778–784.
Fulton, J. F. (1951). "Frontal Lobotomy and Affective Behavior." W. W. Norton, New York.
Gautier, J-P. (1974). *Behaviour 51*, 209–273.
Goodall, J. (1965). *In* "Primate Behavior: Field Studies of Monkeys and Apes" (I. DeVore, ed.), pp. 425–473. Holt, Rinehart and Winston, New York.
Granit, R. (1970). "The Basis of Motor Control." Academic Press, New York.
Green, H. D. and Walker, A. E. (1938). *J. Neurophysiol. 1*, 262–280.
Green, S. (1975a). *In* "Primate Behavior" (L. A. Rosenblum, ed.), Vol. 4, pp. 1–102. Academic Press, New York.
Green, S. (1975b). *Z. Tierpsychol. 38*, 304–314.

Grim, M. (1967). *Folia Morph. 15,* 124–131.

Grimm, R. J. (1967). *J. Zool. Lond. 152,* 361–373.

Guidetti, B. (1957). *Rev. Neurol. 97,* 121–131.

Hall, K. R. L. and DeVore, I. (1965). *In* "Primate Behavior. Field Studies of Monkeys and Apes." (I. DeVore, ed.), pp. 53–110. Holt, Rinehart and Winston, New York.

Hall, K. R. L., Boelkins, R. C. and Goswell, M. J. (1965). *Folia Primatol. 3,* 22–49.

Hast, M. (1967). *Pract. Otorhinolaryngol. 29,* 53–56.

Hast, M. (1969). *Acta Otolaryngol. 67,* 84–92.

Hast, M. H., Fischer, J. M., Wetzel, A. B., and Thompson, V. E. (1974). *Brain Res. 73,* 229–240.

Hayes, K. J. and Hayes, C. (1952) *J. Comp. Physiol. Psychol. 45,* 450–453.

Hill, J. H. (1972). *Amer. Anthropol. 74,* 308–317.

Holloway, R. J., Jr. (1968). *Brain Res. 7,* 121–172.

Itani, J. (1963). *Primates 4,* 11–66.

Jay, P. C. (1962). "The Social Behavior of the Langur Monkey," University of Chicago Doctoral Dissertation.

Jordan, J. (1971). *Folia Morph. 30,* 323–340.

Jürgens, U., Maurus, M., Ploog, D., and Winter, P. (1967). *Exp. Brain Res. 4,* 114–117.

Jürgens, U. and Ploog, D. (1970). *Exp. Brain Res. 10,* 532–554.

Keene, M. F. L. (1961). *J. Anat.* (Lond). *95,* 25–29.

Kellogg, W. N. (1968). *Science. 162,* 423–427.

Keleman, G. (1948). *J. Morph. 82,* 229–256.

Klock, L. E. and Beckwith, J. B. (1970). *In* "Sudden Infant Death Syndrome" (A. B. Bergman, J. B. Beckwith, and C. G. Ray, eds.), pp. 102–104. Univ. Washington Press, Seattle.

Klopfer, P. H. (1976). *In* "Communicative Behavior and Evolution." (E. C. Simmel and M. E. Hahn, eds.), pp. 7–21. Academic Press, New York.

Krnjevic, K. and Miledi, R. (1958). *J. Physiol. 140,* 427–439.

Larson, C. R., Sutton, D. and Lindeman, R. C. (1974). *Folia Primatol. 22,* 315–323.

Leander, J. D., Milan, M. A., Jasper, K. B., and Heaton, K. L. (1972). *J. Exp. Anal. Behav. 17,* 229–235.

LeBeau, J. (1954). *J. Neurosurg. 11,* 268–276.

Lewin, W. and Whitty, C. W. M. (1960). *J. Neurophysiol. 23,* 445–447.

Leyton, A. S. F. and Sherrington, C. S. (1917). *Quarterly J. Exp. Physiol. 11,* 135–222.

Lieberman, P., Crelin, E. S., and Klatt, D. H. (1972). *Amer. Anthropol. 74,* 287–307.

Lorente de No, R. (1949). *In* "Physiology of the Nervous System." (J. F. Fulton, ed.), pp. 288–330. Oxford Univ. Press, London.

Marler, P. (1969). *Recent Adv. Primatol. 1,* 94–100.

Marler, P. (1972). *Behaviour. 42,* 175–197.

Marler, P. (1973). *Z. Tierpsychol. 33,* 223–247.

Marler, P. (1975). *In* "The Role of Speech in Language" (J. E. Kavanagh and J. E. Cutting, eds.), pp. 11–37. MIT Press, Cambridge, Mass.

Marshall, J. E., Jr. and Marshall, E. R. (1976). *Science. 193,* 235–238.

Martensson, A. and Skoglund, C. R. (1964). *Acta Physiol. Scand. 60,* 318–336.

McPhedran, A. M., Wuerker, R. B., and Henneman, E. (1965). *J. Neurophysiol. 28,* 71–84.

Moynihan, M. (1964). *Smithsonian Misc. Publication: 146,* 1–84.

Myers, S. A., Horel, J. A., and Pennypacker, H. S. (1965). *Psychon. Sci. 3,* 389–390.

Negus, V. E. (1962). "The Comparative Anatomy and Physiology of the Larynx." Hafner, New York.

Newman, J. D. and Symmes, D. (1974). *Developmental Psychobiol. 7,* 351–358.

Nielsen, J. M. and Jacobs, L. L. (1951). *Bull. Los Angeles Neurol. Soc. 16,* 231–234.

Nishimura, A. (1973). In *Symp. IVth Int. Congr. Primat.* Vol. 1: pp. 76–87. Karger, Basel.

Passingham, R. E. (1973). *Brain Behav. Evol. 7,* 337–359.

Paulsen, K. (1958). *Zeit. Zellforsch. 48*, 349–355.

Penfield, W. and Welch, K. (1951). *Arch. Neurol. Psychiat. 66*, 289–317.

Penfield, W. and Roberts, L. (1959). "Speech and Brain Mechanisms." Princeton Univ. Press, Princeton.

Peters, C. R. (1972). *Man. 7*, 33–49.

Pola, Y. V. and Snowdon, C. T. (1975). *Anim. Behav. 23*, 826–842.

Ramón y Cajal, S. R. (1911). "Histologie de Systeme Nerveux de l'Homme et des Vertebres." Maloine, Paris.

Randolph, M. C. and Brooks, B. A. (1967). *Folia Primatol. 5*, 70–79.

Reynolds, V. and Reynolds, F. (1965). *In* "Primate Behavior: Field Studies of Monkeys and Apes" (I DeVore, ed.), pp. 368–424. Holt, Rinehart and Winston, New York.

Richman, B. (1976). *J. Acoust. Soc. Am. 60*, 718–724.

Robinson, B. W. (1967). *Physiol. Behav. 2*, 345–354.

Rosabal, F. (1967). *J. Comp. Neurol. 130*, 87–108.

Rossi, G. and Cortesina, G. (1965). *Acta Otolaryng. 59*, 575–592.

Rowell, T. E. and Hinde, R. A. (1962). *J. Zool. 138*, 279–294.

Sanides, F. (1970). *In* "Advances in Primatology" (C. Noback and W. Montagna, eds.), Vol. I, pp. 137–208. Appleton-Century-Crofts, New York.

Sanides, F. (1975). *Brain and Language. 2*, 396–419.

Schaller, G. B. (1963). "The Mountain Gorilla: Ecology and Behavior." Univ. of Chicago Press, Chicago.

Spuhler, J. N. (1959). *In* "The Evolution of Man's Capacity for Culture" (J. N. Spuhler, ed.), pp. 1–13. Wayne State Univ. Press, Detroit.

Starck, D. and Schneider, R. (1960). *In* "Primatologia" (H. Hofer, A. H. Schultz, and D. Starck, eds.), Vol. 3, pp. 423–587. Karger, Basel.

Struhsaker, T. T. (1967). *In* "Social Communication Among Primates" (Altmann, S., ed.), pp. 281–324. Univ. Chicago Press, Chicago.

Struhsaker, T. T. (1970). *In* "Old World Monkeys. Evolution, Systematics and Behavior" (J. R. Napier and P. H. Napier, eds.), pp. 365–445. Academic Press, New York.

Sutton, D., Larson, C., Taylor, E. M., and Lindeman, R. C. (1973). *Brain Res. 53*, 225–231.

Sutton, D., Larson, C. R., and Lindeman, R. C. (1974). *Brain Res. 16*, 61–75.

Sutton, D., Larson, C. R., and Lindeman, R. C. (1974). *Brain Res. 16*, 61–75.

Sutton, D., Taylor, E. M., and Lindeman, R. C. (1977). *Acta Anat. 97*, 57–67.

Sutton, D., Samson, H. H., and Lindeman, R. C. (1978). In "Recent Advances in Primatology" (D. J. Chivers and J. Herbert, eds.), Vol. 1. pp. 764–784, Academic Press, London.

Symmes, D. and Newman, J. D. (1974). *Exp. Brain Res. 19*, 365–376.

Talairach, J., Bancaud, J., Geier, S., Bordas-Ferrer, M., Bonis, A., Szikla, G., and Rusus, M. (1973). *Electroenceph. Clin. Neurophysiol. 34*, 45–52.

Taylor, E. M., Sutton, D., and Lindeman, R. C. (1976). *Growth 40*, 69–74.

Vogel, C. (1973). *Am. J. Phys. Anthropol. 38*, 469–480.

Vogt, C. and Vogt, O. (1919). *J. f. Psychol. u. Neurol. 25*, 277–462.

Walker, A. E. (1940). *J. Comp. Neurol. 73*, 59–86.

Walker, A. E. and Green, H. D. (1938). *J. Neurophysiol. 1*, 152–165.

Wilson, D. H. and Chang, A. E. (1974). *Conf. Neurol. 36*, 61–68.

Wilson, W. A. Jr. (1975). *Animal Behav. 23*, 432–436.

Wind, J. (1970). "On the Phylogeny and the Ontogeny of the Human Larynx." Wolters-Noordhoff, Groningen.

Winter, P., Ploog, D., and Latta, J. (1966). *Exp. Brain Res. 1*, 359–384.

Winter, P., Handley, P., Ploog, D., and Schott, D. (1973). *Behaviour. 47*, 230–239.

Yerkes, R. M. and Learned, D. W. (1925). "Chimpanzee Intelligence and its Vocal Expression." Williams and Wilkins, New York.

Zenker, W. and Anzenbacher, H. (1962). *Acta Anat. 51*, 29–49.

4

Central Nervous System Processing of Sounds in Primates

JOHN D. NEWMAN

Introduction

This chapter is primarily a review of the physiology of sound reception in primates. Major emphasis is placed on the role of the cerebral cortex in the detection and discrimination of sounds, as revealed by single-unit, evoked potential, and selective lesion techniques. Studies of human auditory physiology are for the most part omitted, primarily because the detailed anatomical correlates of physiological phenomena are generally unavailable. It should be remembered, however, that some of the earliest evidence linking the temporal lobe of the cerebrum to hearing came from *postmortem* examination of brains from human patients with hearing (and often other sensory or psychomotor) defects.

The following paragraphs in this section are intended to provide a brief conceptual and historical background of research on primate hearing and its anatomical, physiological, and behavioral correlates.

Neuroanatomy

A strong thread of interest for the phylogeny of language ties together much of the early work on primate cortical anatomy. Although this thread is often difficult to follow, it seems to have arisen with the connection between damage to the first (superior) gyrus of the temporal lobe, often only in the left hemisphere, and an inability to comprehend spoken words ("word deafness").

It is a sobering fact that the basic anatomy of the auditory pathway and the concept of cerebral dominance regarding interpretation of spoken language were already established by 1905, the year Campbell published his classic monograph on the histology and functional parcellation of the cerebral cortex. Campbell recognized that a limited part of the superior temporal gyrus (STG) was the terminus of the auditory pathway. This region, lying chiefly within the transverse gyri, was found by him to have a distinct cytoarchitecture and fiber arrangement. By examining chimpanzee and orangutan brains, Campbell determined that a comparable region of the STG in these species had the same distinctive histology.

There followed from this early work a period of intensive comparative cytoarchitectural study. By the late 1950s, the cortical structure in about 17 primate genera had been described. The history and status of these efforts can be found in Bonin and Bailey (1961). Their review summarizes some interesting differences between primates, which only partly conform with other measures of position on the primate phylogenetic tree: A primate-type cortex (possessing a distinct temporal pole) is found in the tree shrew (*Tupaia*); Lemuriforms do not possess a supratemporal koniocortex; *Callithrix* (New World; marmoset) has a cortex resembling the brains of higher primates; *Cebus* (also New World) and *Macaca* (Old World) have very similar brains; the brain of the chimpanzee (*Pan*) is about four times larger than that of the macaques, but is otherwise essentially the same. From this summary, it is evident that all but the most primitive primates possess a cytoarchitecturally distinct koniocortex within the supratemporal plane (see Anatomy of Auditory Cortex, p. 74). Whether this characteristic bears any functional significance, however, is unclear, although later sections in this chapter will document differences in the functional properties of tissue within the supratemporal plane and elsewhere in the STG.

Physiology

From a historical perspective, studies of the physiology of the auditory pathway have emphasized the boundaries of different auditory centers and the organization of the representation of a given parameter within each component of the auditory pathway. By combining methods for recording electrical activity on the cortical surface with focal electrical stimulation of the cochlear nerve, a topographical projection of the cochlea onto the cortical surface could be demonstrated (Woolsey, 1972). Similar studies, using tones instead of cochlear nerve stimulation, also demonstrated a systematic distribution of tissue maximally activated by specific frequencies. This frequency (tonotopic) "map" in general matched that revealed by electrical stimulation, in the sense that stimulating fibers innervating a part of the basilar membrane known to be sensitive to a particular frequency band activated a sector of cortex likewise activated by the same band of frequencies. Since an overview of the functional organization of auditory cortex has been published by Brugge (1975), further details will be omitted here.

Much of the more recent study of central auditory mechanisms has focused on the representation and coding of acoustic parameters in individual cells at different levels of the auditory pathway (see Figure 4.1 for a schematic overview of the pathway). Various aspects of these efforts have received consideration in other publications (Aitkin, 1976; Webster and Aitkin, 1975). Furthermore, the fine-grained analysis of the functional properties of neurons in auditory cortex has been treated in several reviews (Brugge, 1975; Goldstein and Abeles, 1975). Less well developed is the analysis of mechanisms by which the central auditory system detects and discriminates sounds of biological significance. This particular line of

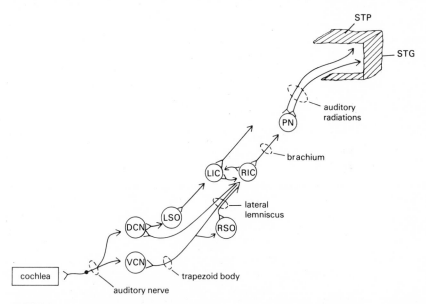

FIGURE 4.1.   *Major constituents of the primate auditory pathway. Nerve fiber bundles are enclosed by broken circles. Cell populations are indicated by solid circles and are abbreviated as follows: DCN, dorsal cochlear nucleus; VCN, ventral cochlear nucleus; LSO and RSO, left and right superior olivary nuclei; LIC and RIC, left and right inferior colliculi; PN, posterior nuclei of the thalamus (medial geniculate body and medial pulvinar nucleus); STG, superior temporal gyrus; STP, supratemporal plane of the STG. Note that the nerve fiber bundles leaving the cochlear nuclei project more strongly to higher stations on the contralateral side of the brain (only the right half of the pathway is completely illustrated).*

research is of general interest due to the adaptive significance of relevant detection mechanisms, and several interesting summaries are now available (Worden and Galambos, 1972; Evans, 1974; Bullock, 1977; Schwartzkopff, 1977).

Behavioral Correlates

The message of this chapter is that a full understanding of the central mechanisms for processing sound requires the determination of how biologically significant sounds are represented and encoded in cortical tissue. This implies, first, identification and physical analysis of biologically significant sounds and, second, identification of specialized central detection mechanisms. Not surprisingly, most of the work in this direction has emphasized the study of sounds used in species-specific communication or echo location (Worden and Galambos, 1972; Newman, 1977; Schwartz-kopff, 1977), where functional and structural analyses of the sounds are most feasible.

PRIMATE VOCALIZATIONS

Although a relatively new subject, the study of vocal communication in primates is growing rapidly, and information on the physical and social characteristics of primate vocalizations is becoming available (e.g., Ploog and Melnechuck, 1969; Peters and Ploog, 1973; Green, 1975; Ploog et al., 1975). However, there remains a serious need for descriptions of sound signals and their contextual significance (Marler, 1977).

From the standpoint of the demands placed on the central acoustic analyzing mechanism, there are several biological levels at which identification of sounds probably operates. Thus, at the level of species specificity, the relevant issues are (a) the characteristics of vocalizations that identify the species and (b) how they differ from the sounds of sympatric or related species serving the same function [e.g., gibbon "song" (Marshall and Marshall, 1976)]. A second level is that of the subspecies, population, or troop [e.g., squirrel monkey isolation peep (Winter, 1969)]. A third level where physical character of vocalizations is significant is that of the individual [e.g., the pant-hoot of chimpanzees (Marler and Hobbett, 1975)]. In addition, there may be structural variations of a single call type that are uttered by specific age classes [e.g., the twitter of squirrel monkeys (Newman et al., 1978)] or in a specific social context [e.g., the coo of Japanese macaques (Green, 1975)]. Added to these different levels of intraspecific social organization are constraints of environment or detection by predators on call structure that transcend species differences (e.g.,

Vencl, 1977). This richness and complexity, then, make primate vocal communication a particular challenge to study.

PRIMATE HEARING

While the elaborate nature of primate vocal communication suggests the necessity of a comparably elaborate neural analyzer, it should be remembered that there are basic attributes of hearing that impose limits on which vocalizations can be perceived and on the degree of resemblance between calls that can still permit discrimination. Laboratory testing of a number of primate species has determined some of these basic attributes, using psychophysical methods designed for nonverbal determination of thresholds [see Stebbins (1971, 1975, 1978) for a review of the methods and a summary of the results]. In general, prosimians have lower absolute hearing thresholds for tones above 8 kHz, whereas anthropoid primates are superior at lower frequencies. Many tests of central auditory function used in clinical evaluation of humans have yet to be applied to nonhuman primates (Berlin and Lowe, 1972). Application of these tests in primate studies will be useful not only in comparing psychoacoustic data but in evaluating the effects of experimental brain lesions.

FEATURE EXTRACTION

Of considerable heuristic value to the study of the central mechanisms for perceiving vocalizations has been the concept of the "trigger feature" or "releaser" (e.g., Scheich, 1977). The basic idea is that certain parameters or elements of a sound are essential for triggering the appropriate behavioral response from the receiving individual. In species with limited bioacoustic abilities, this mechanism may involve a peripheral hearing organ that filters out all but a limited frequency band and is sensitive mainly to a few sounds with specific spectral and temporal characteristics [e.g., anuran amphibians and certain insects (Worden and Galambos, 1972)]. In such organisms, we may think of a central process of *feature detection*, which is activated when the appropriate sounds are transduced by the peripheral auditory apparatus. In animals with a peripheral auditory apparatus sensitive to a wide range of frequencies, a more specialized central mechanism is required; that is, one in which the neural code for specific sound complexes is detected out of the rather noisy transmission coming from the periphery. This mechanism we may call *feature extraction*, because it involves extracting one neural code or set of codes out of many. This condition, of course, is applicable to most mammals. It has been the goal of a growing number of studies to discover the central sites and specific attributes of these specialized mechanisms in a variety of species. Various strategies have been employed, but most studies using methods for recording the activity of single auditory neurons assume "specialization" to be manifested in the exclusive response of a single neuron to a single vocalization. This assumption derives from the belief that somewhere in the brain are neurons that simply signal presence or absence of a sound [see Bullock (1961) for a detailed formulation of this idea]. Although a small number of neurons with high selectivity have been found (see Unit Activity in Primate Auditory Cortex, p. 82), demonstrating response exclusivity is unlikely to be convincing without exhaustive testing. In my opinion, the expectation that a neuron should respond only to a single stimulus is an unnecessary constraint. It seems more likely that even neuronal elements specialized for detection of a specific vocalization will occasionally be triggered by other incoming signals, but that there is an *optimal stimulus* for each neuron. This means that only a limited number of sound configurations will trigger activity consistently, other sounds being less consistently effective. The goal, then, becomes one of demonstrating stimulus effectiveness *quantitatively*, rather than as an all-or-none phenomenon.

At this stage of our knowledge, it would appear that the best strategy is to describe in as much detail as possible the nature of the excitatory and inhibitory response areas controlling the activity of individual neurons. This approach will inevitably link together the information gained from testing with simple artificial sounds and with representative tokens from a species' vocal repertoire.

Anatomy of Auditory Cortex

In keeping with an overall chapter emphasis on cortical mechanisms, this section will review the anatomy of auditory cortex as revealed by a variety of techniques. These techniques include stains for Nissl granules (for revealing neurons and their general shape in cytoarchitectural studies), methods for tracing pathways (anterograde degeneration and labeling with radioactive tracer molecules), and methods for identifying neurons projecting to a specific region (retrograde labeling by horseradish peroxidase). The Golgi methods, which define the nature and extent of dendritic and axonal ramifications arising from individual neurons, have not yet been applied to a detailed study of primate auditory cortex or thalamus.

The ultimate value of anatomical studies, at least for the physiologist, is to identify the substrates upon which physiological mechanisms operate.

To this end, functional studies routinely express findings in terms of anatomically defined parcellations. Numerous examples are cited in other sections of this chapter. Here, I will mention anatomical studies that particularly facilitate functional interpretations (as opposed, for example, to interpretations of phylogeny or ontogeny).

The organization of the superior temporal gyrus (STG) has been studied from several perspectives, including the sources and distribution of thalamic afferent fibers to different subregions and laminae, the distribution of intra- and interhemispheric afferents, and the internal organization of the neuronal network. Each of these topics will be discussed in turn.

Sources and Distribution of Thalamic Afferents

While it has long been accepted that a particular restricted region of the STG receives afferents from the medial geniculate body (MGB), not until recently have many of the details of MGB projection patterns been worked out. To relate these projections to their sources, a brief description of the primate MGB is warranted. The MGB is generally divisible into two or three subdivisions. Where only two divisions are recognized, these consist of a dorsomedial part (situated further rostrally) and a ventrolateral part composed of smaller cells (Walker, 1937). Advocates of three MGB subdivisions recognize a principal nucleus, further divisible into a ventral (MGBv) and dorsal (MGBd) division, and a magnocellular nucleus (MGBm) (Burton and Jones, 1976). The fact that these three subdivisions have different cortical projection patterns supports their functional distinctiveness (Burton and Jones, 1976). The most complete analysis of their projections has been performed on rhesus and squirrel monkey brains, following injections of radioactive tracer molecules into the MGB (Burton and Jones, 1976; Jones and Burton, 1976). The basic projection patterns are as follows: MGBv projects exclusively to a cytoarchitecturally distinct area within the supratemporal plane (STP) called koniocortex (see below), also referred to as AI or "primary auditory cortex"; MGBd projects to cellular fields lying anterior, medial, and lateral to AI, including a substantial part of the laterally exposed STG; MGBm projects diffusely to all of the same cortical fields, including AI. Since all three subdivisions contain neurons with a wide range of best frequencies (see Brainstem, p. 80), it is likely that the MGB transmits some of the same information *in parallel* to different parts of the auditory cortex.

One other source of thalamic afferents to the STG has been identified: the medial part of the pulvinar nucleus (Trojanowski and Jacobson, 1975;

Burton and Jones, 1976). This nucleus projects to part of the lateral surface of the STG and receives an efferent projection from this same cortical area (Trojanowski and Jacobson, 1975). The medial pulvinar also sends a projection to part of the frontal lobe (Trojanowski and Jacobson, 1974). Although the responsiveness of medial pulvinar neurons to acoustic stimuli has not been fully assessed, the evidence for this nucleus projecting to two areas that demonstrate responsiveness to sounds (see Acoustic Activation of Units outside the Auditory Pathway, p. 94) suggests that the medial pulvinar may also be involved in perception of sounds.

Figure 4.2 shows the approximate boundaries of the cortical fields described in this section.

Laminar Distribution of Afferent and Efferent Fibers

The fact that the cerebral cortex consists of layers or laminae of cells is both interesting because of its possible functional significance and of con-

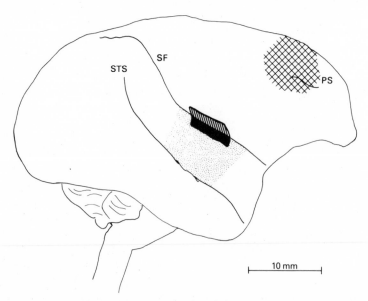

FIGURE 4.2. *Schematic side view of right cerebral hemisphere of squirrel monkey brain. Major auditory zones are indicated: primary auditory cortex, solid black area (extension into buried supratemporal plane indicated by parallel lines); auditory association cortex, stippled area; frontal auditory cortex, cross-hatched lines. Surface features indicated: supratemporal sulcus, STS; Sylvian fissure, SF; principal sulcus, PS.*

siderable assistance in correlating anatomical and physiological data. For example, knowledge of the laminar distribution of afferent fibers from different sources could be of considerable value in interpreting results of single-unit recording experiments. Several relevant anatomical analyses have been attempted. As part of their study of thalamic afferents to the STG, Jones and Burton (1976) analyzed the laminar pattern of thalamic afferent distribution to different parts of the STG. All regions contained a plexus of thalamic terminals in Layer III_B. Only in AI, however, did this plexus extend into adjacent layers (III_A and, particularly, IV). Layer V contained a sparse distribution of terminals in areas surrounding AI. Layer I of all cell fields contained traces of thalamic terminals, apparently arising from MGBm (see above). Another study (Pandya and Sanides, 1973) analyzed the laminar distribution of afferents arising *within* the auditory cortex. Afferents arising from AI pass over the corpus callosum to form exceptionally dense terminations in Layers II–IV, with a prominent concentration in Layer III. However, afferents from elsewhere in the STG send fibers to Layers I and II. A complete "wiring diagram" of auditory cortex would include identification of the neurons that are the source of efferent fibers. While this is not yet available for primates, a recent study (Ravizza *et al.*, 1976) has provided partial information on subcortical efferents for the golden Syrian hamster. Briefly, cells giving rise to cortical efferents projecting to the inferior colliculus are located in the upper part of Layer V, those to the medial geniculate in Layer VI, and those to contralateral auditory cortex in Layers II, III, IV, and V. Intracortical efferents in the primate STG have been localized mainly to pyramidal neurons in Layer III (Jacobson and Trojanowski, 1977), at least in cellular fields other than AI.

To summarize, there is at least a partial segregation of fibers entering and leaving the STG based on different laminar distributions of cells receiving or giving rise to these fibers.

Internal Organization of the Neuronal Network

Within each cortical region neurons may be differentiated according to their shape, size, and (using appropriate methods) arrangement of dendrites. This morphological differentiation permits a comparison of the organization (cytoarchitecture) of different cell populations within the STG. The most distinctive is the primary auditory cortex, designated as *koniocortex*. This region is identified by the cell density and thickness of Layers II, III, and IV (which fuse together), the thick but cell-sparse Layer V, and a denser Layer VI. The cells in Layers II and III are generally small (granule cells), while in Layer IV pyramidal cells predominate and are arranged in vertical columns perpendicular to the layers.

Physiology of the Auditory Pathway in Primates

Auditory Nerve

While this chapter emphasizes cortical mechanisms, it is necessary to preface the analysis with an overview of the properties of neural elements at earlier stations of the auditory pathway. This will not only help put the cortical work into better perspective, but may point out some gaps in our knowledge of basic auditory mechanisms in primates. Since there is no recent overview of central hearing mechanisms in primates in the literature, this section may also serve as a starting point for a more extensive future effort.

The classic study of primate cochlear nerve fibers is that published by Katsuki *et al.* (1962). These authors examined the functional properties of some 700 auditory fibers in three species of macaques in terms of tonotopic organization, response patterns, and threshold of activation by tones. By advancing an electrode through the nerve from a dorsoposterior entry point, units with progressively higher best or characteristic frequencies (CF) were found. Measurement of tonal response areas revealed two types: asymmetrical (in which the response area cut off sharply above the CF) and symmetrical. All neurons with CFs lower than 400 Hz had symmetrical response areas, and all neurons with CFs above 4000 Hz were asymmetrical. Although not emphasized by the authors, examination of their data shows a disproportionate number of fibers with CFs between 9 and 20 kHz. These authors also examined the relation between CF and threshold (lowest sound intensity at the CF producing a change in activity). They found that neurons with CFs to around 6 kHz had a bimodal distribution of thresholds, whereas the thresholds in the group of CFs between 9 and 20 kHz formed a unimodal distribution. They suggested that the high and low threshold groups in the lower CF population innervated the inner and outer hair cells, respectively. By plotting the number of impulses produced by fibers at different intensities at the CF, it was found that the low threshold units exhibited a lower rate of change of firing rate with changing sound intensity than the high threshold units. The plotted function for the low threshold units had a slope similar to that for the entire population of units with CFs in the 9- to 20-kHz range (i.e., the slopes of the regression lines were similar).

In addition to properties depending on the CF, there are several basic properties that are largely independent of CF in primary auditory nerve fibers. These have been worked out in a series of studies on the auditory nerve of squirrel monkeys (Rose *et al.*, 1967, 1969, 1971; Hind *et al.*, 1967;

Brugge *et al.*, 1969; Anderson *et al.*, 1971). The most significant of these properties is the fact that the discharges of a fiber tend to group around the period or its multiple of the tonal frequency. This phenomenon is evident for fibers with CFs of up to about 5 kHz, although a few instances of this "phase locking" were discovered in units with higher CFs. An important attribute of phase locking is that it is a mechanism for coding the frequency of a tone in fibers with a wide range of CFs. The two criteria for transmitting the code by a given fiber is that the stimulus contains spectral energy falling within the response area of the fiber and that it is of some minimal intensity. Indeed, in spontaneously active units, a tone other than the CF may change the distribution of discharges (but not the rate) at an intensity below the CF threshold. Since fibers with different CFs can signal the presence of a particular tonal frequency by the periodicity of their discharges, coding of frequency is divorced from the place on the basilar membrane to which a fiber is most sensitively related. Using two simultaneous tones, it was possible to work out some basic principles governing primary fiber responses to complex sounds (Hind *et al.*, 1967): (*a*) The timing of discharges is a sensitive function of both the intensity of the component sinusoids and the phase relation between them and (*b*) a tone that is ineffective by itself can, in combination with an effective tone of appropriate frequency ratio, greatly influence the distribution of discharges. Careful analysis of the effects of stimulus intensity on the shape of the response area and the spread of responsiveness led to an equally fundamental conclusion (Rose *et al.*, 1971): The representation of frequency along the cochlear partition is not uniform; below 2 kHz the anatomical resolution is about twice as great as that for higher frequencies. This means that the absolute differential frequency limen in squirrel monkeys should rise steeply above 2 kHz. A second conclusion from this analysis derives from the fact that auditory nerve fibers may differ significantly in the number of hair cells they innervate and thus form "acoustic units" of different sizes.

The studies on auditory nerve fibers so far summarized have all used tonal stimuli to characterize response properties. In a study using white noise as a stimulus, Ruggero (1973) found that auditory nerve fibers responded much as they would to tones. Specifically, units with CFs below 2 kHz responded with a periodicity close to their best frequency. There was a major difference, however, in that periodic discharges to white noise dampened out after a few milliseconds. There was a correlation between sharpness of the tuning curve and the number of periodic cycles before becoming fully dampened. As is the case with a linear filter, the sharper the tuning the less dampened the oscillatory response component proved to be. Testing with narrow band noise revealed that increases in band width, within a certain range, led to increases in discharge rate. In particular, a doubling of band width produced about the same discharge increase as a 3-dB increase in intensity of a noise band with constant width, indicating that a summation of spectral components was the underlying mechanism.

The foregoing discussion has demonstrated the emphasis placed on periodicity coding in studies of the primate auditory nerve. Beyond demonstrating its existence, the phase-locked response of auditory nerve fibers is a useful tool in resolving the relative effectiveness of multitone stimuli. Whether it would be appropriate to increase the complexity of stimuli further is open to question. However, the suggestion that there is an auditory mechanism common to many mammals that permits categorization of human speech sounds (Miller, 1977) could lead to a search for this mechanism at the level of the peripheral auditory system. Miller suggests that the first stage of this search would be to map the frequencies of human speech onto the basilar membrane, to demonstrate the extent to which sounds in the appropriate frequency band are conveyed centrally. This could be done, for example, by surveying CFs of fibers in the auditory nerve of different species. I see as an even more interesting line of research, however, the analysis of the representation of steady-state vowel sounds by the auditory nerve fibers. Since vowels consist of simultaneously occurring tones at (usually) nonharmonically related intervals, it would be interesting to know how they are coded by individual fibers.

Brainstem

The detailed studies on specific aspects of the physiology of the cochlear nerve and the auditory cortex are unfortunately not fully matched by a comparable quantity of studies at intermediate stations of the auditory pathway. Only a handful of papers have been published on the physiology of the primate auditory thalamus and midbrain, covering tonotopic organization, binaural interactions, and effects of behavioral performance on unit response patterns.

TONOTOPIC ORGANIZATION

The auditory thalamus and midbrain have each been the focus of a single study on tonotopic organization. Both studies used squirrel monkeys, and in both a comparable level of anesthesia was maintained during unit recording. Both studies also attempted to correlate acoustic characteristics of single units with cellular architecture and used comparable methods for delivering tone bursts and determining CFs of units or unit clusters. In the inferior colliculus (IC), FitzPatrick (1975) was able to reconstruct the representation of frequencies as determined by mapping the location of units with different CFs. In the central nucleus, there is a regular progression of frequencies from low to high when moving from a dorsal to a ventral location. This progression is particularly evident when entering the colliculus dorsocaudally and advancing rostroventrally. Examination of Golgi-stained material revealed dendritic laminae oriented perpendicularly to dorsocaudally oriented penetrations; hence, the dendritic layers may form the anatomical substrate for a series of frequency planes. Reconstruction of CFs from a number of penetrations demonstrated that isofrequency planes were in fact oriented parallel to the dendritic laminae. FitzPatrick also assessed the proportional representation of frequency within the central nucleus of the IC. As a general rule, penetrations encountered units with CFs above about 3 kHz over longer distances than units with lower CFs. By plotting the total collicular linear distance over which units with different CFs were found, it was possible to estimate the relative amount of the central nucleus devoted to each octave over the entire tested frequency range. This analysis revealed that successively higher octaves are represented by successively larger amounts of tissue up to the 8- to 16-kHz band. The relative amount of tissue devoted to frequencies above 16 kHz was less than for the 8- to 16-kHz band, but greater than that devoted to lower octaves. There is a close correlation between proportion of IC devoted to different octaves and the density of innervation of the basilar membrane devoted to each octave, up to and including the 8- to 16-kHz band. Thus, the 8- to 16-kHz band is represented by the greatest amount of IC tissue and receives the densest spiral ganglion innervation at the basilar membrane. Not mentioned by FitzPatrick in this interpretation is the fact that her sound delivery system was calibrated only up to 26 kHz and may have been relatively ineffective at higher frequencies. Thus, the proportion of collicular tissue representing frequencies above 26 kHz (contributing to her 16- to 32-kHz population) may actually be greater than her analysis indicated.

The study of the squirrel monkey thalamus (Gross *et al.*, 1974), like that of the IC, surveyed most of the auditory nucleus for responses to tones. In the present case, the subdivision of the medial geniculate body (MGB) most comparable to the central nucleus of the IC is the small-celled region (MGBv). In this subdivision, CFs of units or unit clusters change progressively from low to high along penetrations entering the lateral border and running perpendicularly to the sagittal plane. This means that isofrequency planes run approximately vertically and parallel to the sagittal plane. In the dorsal subdivision (MGBd), there is a similar tonotopic organization. Here, however, the linear distance over which the complete acoustic range is represented is longer than in the MGBv. In the medial (magnocellular) division (MGBm), there is no apparent tonotopic organization but an intermixing of units with divergent CFs.

BINAURAL INTERACTIONS

A single study (Starr and Don, 1972), using anesthetized squirrel monkeys, reported some effects of interaural differences in click intensity and time of stimulation of units in the MGB and IC. In the more thoroughly studied MGB cells, a significant proportion was sensitive to intensity or time differences but not both. Systematic changes in time interval were used to produce an apparent movement of the sound source, but no units sensitive to the direction (left to right versus right to left) of this "moving" sound were found. The most significant difference between MGB and IC units was in the proportion whose activity was suppressed by monaural clicks to either ear. This class of units accounted for 53% of the MGB total, but only 1% of those tested in the IC.

BEHAVIORAL EFFECTS UPON UNIT RESPONSIVENESS

The final paper to be considered in this section differs from the others both in its conceptual base and in the fact that awake monkeys were used as subjects. This is the only study to consider effects of behavioral performance upon unit responses to sounds at a level below the cortex (Ryan and Miller, 1977). As found in the cortical studies (see Unit Activity in Primate Auditory Cortex, below), the majority of units increased their evoked activity when the subject was performing a reaction-time task. Unlike the cortical results, however, the latency of the units' responses tended to *increase*. Furthermore, excitatory "off" response components dropped out during performance when longer duration stimuli were used. These results suggest that the mechanisms regulating performance-related unit activity in the IC are at least partly different from those regulating cortical unit activity.

Unit Activity in Primate Auditory Cortex

The study of the primate auditory pathway in terms of single-unit activity is still in its infancy. Nevertheless, a significant number of published reports deal with this topic at the level of the superior temporal gyrus. The report by Katsuki *et al.* (1962) pioneered both the study of the primate auditory cortex with microelectrodes and the study of the activity of cortical neurons in the absence of anesthesia. Their cortical studies provided evidence for the complexity of acoustic response properties of units in the

STG and for their influence by nonacoustic input. Furthermore, by testing the effect of a second continuous tone on the tuning curves of cortical neurons, they found that neurons with broad tuning curves showed a decrease in the breadth of their response area, while similar tests of narrowly tuned neurons sometimes caused an *expansion* of the response area, including the appearance of new separated (supplementary) areas. Performing this testing procedure on several neurons with similar response areas produced different effects, indicating that cortical neurons with similar response areas to simple tones could respond differently to complex tonal stimuli. Katsuki and his co-workers concluded that inhibition plays a major role in determining the responses to complex sounds by cortical neurons.

The study by Katsuki *et al.* concentrated on primary auditory cortex, tissue lying within the buried supratemporal plane. A subsequent study (Gross *et al.*, 1967) presented the first single-unit data from auditory association cortex in primates. The main findings of the study by Gross, Schiller, Wells, and Gerstein were that (*a*) only about half of the tested units responded to the acoustic stimuli (click, 500-Hz tone); (*b*) the responses were often weak and predominantly excitatory; (*c*) responses under light Nembutal anesthesia were virtually identical to those obtained from locally anesthetized, muscle-paralyzed preparations; (*d*) spontaneous activity ranged from 50/sec to under 1/sec and bore no relationship to response properties.

Since these pioneering efforts, studies of cortical unit activity have fallen into three main categories: (*a*) representation of the cochlea at the cortex (tonotopic organization); (*b*) control of unit activity under different stimulus conditions, including concomitant operant conditioning; (*c*) coding of species-specific vocalizations.

TONOTOPIC ORGANIZATION

Working from the foundation of earlier anatomical and evoked potential studies, attempts to map the representation of frequency in cortex have been made with microelectrodes. The basic procedure involves determining characteristic frequencies (CFs) (i.e., the frequency that produces detectable rate change at lowest intensity) for individual neurons or neuron clusters in subsections of auditory cortex and plotting the distribution of CFs over the cortical surface. The clearest maps have been obtained using anesthetized preparations and plotting the frequency data for individual monkeys, rather than pooling the data from several monkeys. Tonotopic organization in the STG has been worked out for *Macaca arctoides* (stump-tailed macaque), *M. mulatta* (rhesus macaque) (Merzenich and Brugge, 1973), and *Aotus trivirgatus* (owl monkey) (Imig *et al.*, 1977) and correlated with cytoarchitecture of different cellular fields. Using somewhat different methods, Funkenstein and Winter (1973) found that frequency is represented in squirrel monkey STG in a manner approximating that for the owl monkey: low frequencies rostrally and higher frequencies progressively more caudally. Units with response areas restricted to frequencies above 10 kHz were found only in the posterior parts of the explored region, both at the lateral extent of the cortex and within the supratemporal plane. The more detailed studies in *Macaca* and *Aotus* reveal a plan of functional organization that may be a general scheme for primate auditory cortex, and which is compatible with anatomical studies (see Sources and Distribution of Thalamic Afferents, p. 75). This scheme has the following basic properties:

1. Fields AI and R (rostral to AI within the STP) form the central core of the auditory cortex, are similar in organization, both having complete representation of the cochlea (low frequencies being represented rostrolaterally and high frequencies caudomedially), and contain mostly units responding over only a narrow frequency range.
2. The fields surrounding AI and R show a greater prevalence of units poorly responsive to tones, more units with broad tone response areas, and a less orderly progression of frequency representation.

The studies in *Macaca* and *Aotus* demonstrate that for a given sector of auditory cortex a limited range of frequencies produces optimal responses. However, it is worthwhile noting several points that are relevant to correlating these findings to studies of the brain in unanesthetized monkeys. First, the determination of frequency representation has concentrated on units within a limited part of the cortical laminae (Layers III_B and IV), where the most vigorous discharges occur in the anesthetized preparation (Imig *et al.*, 1977). Since these laminae also contain the terminals of the auditory radiations (see Laminar Distribution of Afferent and Efferent Fibers, p. 76), the data represent mainly the initial thalamocortical synaptic drive and miss the synaptic effects of intracortical interactions. Second, the involvement of nonacoustic factors in controlling responsiveness of cortical cells in awake monkeys raises questions about the validity of studying cortical neuronal physiology in anesthetized preparations. Third, the complexity of excitatory and inhibitory inputs on cortical neurons is generally less apparent in anesthetized preparations.

FIRING PATTERN CHANGES

In a study in awake macaques, Brugge and Merzenich (1973) examined the effect of changing intensity (sound pressure level, SPL) on unit firing patterns. They found that units commonly increased their discharge rate up to a maximum and then decreased this rate with further increases in SPL. This nonmonotonic characteristic was evident in neurons with widely different CFs, and in some cases it was quite sharply peaked (at the extreme, a shift of 10 dB above or below the most effective SPL greatly reduced the spike count). In one population of units lateral to AI, the optimal intensity at the units' CF varied from 15 to 95 dB SPL (40 to 80 dB in a single monkey). However, other neurons were relatively insensitive to changes in SPL once a maximum value had been reached. The frequency at which the maximum spike count occurred was not necessarily the CF, a characteristic of possible relevance to detection of complex sounds.

CODING OF SPECIES-SPECIFIC VOCALIZATIONS

An interest in the representation and coding of species-specific vocalizations has been a major incentive in our studies on squirrel monkey cortex. One of the hypotheses that led to the initial studies was that cortical mechanisms would favor detection of biologically significant sounds. Therefore, at the level of the cerebral cortex, vocalizations would be more effective stimuli than artificial sounds such as steady tones, clicks, and white noise. This expectation was not fulfilled, however, since as large a percentage of cortical units was activated by artificial sounds (larger in the case of tones) as by vocalizations (Funkenstein, *et al.*, 1971; Winter and Funkenstein, 1973). A second hypothesis was that particular types of vocalizations were represented at the level of individual neurons, leading to the prediction that a cortical unit would be very selective in its responses to vocalizations. This prediction was fulfilled by the demonstration that a majority of units responded to only one or a limited number of structurally similar vocalizations (Winter and Funkenstein, 1973).

Subsequent studies have indicated that these initial reports underestimated the extent of vocal representation by single cortical neurons, due to the method of assessing unit responsiveness and to the sedation of the experimental animal. Units were tested with six repetitions each of different tape-recorded vocalizations. A vocalization was considered effective if it produced a noticeable change in discharge rate in five of the six trials. This means that only secure, consistent responses were counted. The use of a barbiturate to sedate the animal depressed both the spontaneous and driven activity of most neurons. This made it more difficult to detect significant changes in unit activity during testing. Subsequent experiments (Newman and Wollberg, 1973a; Wollberg and Newman, 1972), using a greater number of repetitions of each stimulus call and no barbiturates, demonstrated both a much greater percentage of cortical units influenced by vocal input and a greater representation of different vocal types by individual units. Taken together, these studies lead to the conclusion that many vocalizations may influence the activity of a given cortical unit, but only one call, or a restricted number of calls, drives many units in a vigorous, consistent manner.

STABILITY OF RESPONSIVENESS IN
NEURONS WITH HIGH SELECTIVITY

In our experiments, we have been concerned about the security with which acoustic inputs drive cortical neurons. This question is significant in a practical way because it influences our ability to detect neural responses and also theoretically because it relates to assumptions about how a neural network functions in the process of detecting natural sounds. The question " Does a particular input influence the neuron under study?" is measured by us as a significant change in the neuron's discharge rate. Since we record extracellular action potentials only, we do not know when particular stimuli produce subthreshold changes in a neuron's membrane potential. Nevertheless, neurons exist that are virtually silent (no discharges) in the absence of a single effective vocalization. The most dramatic examples produce action potentials at rates of one every 10 sec or less, yet discharge vigorously to nearly every presentation of one or two specific sounds. These units also do not respond to steady tone bursts tested over a wide frequency range and therefore appear to be specialized for the detection of a specific type of vocalization. For this reason, I refer to them as extreme examples of Specialist neurons (Newman, 1978b). Such dramatic cases have been found only a handful of times. Interestingly, in each case the effective vocalization was the isolation peep, and in each there was a suggestion of a response (in a few trials) to the chirp. In two cases, the unit (isolated in the cortex of a "Gothic-arch" monkey) responded to two or more variants of the "Gothic-arch" isolation peep, but gave no discharges during testing with a "Roman-arch" isolation peep. These units, although frustratingly rare, are intriguing because their selectivity in responsiveness follows a meaningful pattern: When tested with variants from two subspecies, they respond only to the appropriate subspecific variant. For such neurons, it seems plausible that a restricted set of inputs, carrying the code for one type of vocalization, has exclusive ability to trigger activity. One is provoked to think of a highly specified labeled line that is "wired" into the circuitry of the neuron. Yet, to make this assumption more plausible, it is necessary to demonstrate both that the input remains effective over time and that other inputs do not become effective at certain times.

Rigorous proof of this assumption would require monitoring the activity of the neuron over a long time, while testing with many stimuli under a variety of conditions. This has not yet been fully accomplished, but some degree of response stability has been demonstrated. In the best tested case, the entire set of vocal stimuli was presented twice (about 30 min intervening) and in both cases identical results were obtained.

Since these few cases are not the general rule, we must examine the properties of other selective neurons to obtain a more representative picture. These form a more disparate class in terms of the consistency with which they respond to the most effective vocalization and the extent to which other sounds also trigger activity changes. They share the common feature of responding more vigorously to a single vocalization or call type. All of our standard test vocalizations have been effective stimuli for Specialist neurons. The following comments pertain to units studied in primary auditory cortex (PAC), equivalent to AI in *Aotus* (Imig *et al.*, 1977). As a group, PAC Specialist neurons have lower spontaneous activity rates (mean rate is 2.7 spikes/sec) than the whole population of PAC neurons responsive to vocalizations (mean rate is 6.8 spikes/sec). Neurons that fail to respond to *any* test vocalization (about 20% of the PAC neurons we have tested) also have low spontaneous rates (2.2 spikes/sec). This may mean that nonresponsive neurons are really selectively responsive to vocalizations other than those in our standard series. This possibility has been confirmed in some cases by testing with additional vocalizations. The quantitative determination of responsiveness applied to data collected over the past several years has revealed a greater degree of selectivity in PAC neurons than was apparent by inspection of scatter plots and peristimulus histograms. Our quantitative analyses have disregarded weak "responses." Using a more rigorous criterion for responsiveness, 33% of the tested PAC neurons are Specialists. This percentage is slightly higher than the 27% reported by Winter and Funkenstein (see above) and is much higher than our original assessment (Newman and Wollberg, 1973a). The fact that our original data included units outside the PAC does not account for this disparity in the incidence of selective units, since the lateral surface of the STG also contains many selective neurons (Symmes *et al.*, 1976).

Of this population, almost one-third also responds to a narrow (less than 1 octave) frequency band of tones in the same intensity range used to test vocalizations. In each case, this frequency band is compatible with the distribution of frequencies in the effective vocalization (below 5 kHz for shriek and err cackle and between 5 and 10 kHz for the tonal calls). Because of the selectivity of these neurons, it is inaccurate to think of them simply as narrow band filters (like cochlear nerve fibers), responding to any sound with sufficient energy in the appropriate frequency range. Yet, it seems likely that the distribution of frequencies within a vocalization is a factor in determining its effectiveness in triggering neuron activity. Figure 4.3 illustrates a neuron with these basic characteristics.

NEURONS RESPONDING TO
STRUCTURALLY SIMILAR CALLS

Slightly more than 20% of the PAC neurons we have studied respond only to a small number of vocalizations with sound energy distributed in the same frequency range. Part of this category consists of neurons that respond to shriek and err cackle or to shriek, err cackle, and twitter. Shriek

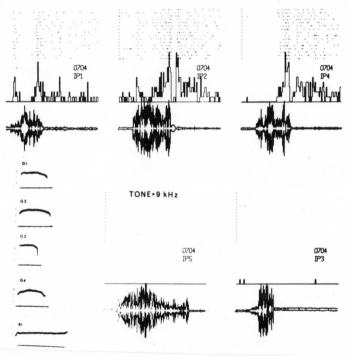

FIGURE 4.3. *A selectively responsive unit from squirrel monkey auditory cortex. This unit responds to a narrow range of tones around 9 kHz, but fails to respond to all but a few vocalizations (Gothic-type isolation peeps 1, 2, and 4) with energy in that frequency band. Unit discharges are indicated by dots in the scatter plots, and the associated vocal stimulus is indicated by the average amplitude plot beneath each peristimulus time histogram. The leftmost rows of vertical dots indicate the start of each trial. The spectrograms of test vocalizations are shown at a smaller scale, frequency marks indicating 4, 8, and 12 kHz. Note that the unit is virtually silent during testing with IP3 (G3) and IP5 (R1). [Adapted from Newman, 1978a.]*

and err cackle contain substantial sound energy below 5 kHz. While most of the sound energy of the twitter is above 5 kHz, the latter part of this call does dip to around 4 kHz. In each case, neurons responding only to this group of calls also respond to tones no higher in frequency than 5 kHz. The rest of the neurons in this class respond only to two or more of the tonal calls, for example, isolation peep and peep or chirp and twitter. Neurons responding to *all* of the tonal calls also respond to a limited frequency range of tones, this frequency band falling within the octave from 5 to 10 kHz. Two units of this type had separated tone response areas, at 6, 8, and between 12 and 14 kHz (Newman and Wollberg, 1973b; Newman, 1978a). Taken as a group, the neurons described in this section respond only to vocalizations of a particular structural class. Thus, it appears appropriate to refer to them as Class Detectors (Newman, 1978b). In the data published by Winter and Funkenstein, more than half of their neurons responded to two or more test vocalizations in either the tonal or noisy/low frequency groups, although none of their units responded to *all* vocalizations in one or the other of these categories.

NONACOUSTIC CONTROL OF CORTICAL ACTIVITY

A growing amount of evidence suggests that nonacoustic factors influence the activity of cortical units; hence, these factors need to be considered in evaluating the role of cortical neurons in auditory perception. In this section, I will consider the evidence that the activity of cortical neurons is changed by nonacoustic factors and summarize the state of our knowledge concerning the underlying mechanisms mediating these changes.

It has been a common observation that neuronal activity changes concomitant with the presentation of a fixed unchanging stimulus are not always constant. This variability may take the form of changes in response strength or changes in the timing or polarity of the response. These changes may be transient, whereby the response changes or fluctuates during testing of a repetitive stimulus, each trial interrupted by a few seconds. However, the changes may also have a longer time constant, whereby the response to a repeated stimulus is relatively stable over a period of a few minutes, but changes over longer periods of time. Changes in unit response strength have been observed in our studies, although, in our data on PAC neurons, constancy of response strength and pattern to an effective vocalization has been the general rule (Newman and Symmes, 1974), at least over the relatively short time periods used when testing most neurons (averaging 20–30 min). Indeed, the response stability of auditory cortex neurons to vocalizations is sufficiently stable to study the effects of experimental manipulations of neuronal excitability, such as electrical stimulation of the

midbrain reticular formation (Newman and Symmes, 1974) and ion-
tophoretic application of putative neurotransmitters (Foote *et al.*, 1975).

Studies on other species, using a variety of stimuli and methods, have
reported more conspicuous changes in the response properties of auditory
cortex neurons. Manley (1976, Manley and Milner, 1975) found that the
neural activity of unit clusters recorded from the cortex of freely moving
tree shrews (*Tupaia*) was highly labile, in that the set of species-specific
vocalizations that were effective in eliciting responses was constantly
changing on both a long-term and a short-term basis. These data are subject to difficulties of interpretation, however, since the contribution of individual units to the measured activity of a cluster is likely to change due to
movements of the microelectrode in these chronic preparations. Consequently, response instability in this case may be more apparent than real.
Other studies with similar results but testing single units are less subject to
this criticism (Manley and Mueller-Preuss, 1978; Glass and Wollberg,
1979). Studies of cortical unit activity in awake but restrained macaques
have reported changes in responsiveness to tones correlated with the
sleep/wake cycle (Brugge and Merzenich, 1973; Pfingst *et al.*, 1977) and
with spontaneous movements of the subject (Brugge and Merzenich, 1973).
Both movement and sleep are correlated with decreased responsiveness.
Since the middle ear muscles are more active during movement, this may
explain the reduced response under such circumstances. The mechanism by
which neuronal responsiveness decreases during sleep is less certain, although the fact that the thalamocortical transmission is enhanced during
arousal may mean that fewer inpulses from lower auditory stations reach
the cortex during periods of low arousal. Manipulation of arousal has been
directly tested for arousal effect on changing responsiveness in cortical
neurons. After implanting bipolar electrodes in the midbrain reticular formation (MRF), we looked for changes in response profiles (i.e., which
specific vocal stimuli were effective) and patterns in STG units concurrent
with MRF stimulation (Newman and Symmes, 1974). We tested this in two
ways. In the first method, we presented the entire set of vocal stimuli and
recorded the unit's response; we then repeated the entire vocal tape, but
preceded each sound with a short train of electrical pulses of the MRF.
These pulse trains reliably induced cortical arousal (as measured by EEG
activation), and using longer stimulus durations induced behavioral
arousal as well. A second testing method involved alternating test trials in
which one consisted of the vocal stimulus alone and the next followed with
the MRF stimulation preceding the sound. The results obtained with either
method were the same: 36% of the tested units showed pattern changes
during MRF stimulation, but only 10% changed their response profiles.
This is the only report in which the effect of induced arousal changes on
unit activity in the primate auditory cortex has been presented in quan-

titative terms. In an earlier documented example of arousal affecting unit responsiveness induced by ammonia fumes (Katsuki *et al.*, 1962), the subject struggled during arousal. In that case, arousal-related activation of the middle ear muscles may have been a factor in the decreased unit response. In our study, however, changes in responsiveness included the appearance of responses to previously ineffective vocalizations; hence, middle ear activation is a less satisfactory explanation. It should also be mentioned that 25% of the neurons tested by us gave time-locked responses to successive trains of reticular formation stimulation. This percentage is of the same order of magnitude as the 20% found to respond to 10/sec stimulation of the intralaminar nuclei (nonspecific thalamus, Nelson & Bignall, 1973).

Of all the reports of lability in auditory cortex unit responses, perhaps the most intriguing are those correlated with training monkeys in operant conditioning tasks. Data using two different training methods (Miller *et al.*, 1972; Hocherman *et al.*, 1976) agree with regard to showing increased levels of driven activity in units recorded from the auditory cortex in trained monkeys. Furthermore, even in trained monkeys, the differences in the strength of driven activity in the same unit correlate with the subject actively performing a discrimination task (Beaton and Miller, 1975) or expecting a reward (Hocherman, *et al.*, 1976), rather than passively sitting.

CENTRAL MECHANISMS AFFECTING RESPONSIVENESS OF AUDITORY CORTEX NEURONS

Having surveyed the evidence for lability in response properties of auditory cortex neurons, it is appropriate to examine the possible mediating mechanisms. Peripheral phenomena, especially activation of the middle ear muscles (MEM), are viewed as a minor contributing factor by most authors for several reasons: (*a*) The latency for a MEM effect is longer than the latency of the response components most sensitive to change; (*b*) the level of attenuation of sound induced by MEM contraction has been shown to be insufficient to account for differences in response strength; and (*c*) the changes may be either an increase or a decrease in response strength during the same testing conditions, whereas attenuation via MEM activation would be expected to produce only decreases. Given the unlikely participation of the MEM, then, what *central* mechanisms might be involved?

Potential central mechanisms are of two kinds: local and distant. Local mechanisms include persistent changes in membrane excitability or shifts in synaptic interactions to facilitate synaptic transmission under the appropriate conditions. Distant mechanisms include modulating influences of pathways arising in other parts of the CNS. Let us consider local mechanisms first. In 1949, D. O. Hebb proposed the concept of the "cell assembly," which he defined as a diffuse structure composed of cells in the cortex, diencephalon, and possibly the basal ganglia that could act briefly as a closed system. The assembly could facilitate other such systems, in particular systems controlling specific motor responses. Hebb further proposed that each assembly action was aroused by a preceding assembly (in a chain of such systems), by a sensory event, or (normally) by both. Hebb proposed this theory in an attempt to explain the neural basis for concept formation and perceptual learning. He suggested that the process of forming a cell assembly depended initially on mutual facilitation of two cells or groups of cells activated at the same time by the same stimulus and that over many such facilitations the operation of the assembly as a whole would improve. Several important conceptual assumptions followed from this model. One was that, while at *some* point the perceptual process must act on specific cells in order to determine a specific response, an individual cell could enter into more than one assembly at different times. A second assumption was that cell assemblies provided the basis for perceiving *contrasts* and *relative properties* of stimulation, such as differences in size or pitch. The role of sensory mechanisms in Hebb's theory was restricted to the perception of the "primitive unity" of an entity; that is, the physical attributes of the stimulus that permit it to be segregated from the background "noise" of the environment. Learning, on the other hand, established the *boundaries* that separate entities or parts of entities into identifiable units. Although not specifically addressed by Hebb, it should be apparent that his theory is relevant to the issue of categorical versus continual perception of speech and other species-specific sounds, a topic to be considered further under General Conclusions (p. 105). At this point, however, I would like to explore the extent to which Hebb's model relates to the experimental data on unit response lability in the STG. According to Hebb, we should expect repetition of the same stimulus in a learning context to lead to facilitation or to increasingly reliable responses at the single-unit level. This is, in fact, what has been reported (Miller *et al.*, 1972). A paradox exists, however, in that the response components experiencing the greatest change with training are those with the shortest latency (Pfingst *et al.*, 1977); these response components increase in consistency but not duration. Hebb's theory predicts that, during learning, cells participating in cell assemblies will fire for *longer* periods; hence, the major change expected would be in the duration of a unit's response. Thus, it is questionable whether the data presented by Miller and his colleagues represent the formation of cell assemblies in the sense proposed by Hebb.

With this brief survey of some local mechanisms controlling unit responsiveness, we turn here to consider the possible involvement of neural tissue outside the central auditory pathway. I have already briefly mentioned our attempts to influence STG unit responses by stimulating the midbrain reticular formation (RF). The RF and its thalamic extension the nonspecific thalamic nuclei (intralaminar complex) have long been implicated as the mechanism of cortical "activation"; that is, conversion of EEG brain wave patterns from slow wave activity to the low voltage, fast activity of the awake, aroused individual. Defining the role of the RF in sensory perception has been a more elusive task, but considerable evidence supports the notion that the RF enhances cortical processing of sensory input (e.g., Symmes and Anderson, 1967). At the level of the single unit, however, the evidence is less clear. I have cited our own study and the fact that RF stimulation changed the response profiles of only 10% of the tested units. However, in spite of this relatively weak short-term effect, one may ask whether the RF is important in mediating unit changes over a longer time span. Specifically, is the RF involved in the increased firing strength of STG units in trained monkeys? One line of evidence supporting this possibility is that the reported changes in unit responses are mainly in the early ("on") components. Since the RF has been shown to be sensitive to brief, transient changes in sensory input (Starr and Livingston, 1963), and short latency responses to electrical stimulation of the intralaminar system have been recorded in association cortex also activated by acoustic imputs (Nelson and Bignall, 1973), the implication of the RF becomes a real possibility. It is puzzling, however, that some STG neurons show training-correlated response increases and some do not. Furthermore, the evidence to date does not suggest that cortical neurons receiving direct intralaminar input [i.e., in Layer I (Jones, 1975)] are any more likely to show changes in response strength than neurons in other layers.

In addition to nonspecific thalamus and the RF, pathways from other nonauditory sources are known to enter the superior temporal gyrus. Perhaps most interesting are the afferents arising from the frontal cortex and from areas in and near the cingulate gyrus. The first of these arises mainly from the dorsolateral prefrontal cortex (Pandya *et al.*, 1971), whereas the "cingulate" pathway arises from the cingulate and subcallosal gyri (Müller-Preuss and Jürgens, 1976). Both pathways innervate the lateral surface of the STG. A direct test of the functional significance of the frontotemporal pathway was attempted in squirrel monkeys (Alexander *et al.*, 1976) by implanting subdural electrodes over the prefrontal convexity and monitoring the effect of stimulating these sites on the activity of units in the STG. Prefrontal stimulation induced activity changes in 18% of the STG units studied. All but one of these units responded to acoustic stimulation as well. Different stimulation sites were effective for different STG units, but sites along the dorsal convexity were generally more effective than sites along the ventral convexity. The most common effect of prefrontal stimulation was prompt inhibition of spontaneous activity lasting throughout, and for 12–500 msec beyond, the stimulus train. The mean latency for the onset of this inhibition was 16.5 msec (range, 10–40 msec). In a smaller number of STG units, prefrontal stimulation affected only the driven activity resulting from cotemporaneous acoustic stimulation. This effect was always to reduce or obliterate the acoustic response. Physiological investigations of the cingular projection have yet to be published. The interest in this pathway derives from the documented role of the cingulate region in the production of vocalizations (Jürgens and Mueller-Preuss, 1977; Sutton *et al.*, 1974). The pathway to the STG could be an example of an "efference copy" mechanism, providing to central auditory neurons a mechanism for differentiating self-generated sounds.

Acoustic Activation of Units outside the Auditory Pathway

The anatomical tracing of connections originating at the cochlea and projecting centrally has led to what is today a well-documented picture of the central auditory pathway. Since this picture has depended largely on the degree of resolution of staining procedures, it is not surprising that many connections have been missed. This section will document the representation of sounds in structures outside the classic auditory pathway. The criterion for inclusion here will be demonstrable evidence of single-unit responses to sounds. In some cases, the pathway over which the acoustic input travels is uncertain.

Although the superior temporal gyrus is the terminus of the classic auditory pathway, other parts of the cerebral cortex are also responsive to sounds. Some of the evidence for this comes from recordings of evoked potentials; hence, the precise locus of the sound-evoked activity has been unclear. However, one large cortical region where sounds are effective stimuli is in the frontal lobe. The major part of this zone lies dorsally, between the central and arcuate sulci. This region, called the "postarcuate polysensory area" (PPA) (Bignall and Imbert, 1969), widely overlaps with cortical tissue sensitive to other sensory modalities. True polysensory neurons (responsive to more than one modality) are rare, most neurons within this region responding to a single modality (at least within the limits of the testing protocols used). Direct projections from STG to the frontal lobe have been described in squirrel monkeys (Forbes and Moskowitz, 1977; Bignall, 1969) and macaques (Pandya *et al.*, 1969; Jacobson and Trojanowski, 1977).

The pathway from STG to the frontal lobe projects to a region ventral and caudal to the arcuate sulcus, although in macaques the projection extends into the frontal pole. Studies of unit responsiveness to sounds in this frontal region confirm that acoustic input can alter the activity of single units (Nelson and Bignall, 1973; Schechter and Murphy, 1975; Newman and Lindsley, 1976). These studies generally agree that units activated by sounds respond with relatively short latencies. Tuning of frontal units to tones typically is broad (Newman and Lindsley, 1976), and activity changes to a variety of sounds are similar, being a poorly timed, brief burst of activity, varying in strength from trial to trial (Newman and Lindsley, 1976). Species-specific vocalizations used as test stimuli in a study of frontal units in squirrel monkeys did not reveal any response characteristics not produced by artificial sounds (Newman and Lindsley, 1976). However, one finding of interest is that vocalizations containing a broad range of frequencies (err cackle, shriek, and chuck) are more effective in activating frontal units than tonal vocalizations. Testing with vocalizations and artificial noise bursts (Newman *et al.*, 1976) demonstrates that these "noisy" vocalizations trigger more frequent and consistent activity changes in frontal units than the noise bursts, suggesting that the relative effectiveness of this group of vocalizations is due to a more subtle characteristic than possessing sound energy widely distributed in frequency. Although only a speculation at this time, it is intriguing to consider that the frontal lobe is part of a mechanism that regulates responsiveness to one class of vocalizations sharing both structural and communicative characteristics.

Whether the acoustic responsiveness of frontal units depends on input from the STG is open to question. Click-evoked potentials can still be recorded in the PPA following STG ablation (Bignall and Imbert, 1969). Furthermore, part of the thalamic input to the frontal cortex, the pulvinar nucleus (Trojanowski and Jacobson, 1975), contains acoustically responsive units (Mathers and Repisardi, 1973; Casey, 1966), as well as neurons that are directly connected to the STG (Trojanowski and Jacobson, 1974).

Another cortical region receiving a projection from the STG is the so-called "posterior association cortex" (Graybiel, 1974), where the temporal, parietal, and occipital regions meet. Testing with sounds in this region is warranted for the possibility of discovering units sensitive to natural sounds, and to the spatial location of the sound source.

Studies of unit activity in limbic cortex (allocortex) and related limbic areas have reported cases of sound-induced activity. Presentation of tones or whistles while recording from units in macaque piriform cortex, hippocampus, and amygdala caused significant activity changes in about 10% of the tested units (Fuster and Uyeda, 1971). No difference was noted in the incidence of acoustic responsiveness of units recorded from naive monkeys or from monkeys trained in a visual discrimination task. In another limbic area, insular cortex overlying the claustrum ("claustrocortex"), about 10%

of the units responded to clicks or tones (Sudakov *et al.*, 1971). More acoustic units were found in the caudal part of the explored tissue. Two types of acoustic responses were found: transient, short latency, high probability responses, and long latency responses, with prolonged discharge and a lower probability of response. A few units responded to tones only over a restricted frequency range below 2 kHz. Studies of cingulate cortex have discovered acoustically responsive units also (Bachman and MacLean, 1971; Sutton *et al.*, 1978), but the incidence of responsiveness to a variety of acoustic stimuli and the detailed response properties of acoustic units in the cingulate gyrus await further study.

The other major forebrain structure in which auditory activity has been demonstrated is the globus pallidus (Travis and Sparks, 1967, 1968). In this case, the auditory stimuli (1.0- and 1.5-kHz tones) were the cue for delivery of a food pellet or an electric shock. About 40% of the units in the globus pallidus changed their activity during presentation of one or both tones. Units sampled in other parts of the corpus striatum (caudate, putamen) were generally unresponsive to the same stimuli. Clarification of the significance of these data awaits further testing with a wider range of acoustic stimuli in naive and trained subjects.

Evoked Potential Mapping of Primate Auditory Cortex

Prior to the advent of the microelectrode, recording of electrical activity in the brain was possible only from large groups of neurons. The "evoked potential" technique slowly faded from active use as more and more investigators turned to record the activity of single units or small clusters of units. Hence, discussion of the evoked potential technique today is often relegated to historical overviews (see Physiology, p. 70). However, evoked potential recording is experiencing a period of renewed interest, because of its value in the clinic and as an electrophysiological monitor of psychoacoustic performance tests in human subjects. The improvement of the signal to noise ratio through averaging techniques has been of great value in the recent use of slow wave records derived from electrodes placed on the scalp or cranial surface [see MacKay *et al.* (1969) for a review of the methods and conceptual issues regarding evoked brain potentials as indicators of sensory information processing].

In this section, only studies relevant to the question of localizing the cortical sources of auditory evoked activity will be presented, that is, results based on records using penetrating electrodes and taken at different depths within the cortex or studies that systematically mapped changes in the distribution and configuration of the acoustically evoked potential (AEP) over the cortex.

Several studies have mapped the distribution of AEPs on the primate cortex. Mapping studies of the STG attempted to define the primary auditory cortex on the basis of the site of the shortest latency and the AEP with the largest amplitude. In squirrel monkeys (Massopust *et al.* 1968), the primary auditory cortex (P1) was thus located within the insular cortex and a second primary area (P2) was located at the lip of the STP bordering the exposed STG. However, since this study used brief low frequency tone bursts (usually 4 cycles of a 500-Hz tone) instead of clicks or white noise, tuning of cortical neurons and incoming thalamic afferents could have influenced the distribution of maximal evoked potentials. Similar results were found in the STG of the cebus monkey (Hoffman *et al.*, 1969) using the same stimuli. In macaques and mangebey (*Cercocebus*) monkeys, Pribram *et al.* (1954) found a larger zone of responsive tissue, extending over the posterior half of the STG, STP, posterior insula, and overlying parietal cortex, in these cases using clicks. All of the responsive tissue in one hemisphere became unresponsive following ablation of the ipsilateral STP and adjacent STG if sufficient time elapsed to permit retrograde degeneration in the MGB before electrophysiological testing. The authors interpreted this to mean that collaterals arising from MGB afferents were the source of auditory input to parietal cortex. However, a more recent study (Arezzo *et al.*, 1975) of awake macaques demonstrated that the *only* active source of AEPs in the buried parts of the STG, insula, and overlying parietal cortex was within the STP. All other AEPs in this region were volume-conducted from the STP. The center of this active zone lay within the cytoarchitecturally distinct koniocortex. Smaller and more variable AEPs were recorded from the STG surface. Results similar to those of Arezzo *et al.* were obtained from humans (Celesia and Puletti, 1969), in the sense that the source of AEPs was localized to a small part of the STP (in this case within the transverse temporal gyri) and that AEPs from the STG (first temporal convolution) were much smaller and more variable.

One other cortical area, outside the STG, has been identified as an active source of AEPs (Arezzo *et al.*, 1975). This lies within frontal agranular cortex, between the central and arcuate sulci. AEPs have been demonstrated in both awake macaques (Arezzo *et al.*, 1975) and squirrel monkeys (Goldring, *et al.*, 1970; Newman *et al.*, 1976). In both species, the largest AEPs come from the rostral parts of the region ("premotor" cortex). Barbiturate anesthesia largely abolishes these AEPs, but they are clearly demonstrable under chloralose (Bignall and Imbert, 1969). Frontal AEPs persist following ablation of the STG (Bignall, 1970) or destruction of the reticular formation and midline nonspecific thalamus (Bignall and Singer, 1967). Electrical stimulation of the MGB evoked activity over the same region activated by clicks, but so did electrical stimulation of nonspecific thalamus (Nelson and Bignall, 1973). These somewhat confusing results were partially clarified in a series of experiments by Bignall (1970). Bignall found that a lesion in cortical white matter abolishing click-evoked potentials in the frontal lobe did not affect frontal response to direct stimulation of the STG. Bignall then sought to determine the thalamic source of the click-evoked responses. Lesions of thalamus medial to the MGB (including nucleus pulvinaris, nucleus medialis dorsalis, nucleus ventralis lateralis, nucleus anterior, and midline nonspecific nuclei) were without effect in abolishing frontal AEPs. In contrast, destruction of the MGB abolished the AEPs. Destruction of the MGB had to be virtually complete for complete abolition of the frontal responses. Taking these results together leads to the conclusion that there are multiple sources of auditory input to the frontal cortex, including, at the least, one passing through the STG, one arising from the MGB and bypassing the STG (via an as yet unknown path), and perhaps one from the medial pulvinar. Given the lack of knowledge concerning the function of the frontal agranular cortex, it would be unwise to speculate on the significance of these multiple inputs. On the other hand, the fact that two or more pathways presumed to carry auditory information converge on this relatively restricted part of nonprimary cortex is ample reason to study further the role of this region in auditory function.

Effects of Auditory Cortex Ablations on Sound Perception

The relative contribution of different levels of the auditory pathway to hearing processes is still not well known. A traditional approach to this question, particularly before the advent of modern electrophysiological techniques, was to destroy part of the auditory pathway surgically and attempt to describe the resulting deficits. The cerebral cortex has been the focus of much of this work, partly because of the early evidence relating the temporal lobe to hearing deficits and partly because of its accessibility for experimental neurosurgery. Consequently, this section will review the literature on the effects of cortical ablations on primate hearing. Most of the cortical ablation studies pertaining to primate hearing consist of destruction of all or parts of the superior temporal gyrus. These studies relate to a variety of hearing capacities: frequency resolution; pattern discrimination; perception of natural sounds; localization of a sound source.

Frequency Resolution

The results of a number of studies, some now quite old, have left as a generalization the conclusion that the STG is not essential for making frequency discriminations. Before presenting a summary of these results, a prefatory comment pertaining to the method of assessing deficits in performance is in order. Although the behavioral audiogram of several primate species has been determined (see Behavioral Correlates, p. 73), no study has compared the complete behavioral audiogram of a primate before and after selective brain lesions. Since audiometric testing is a routine part of human clinical evaluation, it is thus difficult to compare in a precise way the effects of cortical lesions in nonhuman primates with clinical data. Instead, the practice has been to train primate subjects to signal by some operant behavior their ability to discriminate between sounds differing in frequency (or some other parameter) and then measure the retention of this capacity following surgery. As a consequence, the performance of a learned act has been linked inextricably to tests of the animal's hearing capabilities following brain damage. These studies, then, are not so much tests of frequency resolution as they are tests of the retention of the association between sound identification and behavioral performance.

In nearly every case in which the ability to discriminate between two sounds was tested, extensive ablations of the STG (with varying amounts of STP damage) resulted in a deficit in performance from the highly trained preoperative state. This deficit is in terms of a loss of association between conditional stimuli and response: After adequate retraining, relearning to the preoperative criterion of performance is generally demonstrable. Thus, relearning of a noise versus tone discrimination task (Weiskrantz and Mishkin, 1958) was necessary (and possible) following bilateral lesions of the posterior STG of macaques (sparing the STP). Deficits were greater in a study evaluating combined STG–STP lesions (Iversen and Mishkin, 1973) on discrimination of two tones (300 versus 2400 Hz), although all subjects eventually regained the criterion level of performance. In the latter study, lesions restricted to *either* STP or STG resulted in a much milder deficit. Iversen and Mishkin also tested the effects of the same lesions on resolving a conditional tone from tones of other frequencies. Even monkeys with combined lesions were able to achieve preoperative levels of performance. However, the authors did not test their subjects with conditional tones differing by less than about 60 Hz. Another group of investigators found that similarly placed lesions resulted in raised frequency difference thresholds, but the postoperative increase was only 25–30 Hz (against a fixed conditional tone of 500 Hz) over the preoperative difference threshold of 5 Hz (Massopust *et al.*, 1967). This group also did a similar study using tones of higher frequencies (4000 and 5000 Hz) as the conditional stimuli (Massopust *et al.*, 1970). In this case, lesions of the middle third of the STG, plus STP and insula, resulted in a more profound deficit in frequency resolution (from 30–50 Hz preoperatively to 400–900 Hz postoperatively). Finally, a study primarily focused on sound localization deficits (see below) found that extensive STG–STP lesions produced an elevation in discrimination thresholds for certain higher frequencies, although the entire audible frequency range was not tested (Heffner and Masterton, 1978). These last three studies thus stand as the only attempts to define in psychophysical terms the effects of STG lesions on resolution of frequency differences, and the only ones in which a clear deficit in perception of tonal frequency is demonstrable following STG–STP ablations.

Pattern Discrimination

Several studies have tested the ability of monkeys to discriminate electronically generated sound sequences. The most direct tests, and most profound deficits following surgery, were demonstrated in studies using two different sequences of tones [1000–800–1000 versus 800–1000–800 Hz (Jerison and Neff, 1953)] and trains of chopped noise differing in chopping rate (Symmes, 1966). In both studies, extensive bilateral ablations of the STG and STP were made and permanent deficits occurred postoperatively in that subjects failed to relearn the task. In both studies, the same subjects were trained to perform other, presumably simpler, auditory discrimination tasks postoperatively. In a separate study (Wegener, 1976), discrimination of different tone sequences was tested following ablations restricted to the *rostral* STP. Severe performance deficits were found following surgery, but in each case the subjects were able to relearn the task.

Two studies have tested performance on auditory sequence discrimination tasks following ablations restricted to the STG (sparing the STP). In one (Stepien *et al.*, 1960), *Cercopithecus* monkeys were trained to differentiate between two conditions using a pair of click trains. The two conditions differed by whether the paired trains were identical or not, a delay being interposed between the two members of a pair. Postoperatively, monkeys trained on this task failed to distinguish between the two conditions. Further testing revealed that the deficit was not primarily in discriminating the two click trains per se, because elimination of the delay separating the trains of a given pair resulted in greatly improved performance (although not to preoperative levels). Rather, the authors interpreted their results as showing a deficit in short-term memory, by which the monkey forgets the stimulus last heard in stimulus pairs separated by some time interval. A second study (Dewson *et al.*, 1970) trained macaques to signal discrimination between 1000-Hz tones and white noise bursts paired in four different sequences (T–T, T–N, N–N, N–T). Postoperative testing showed deficits in performance after varying the duration of each sound segment (tone or noise) or the interval between them to determine the limits of each dimension at which the monkeys continued to perform at criterion level. The deficits were in the form of shorter silent intervals and shorter segment durations required postoperatively to perform at criterion. Interestingly, even *unilateral* ablations resulted in some lasting deficit in these dimensions (although less than with bilateral ablations). Further analysis of this unilateral effect (Cowey and Dewson, 1972), in which one or the other ear was occluded during testing, showed that postoperative deficits were greater when the animal was listening with the ear (in this case the right) contralateral to the ablation. It is worth noting that these last results derive from only a single monkey and that this animal showed better performance to left ear occlusion prior to ablation of the left STG. More recently, Dewson (1978) has reviewed his past findings and presented further evidence of unilateral STG ablation effects. In these studies, a series of monkeys was trained on a more complex task (nonspatial, conditioned delayed matching to sample), in which the monkey must learn to associate a color (red or green) with a particular sound (tone or noise). Trained in this task, the most persistent deficits were obtained following lesions to the left STG (and sparing the STP). The deficit is in the form of the length of delay that can be interposed between a colored light and sound and still maintain criterion performance. The errors are in the form of random choices of the response key (illuminated by the red or green light) when detectable delays are interposed. Removal of the delay, or adding a spatial attribute, results in the same monkeys performing at almost preoperative levels.

Perception of Natural Sounds

Three studies in the primate literature document an effect of auditory cortex ablations on perception of natural sounds. In one of these, the "natural sounds" were human speech sounds (vowels), whereas in the others the stimuli were natural species-specific vocalizations. Dewson *et al.* (1969) trained rhesus macaques to discriminate between the vowels i and u as well as to perform other auditory discrimination tasks. The most significant finding in this study was that bilateral ablation of the STP resulted in an apparently total inability to relearn a conditional discrimination between the vowels. The two vowels were of equal duration and overall intensity, differing most conspicuously in the frequency position of the second formant.

As part of a larger series of auditory discrimination tasks, Pratt (Pratt and Iversen, 1978) trained baboons (*Papio cynocephalus*) to signal discrimination between the "yak" vocalization and the same call presented backward. Animals with bilateral STP lesions performed at significantly lower levels postoperatively, whereas animals with bilateral lesions sparing the STP showed no significant performance deterioration. Similar results occurred when testing for retention of tone sequence discrimination or identification of the spatial position of a noise burst source.

The most comprehensive and significant ablation study using natural sounds is that by Hupfer *et al.* (1976, 1977). Squirrel monkeys were the subjects, and the test stimuli were an extensive series of species-specific vocalizations and nonprimate sounds. The task consisted of jumping from a perch when hearing *any* species-specific vocalization (Go stimulus) and refraining from jumping to any other sound (No-Go stimulus). Correct selection of Go stimuli resulted in a period of limited access to another monkey (i.e., social reinforcement). Thus, the basic task was to signal discrimination between squirrel monkey vocalizations and other natural sounds, which all subjects readily learned to do. Postoperative testing following lesions variously placed in the STP and STG suggested a major role of the STP in performing this task. Animals with large bilateral STG–STP lesions (with nearly total retrograde degeneration in the rostral and intermediate MGB) failed to relearn the task. In addition, these subjects could not learn to discriminate a single vocalization (yap) from white noise. Animals with lesions sparing more of the STP showed considerable postoperative retention losses, but some relearning was demonstrable. No significant effect of unilateral ablations was evident with these tests.

Sound Localization

Although less directly connected to the main themes of this chapter, locating a sound source in space is an important acoustic function and is one for which significant deficits are demonstrable following cortical ablations. Study of this problem in primates has been carried out by Heffner and Masterton (1975), using rhesus monkeys. Earlier work on nonprimates suggested that ablations of the auditory cortex did not disrupt all responses to a localizable sound source; hence, the question of what the auditory cortex contributes to this function remained. Two types of localization tests were used, the stimuli always being single or grouped clicks. Monkeys with significant amounts of bilaterally ablated STG and STP continued to signal the source (left versus right) of the stimuli by pressing one of two keys. However, the same monkeys failed to identify the source of the same stimuli if required to walk from some intermediate location to the appropriate speaker. This finding led the authors to conclude that the effect of the ablations was not primarily sensory but instead either disrupted the connection between perception and the motor response or interfered with short-term memory (see the discussion of the study by Stepien *et al.*, p. 100). In a subsequent series of experiments (Heffner and Masterton, 1978), a further attempt was made to define the nature of the deficit in localization tasks. Out of these experiments came the discovery that STG–STP ablations increase the frequency difference limen at higher frequencies (see above) and, in addition, decrease the ability to localize high frequency (16-kHz) tones.

Effects of Lesions in Other Cortical Areas

The effects on postoperative performance in auditory discrimination tasks of lesions in several nontemporal cortical regions have been published over the years. Many of the results of the older studies have been summarized in the excellent review by Wegener (1964). More recently, attention has focused on one of these areas, the frontal granular cortex ("prefrontal cortex"), rostral to the arcuate sulcus. This area is rostral to the premotor area, where most of the auditory unit and evoked potential studies have concentrated their efforts (see Acoustic Activation of Units outside the Auditory Pathway, p. 94, and Evoked Potential Mapping of Primate Auditory Cortex, p. 96). It has long been suspected that the deficits following prefrontal lesions are primarily nonsensory, that is, related to some aspect of the task other than resolution of acoustic differences between the conditional stimuli. Some evidence favors perseverative errors as a major factor in prefrontal deficits, at least for lesions in the inferior portions of this region (Iversen and Mishkin, 1970). Symmes (1967), however, found little to differentiate prefrontal and STG lesion effects, in terms of degree of deficit or nature of errors, in a chopped versus steady noise discrimination task.

To conclude this section, it may be stated that ablation methods are still of significant value in filling gaps regarding the specific contributions of different sectors of auditory cortex to sound perception. An improved understanding of anatomically and electrophysiologically determined STG boundaries (see Sources and Distribution of Thalamic Afferents, p. 75, and Unit Activity in Primate Auditory Cortex, p. 82) may facilitate further work. The evident parallel pathways from thalamus to cortex (again, Sources and Distribution of Thalamic Afferents, plus Acoustic Activation of Units outside the Auditory Pathway, p. 94) suggest experiments in which the effects of disrupting different parts of the afferent path are assessed. For example, the possibly different effects of selective lesions of MGBv or other segments of the auditory thalamus (p. 75) remain to be examined. Finally, the role of frontal agranular cortex in mediating auditory behavior needs to be tested, in view of the inputs from the auditory thalamus and the STG to this region (Evoked Potential Mapping of Primate Auditory Cortex, p. 96).

General Conclusions

The various sections in this chapter have treated a limited number of topics in some detail. Here, I hope to conclude with a more general overview of primate hearing.

Evolution of the Primate Auditory Analyzer

Statements about the evolution of brain mechanisms must be largely speculative, since behavior and neural circuitry do not leave fossil records. Even the significance of fossilized traces that may be found (cranial endocasts) is unclear, given the uncertain relationship between brain shape and size to specific cerebral functions. However, a comparison of certain attributes among living species suggests some generalities about the phylogenetic history of hearing and may lead to clues regarding the adaptive significance of known central auditory mechanisms.

Two of these attributes are the size and the shape of the audible field. The size of an individual's audible field is determined by the range of frequencies to which the individual is sensitive (tested, one frequency at a time, with single pure tones). The shape of the audible field is determined by the lowest intensity (threshold) at which each specific frequency is detected. By plotting the size and shape of the audible field of different species, one can compare the extent to which species differ in their ability to hear different parts of the sound spectrum. Comparisons of audible fields have been made (Masterton and Diamond, 1973; Masterton *et al.*, 1969), and the results lead to several interesting conclusions:

1. Birds and mammals are generally superior to reptiles and amphibians in both absolute sensitivity and range of audible frequencies.
2. Mammals are superior to birds in their capacity to hear high (above about 15 kHz) frequencies, due largely to the mechanical properties of the three-ossicled mammal middle ear.

3. Hominoids (man and chimp) are unique among the primates in the low upper limit to their hearing range.
4. Low frequency (below 1 kHz) hearing in mammals is a secondary specialization, usually at the expense of detecting higher frequencies. Within mammals, the relative bias toward high or low frequencies generally correlates with head size and, consequently, dictates the cue (interaural intensity versus time differences) most useful in sound localization. Thus, small species (with ears closely spaced) produce, and are best able to localize, high frequency sounds. Since low frequency sounds are less attenuated over distance, low-frequency hearing facilitates long-distance communication as well.

Specializations for Species-Specific Communication

Measures of species differences based on audibility threshold curves, or on neuroanatomical peculiarities, are useful in understanding the adaptability of a species to a particular habitat or ecological niche. However, they are usually less successful as guides to the perceptual mechanisms underlying recognition and differentiation of intraspecific vocal signals. To understand these perceptual mechanisms requires probing into the physiology of underlying cerebral mechanisms. This chapter reviewed the recent progress of such probes, initiated from a variety of vantage points. The information from these approaches is still too fragmented to be synthesized into a general model of how sounds are processed by the central auditory system. Further progress in this direction will likely depend on discovering the rules governing the activation of auditory neurons by a great variety of sounds.

Given the considerable overlap in frequency sensitivity and vocal characteristics of different species, we may ask what mechanisms facilitate species-specific communication. Behavioral information regarding response biases to different vocalizations is largely lacking for most primate species, but some evidence now favors a special perceptual mechanism for differentiating species-specific sounds. In one study, Japanese macaques (*Macaca fuscata*) were trained to signal discrimination of two variants of the coo call differing in one distinctive feature, the location of an upward pitch inflection (Beecher *et al.*, 1977; Zoloth, *et al.*, 1979). When tested for their ability to perform this discrimination using new sets of variants, the subjects continued to perform as though there were still the two classes of tokens. Thus, they demonstrated a capacity to *categorize* one type of species-specific vocalization according to the details of a structural attribute known to have social significance (Green, 1975). This capacity was lacking in other monkey species (*M. nemestrina* and *Cercopithecus aethiops*) tested in the same way with the same tokens. This difference in performance between species suggests that the Japanese macaque may possess a perceptual mechanism specialized for detecting its vocalizations, which the other monkeys lack. It is appealing to speculate that for each primate species the acoustic characteristics imparting social significance to different call variants are species specific and that a species-specific perceptual mechanism exists for recognizing these socially meaningful characteristics. Further behavioral tests with other species are required to explore the extent to which the findings just outlined can be demonstrated more widely. It seems clear, in any case, that "categorical perception" (Macmillan *et al.,* 1977) must have its neural correlate, and exploration at the neuronal level of this perceptual phenomenon has only just begun.

References

Aitkin, L. M. (1976). *In* "International Review of Physiology" (R. Porter, ed.), Vol. X, pp. 249–279. University Park Press, Baltimore.

Anderson, D. J., Rose, J. E., Hind, J. E., and Brugge, J. F. (1971). *J. Acoust. Soc. Amer. 49,* 1131–1139.

Arezzo, J., Pickoff, A., and Vaughan, H. G. Jr. (1975). *Brain Res. 90,* 57–73.

Alexander, G. E., Newman, J. D., and Symmes, D. (1976). *Brain Res. 116,* 334–338.

Bachman, D. S. and MacLean, P. D. (1971). *Intern. J. Neuroscience 2,* 109–112.

Beaton, R. and Miller, J. M. (1975). *Brain Res. 100,* 543–562.

Beecher, M. D., Zoloth, S. R., Petersen, M. R., Moody, D. B., and Stebbins, W. C. (1977). *J. Acoust. Soc. Amer. 62,* 5101–5102.

Berlin, C. I. and Lowe, S. S. (1972). *In* "Handbook of Clinical Audiology" (J. Katz, ed.), pp. 280–312. Williams and Wilkins, Baltimore.

Bignall, K. E. (1969). *Brain Res. 13,* 319–327.

Bignall, K. E. (1970). *Brain Res. 19,* 77–86.

Bignall, K. E. and Imbert, M. (1969). *EEG clin. Neurophysiol. 26,* 206–215.

Bignall, K. E. and Singer, P. (1967). *Exp. Neur. 18,* 300–312.

Bonin, G. von and Bailey, P. (1961). *In* "Primatologia: Handbook of Primatology" (H. Hofer, A. H. Schultz and D. Starck, eds.), Vol II, Pt. 2, Section 10, pp. 1–42. Karger, Basel.

Brugge, J. F. (1975). *In* "The Nervous System" (D. B. Tower, ed.), Vol. III, pp. 97–111. Raven, New York.

Brugge, J. F., Anderson, D. J., Hind, J. E., and Rose, J. (1969). *J. Neurophysiol. 32,* 386–401.

Brugge, J. F. and Merzenich, M. M. (1973). *J. Neurophysiol. 36,* 1138–1158.

Bullock, T. H. (1961). *In* "Sensory Communication" (W. A. Rosenblith, ed.) pp. 717–724. MIT Press, Cambridge, Mass. and Wiley, New York.

Bullock, T. H. (1977). "Recognition of Complex Acoustic Signals". Dahlem Konferenzen, Berlin.

Burton, H. and Jones, E. G. (1976). *J. Comp. Neur. 168,* 249–302.

Campbell, A. W. (1905). "Histological Studies of the Localization of Cerebral Function". Cambridge Univ. Press, Cambridge.

Casey, K. L. (1966). *J. Neurophysiol. 29,* 727–750.

Celesia, G. C., and Puletti, F. (1969). *Neurology 19*, 211–200.

Cowey, A. and Dewson, J. H. (1972). *Neuropsychologia 10*, 279–289.

Dewson, J. H. III (1978). *In* "Recent Advances in Primatology" (D. Chivers and J. Herbert, eds.), Vol. I, pp. 763–768. Academic Press, London.

Dewson, J. H. III, Pribram, K. H., and Lynch, J. C. (1969). *Exp. Neur. 24*, 579–591.

Dewson, J. H. III, Cowey, A., and Weiskrantz, L. (1970). *Exp. Neur. 28*, 529–548.

Evans, E. F. (1974). *In* "The Neurosciences: Third Study Program" (F. O. Schmidt and G. Quarton, eds.), pp. 131–145. MIT Press, Cambridge, Mass.

FitzPatrick, K. A. (1975). *J. Comp. Neur. 164*, 185–208.

Foote, S. L., Freedman, R., and Oliver, A. P. (1975). *Brain Res. 86*, 229–242.

Forbes, B. F. and Moskowitz, N. (1977). *Brain Res. 136*, 547–552.

Funkenstein, H. H., Nelson, P. G., Winter, P., Wollberg, Z., and Newman, J. D. (1971). *In* "The Auditory System; A Workshop" (M. Sachs, ed.), pp. 307–315. National Educational Consultants, Baltimore.

Funkenstein, H. H. and Winter, P. (1973). *Exp. Brain Res. 18*, 464–488.

Fuster, J. M. and Uyeda, A. A. (1971). *EEG clin. Neurophysiol. 30*, 281–293.

Glass, I. and Wollberg, Z. (1979). *Exp. Brain Res. 34*, 489–498.

Goldring, S., Aras, E., and Weber, P. C. (1970). *EEG clin. Neurophysiol. 29*, 537–550.

Goldstein, M. H. Jr. and Abeles, M. (1975). *In* "Handbook of Sensory Physiology" (W. D. Keidel and W. D. Neff, eds.), Vol V, Pt. 2, pp. 199–218. Springer-Verlag, Berlin.

Graybiel, A. M. (1974). *In* "The Neurosciences: Third Study Program" (F. O. Schmidt and F. G. Worden, eds.), pp. 205–214. MIT Press, Cambridge, Mass.

Green, S. (1975). *In* "Primate Behavior" (L. Rosenblum, ed.), Vol. IV, pp. 1–102. Academic Press, New York.

Gross, C. G., Schiller, P. H., Wells, C., and Gerstein, G. L. (1967). *J. Neurophysiol. 30*, 833–843.

Gross, N. B., Lifschitz, W. S., and Anderson, D. J. (1974). *Brain Res. 65*, 323–332.

Hebb, D. O. (1949). "The Organization of Behavior." Wiley, New York.

Heffner, H. and Masterton, B. (1975). *J. Neurophysiol. 38*, 1340–1358.

Heffner, H. and Masterton, B. (1978). *In* "Recent Advances in Primatology" (D. Chivers and J. Herbert, eds.), Vol. I, pp. 735–754. Academic Press, London.

Hind, J. E., Anderson, D. J., Brugge, J. F., and Rose, J. E. (1967). *J. Neurophysiol. 30*, 794–816.

Hocherman, S., Benson, D. A., Goldstein, M. H. Jr., Heffner, H. E., and Heinz, R. D. (1976). *Brain Res. 117*, 51–68.

Hoffman, J. P., Walker, J. V., Wolin, L. R., Kadoya, S., and Massopust, L. C. Jr. (1969). *J. Auditory Res. 9*, 89–99.

Hupfer, K. (1976). "Erkennung arteigener Laute des Totenkopfaffen (*Saimiri sciureus*) nach zerstörung der Hörrinde." Doctoral dissertation, Ludwig - Maximilians University, Munich.

Hupfer, K., Jürgens, U., and Ploog, D. (1977). *Exp. Brain Res. 30*, 75–88.

Imig, T. J., Ruggero, M. A., Kitzes, L. M., Kavel, E., and Brugge, J. F. (1977). *J. Comp. Neur. 171*, 111–128.

Iversen, S. D. and Mishkin, M. (1970). *Exp. Brain Res. 11*, 376–386.

Iversen, S. D., and Mishkin, M. (1973). *Brain Res. 55*, 355–367.

Jacobson, S. and Trojanowski, J. Q. (1977) *Brain Res. 132*, 209–233.

Jerison, H. J. and Neff, W. D. (1953). *Fed. Proc. 12*, 73–74.

Jones, E. G. (1975). *J. Comp. Neur. 160*, 167–204.

Jones, E. G. and Burton, H. (1976). *J. Comp. Neur. 168*, 197–248.

Jürgens, U. and Mueller-Preuss, P. (1977). *Exp. Brain Res. 29*, 75–83.

Katsuki, Y., Suga, N., and Kanno, Y. (1962). *J. Acoust. Soc. Amer. 34*, 1396–1410.

Mackay, D. M., Evans, E. F., Hammond, P., Jeffreys, D. A., and Regan, D. (1969). *Neurosciences Res. Prog. Bull. 7*, 181–276.

Macmillan, N. A., Kaplan, H. L., and Creelman, C. D. (1977). *Psych. Rev. 84*, 452–471.

Manley, J. A. (1976). "Multiple-unit Recording from the Auditory Cortex of Tree Shrews". Doctoral dissertation, McGill University, Montreal.

Manley, J. A., and Milner, P. M. (1975). *J. Acoustic Soc. Amer. 57*, S54.

Manley, J. A. and Mueller-Preuss, P. (1978). *Exp. Brain Res. 32*, 171–180.

Marler, P. (1977). *In* "Recognition of Complex Acoustic Signals". (T. H. Bullock, ed.), pp. 17–35. Dahlem Konferenzen, Berlin.

Marler, P. and Hobbett, L. (1975). *Z. Tierpsychol. 38*, 97–109.

Marshall, J. T. Jr. and Marshall, E. R. (1976). *Science 193*, 235–237.

Massopust, L. C. Jr., Wolin, L. R., Meder, R., and Frost, V. (1967). *Exp. Neur. 19*, 245–255.

Massopust, L. C. Jr., Wolin, L. R., and Kadoya, S. (1968). *Exp. Neur. 21*, 35–40.

Massopust, L. D. Jr., Wolin, L. R., and Frost, V. (1970). *Exp. Neur. 28*, 299–307.

Masterton, B. and Diamond, I. T. (1973). *In* "Handbook of Perception" (E. G. Carterette, ed.), Vol. III, pp. 407–448. Academic Press, New York.

Masterton, B., Heffner, H., and Ravizza, R. (1969). *J. Acoust. Soc. Amer. 45*, 966–985.

Mathers, L. H. and Repisardi, S. C. (1973). *Brain Res. 64*, 65–83.

Merzenich, M. M. and Brugge, J. F. (1973). *Brain Res. 50*, 275–296.

Miller, J. M., Sutton, D., Pfingst, B., Ryan, A., Beaton, R., and Gourevitch, G. (1972). *Science 177*, 449–451.

Miller, J. D. (1977). *In* "Recognition of Complex Acoustic Signals (T. H. Bullock, ed.), pp. 49–58. Dahlem Konferenzen, Berlin.

Müller-Preuss, P., and Jürgens, U. (1976). *Brain Res. 103*, 29–43.

Nelson, C. N. and Bignall, K. E. (1973). *Exp. Neur. 40*, 189–206.

Newman, J. D. (1978a). *In* "Recent Advances in Primatology" (D. Chivers and J. Herbert, eds.), Vol. I, pp. 755–762. Academic Press, London.

Newman, J. D. (1978b). *J. Med. Primat. 7*, 98–105.

Newman, J. D. and Lindsley, D. F. (1976). *Exp. Brain Res. 25*, 169–181.

Newman, J. D. and Symmes, D. (1974). *Brain Res. 78*, 125–138.

Newman, J. D. and Wollberg, Z. (1973a). *Brain Res. 54*, 287–304.

Newman, J. D. and Wollberg, Z. (1973b). *Exp. Neur. 40*, 821–824.

Newman, J. D., Symmes, D., and Alexander, G. E. (1976). *Proc. 6th Soc. Neuroscience Meeting, Toronto 1976*, Vol. II, p. 695. Soc. Neuroscience, Bethesda.

Newman, J. D., ed. (1977). *In* "Recognition of Complex Acoustic Signals" (T. H. Bullock, ed.), pp. 279–306. Dahlem Konferenzen, Berlin.

Newman, J. D., Lieblich, A., Talmage-Riggs, G., and Symmes, D. (1978). *Z. Tierpsychol. 47*, 77–78.

Pandya, D. N., Hallett, M., and Mukherjee, S. K. (1969). *Brain Res. 14*, 49–65.

Pandya, D. N., Dye, P., and Butters, N. (1971). *Brain Res. 31*, 35–46.

Pandya, D. N. and Sanides, F. (1973). *Z. Anat. Entwickl.-Gesch. 139*, 127–161.

Peters, M. and Ploog, D. (1973). *Ann. Rev. Physiol. 35*, 221–242.

Pfingst, B. E., O'Connor, T. A., and Miller, J. M. (1977). *Exp. Brain Res. 29*, 393–404.

Ploog, D. and Melnechuk, T. (1969). *Neurosciences Res. Prog. Bull. 7*, 419–510.

Ploog, D., Hupfer, K., Jürgens, U., and Newman, J. D. (1975). *In* "Growth and Development of the Brain" (M. A. B. Brazier, ed.), pp. 231–254. Raven, New York.

Pratt, S. R. and Iversen, S. D. (1978). *In* "Recent Advances in Primatology" (D. Chivers and J. Herbert, eds.), Vol. I, pp. 807–809. Academic Press, London.

Pribram, K. H., Rosner, B. S., and Rosenblith, W. A. (1954). *J. Neurophysiol. 17*, 336–344.

Ravizza, R. J., Straw, R. B., and Long, P. D. (1976). *Brain Res. 114*, 497–500.

Rose, J. E., Brugge, J. F., Anderson, D. J., and Hind, J. E. (1967). *J. Neurophysiol. 30*, 769–793.

Rose, J. E., Brugge, J. F., Anderson, D. J., and Hind, J. E. (1969). *J. Neurophysiol. 32*, 402–423.

Rose, J. E., Hind, J. E., Anderson, D. J., and Brugge, J. F. (1971). *J. Neurophysiol. 34*, 685–699.

Ruggero, M. A. (1973). *J. Neurophysiol. 36*, 569–587.

Ryan, A. and Miller, J. (1977). *J. Neurophysiol. 40*, 943–956.

Schechter, P. B. and Murphy, E. H. (1975). *Brain Res. 96*, 66–70.

Scheich, H. (1977). *In* "Recognition of Complex Acoustic Signals" (T. H. Bullock, ed.), pp. 161–182. Dahlem Konferenzen, Berlin.

Schwartzkopff, J. (1977). *Ann. Rev. Psychol. 28*, 61–84.

Starr, A. and Livingston, R. B. (1963). *J. Neurophysiol. 26*, 416–431.

Starr, A. and Don, M. (1972). *J. Neurophysiol. 35*, 501–517.

Stebbins, W. C. (1971). *In* "Behavior of Nonhuman Primates" (A. M. Schrier and F. Stollnitz, eds.), Vol. III, pp. 159–192. Academic Press, New York.

Stebbins, W. C. (1975). *In* "The Nervous System" (D. B. Tower, ed.), Vol. III, pp. 113–124. Raven, New York.

Stebbins, W. C. (1978) *In* "Recent Advances in Primatology" (D. Chivers and J. Herbert, eds.), Vol. I, pp. 705–720. Academic Press, London.

Stepien, L. S., Cordeau, J. P., and Rasmussen, T. (1960). *Brain 83*, 470–489.

Sudakov, K., Maclean, P. D., Reeves, A., and Marino, R. (1971). *Brain Res. 28*, 19–34.

Sutton, D., Larson, C. R., and Lindeman, R. C. (1974). *Brain Res. 71*, 61–75.

Sutton, D., Samson, H. H., and Larson, C. R. (1978). *In* "Recent Advances in Primatology" (D. Chivers and J. Herbert, eds.), Vol. I, pp. 769–784. Academic Press, London.

Symmes, D. (1966). *Exp. Neur. 16*, 201–214.

Symmes, D. (1967). *J. Auditory Res. 7*, 335–351.

Symmes, D. and Anderson, K. V. (1967). *Exp. Neur. 18*, 161–176.

Symmes, D., Newman, J. D., and Alexander, G. E. (1976). *Proc. 6th Soc. Neuroscience Meeting Toronto 1976*, Vol. II, P. 696. Soc. Neuroscience, Bethesda.

Travis, R. P. and Sparks, D. L. (1967). *Physiol. Behav. 2*, 171–177.

Travis, R. P. and Sparks, D. L. (1968). *Physiol. Behav. 3*, 187–196.

Trojanowski, J. Q. and Jacobson, S. (1974). *Brain Res. 80*, 395–411.

Trojanowski, J. Q. and Jacobson, S. (1975). *Brain Res. 85*, 347–353.

Vencl, F. (1977). *Amer. Naturalist 111*, 777–782.

Walker, A. E. (1937). *J. Anat. 71*, 319–331.

Webster, W. R. and Aitkin, L. M. (1975). *In* "Handbook of Psychobiology" (M. S. Gazzaniga and C. Blakemore, eds.), pp. 325–364. Academic Press, New York.

Wegener, J. G. (1964). *J. Auditory Res. 4*, 227–254.

Wegener, J. G. (1976). *Neuropsychologia 14*, 161–173.

Weiskrantz, L. and Mishkin, M. (1958). *Brain 81*, 406–414.

Winter, P. (1969). *Folia primat. 10*, 216–229.

Winter, P. and Funkenstein, H. H. (1973). *Exp. Brain Res. 18*, 489–504.

Wollberg, Z. and Newman, J. D. (1972). *Science 175*, 212–214.

Woolsey, C. N. (1972). *In* "Physiology of the Auditory System" (M. B. Sachs, ed.), pp. 271–282. National Educational Consultants, Baltimore.

Worden, F. G. and Galambos, R. (1972). *Neurosciences Res. Prog. Bull. 10*, 1–119.

Zoloth, S. R., Petersen, M. R., Beecher, M. D., Green, S., Marler, P., Moody, D. B., and Stebbins, W. (1979). *Science 204*, 870–873.

5

Cortical and Subcortical Organization of Human Communication: Evidence from Stimulation Studies[1]

GEORGE A. OJEMANN   CATHERINE MATEER

Electrical Stimulation as a Technique in Behavioral Investigation

Advantages and Limitations

An understanding of the evolution of the biological aspects of social communication is greatly facilitated by a clear model of the neurology of human communication. The information to construct such a model has most often come from studying the changes in communication behavior following spontaneous brain lesions in man: stroke, hemorrhage, focal trauma, or focal seizure. This has been supplemented with information gathered using other methods, including lateralization of function in the intact "normal" brain derived from ingenious test techniques that apparently limit input to one cerebral hemisphere or the other and the identification of areas of intact brain that are metabolically active during specific behaviors (e.g., isotopic blood flow studies). Some models of the neurology of human communication that result from this information are reviewed in other chapters of this volume. Here we present a model for the neurological basis of human communication based on information obtained from another technique: measuring the behavioral change occurring during local elec-

[1] This research was supported by NIH Grant NS 04053, NINCDS, USPHS/DHEW.

NEUROBIOLOGY OF
SOCIAL COMMUNICATION IN PRIMATES

trical stimulation of different areas of the brain. These observations were made during neurosurgical operations under local anesthesia in awake patients.

The model of brain basis for human communication derived from these stimulation studies differs some from the classic models derived from lesion studies. This reflects the different advantages and limitations of each technique for studying the role of human brain organization in communication. The advantages of the stimulation technique include the ability to sample the same behavior during functional changes at multiple brain sites (stimulation mapping), which is limited in a given patient only by the therapeutically determined exposure; the ability to turn the functional alteration on or off so that it may be used as a probe during different parts of a behavioral test; and the apparent lack of behavioral adaptation, at least during the brief period customarily used in stimulation studies. This avoids the problem of recovery of function, which to some degree complicates the interpretation of behavioral changes after almost all brain lesions. Finally, stimulation effects often change over relatively short distances, .5 to 2 cm or less along the same gyrus (Ojemann and Whitaker, 1978a,b), providing a degree of resolution of functional localization unavailable with any other method currently used.

A major limitation of the stimulation technique is that it can be used only in select patient populations—that is, patients undergoing neurosurgery utilizing local anesthesia. In this chapter, we will discuss two such populations: one providing information on cortical function from patients with medically intractable epilepsy undergoing resection of the epileptic focus and the other providing information on thalamic function from patients with dyskinesias (i.e., Parkinsonism or dystonia) undergoing ventral lateral thalamotomy. Whether the different models of brain organization developed from stimulation data apply only to the select patient populations or can be generalized to normal individuals is unknown, for no other technique presently available provides such discrete functional localization. Morphological localization in the normal brain can be achieved on a similar and much more refined scale and, when parallels can be drawn between stimulation data and morphological localization data (as in the question of the degree of individual variability in cortical organization), the results from these two techniques point to the same conclusion. On the other hand, there is evidence of an unusual pattern of brain organization, including the organization of the communications system, in patients with extreme cases of medically intractable epilepsy that began at an early age (Rasmussen and Milner 1977). Therefore, when comparably discrete data obtained with another technique are unavailable, the question of how widely stimulation data can be generalized to other populations remains open.

Historical Background

The human behavioral changes previously reported using brain stimulation techniques are of two general types. Positive responses are one type, for example, a sensation or movement from Rolandic cortex (Penfield and Jasper, 1954), spontaneous speech from the anterior lateral thalamus (Schaltenbrand, 1975), or a specific experience from the lateral temporal cortex (Penfield and Perot, 1963). Positive memory and language responses are rare: rather than being systematically mapped, they have usually been anecdotally reported as isolated cases. Much more common are reports of stimulation disturbing ongoing complex behaviors such as language. Naming common objects is the language behavior most extensively studied in series of cases at a number of brain sites (Penfield and Roberts, 1959; Ojemann and Ward, 1971; Fedio and VanBuren, 1974; Ojemann, 1975, 1977; Whitaker and Ojemann, 1977; Ojemann and Whitaker, 1978a,b; VanBuren et al., 1978). Several other studies have evaluated the effects of stimulation on ongoing verbal memory tasks (Ojemann and Fedio, 1968; Ojemann et al., 1971; Fedio and VanBuren, 1974; Ojemann, 1978a). In this chapter we will consider the information derived from these studies and from new studies of ours concerning the effects of stimulating various brain sites on other aspects of ongoing communication behavior, including reading, nonverbal oral-facial movements, and phoneme identification.

Methodology

Both the previously reported studies and the new stimulation studies from our group used standard tests of specific behaviors. These tests were administered to the patient as slides or from prerecorded tapes; all tests took place during the neurosurgical operation under local anesthesia. Repetitive trials are an essential feature of our standard tests so that some trials can be used to measure the control performance; these control trials are randomly interspersed with the stimulation trials. The alterations in performance on trials under various stimulation conditions are then compared to the performance on the interpolated trials without stimulation. In this way, the effects of stimulation can be separated from those of the unusual operating room testing environment, including any effects of opening the cranium. Special care is taken to ensure that the tests include controls for perceptual alteration and motor disturbance. Care must also be taken to keep the stimulating currents localized. In our studies, current localization has been accomplished in two ways: (a) by using current levels below the thresholds for EEG afterdischarge and (b) by using physiological guides, for example, maintaining the current below the threshold for behavioral changes that would represent spread to nearby structures (such as motor effects that represent the spread of ventral lateral thalamic stimulation to the immediately adjacent internal capsule). All of our studies used 60-Hz biphasic square-wave pulses of 2.5 msec. total duration from a constant-current stimulator. Cortical stimulation is bipolar between electrodes 5 mm apart; thalamic stimulation is delivered in a monopolar manner through an electrode measuring 1 / 5 mm. Stimulation sites on the surface are localized by their relation to the cortical veins, which are photographed at operation; the stimulation sites within the brain are determined using radiological landmarks. All data are recorded on tape, which allows for subsequent analysis.

In many cases, these stimulation studies only substituted a standard test situation for informal stimulation. For example, some stimulation identification of motor or language cortex is an integral part of safely conducting a cortical resection for epilepsy. In addition, standard testing has proven to be a more reliable method for obtaining this therapeutically necessary information as well as providing new knowledge on brain organization (Ojemann, Hoyenga, and Ward 1971; Ojemann, 1978b). Nevertheless, in all of these studies, the information presented to the patient regarding the methods to be used during the operation, from which the patient gives his informed consent, was always reviewed first by the University of Washington Committee for the Protection of Human Subjects. Informed consent for the specific stimulation study was obtained from each subject. Only adult patients were used as subjects.

Scope of Data Presented

The information derived from these stimulation studies bears on three major aspects of the neurological organization of human communication. The first is the general localization of the cortex concerned with language. Object naming is used as a global measure of language behavior since deficits in object naming are a ubiquitous characteristic of lesions anywhere in the classic language cortex. We address such questions as the extent of cortex in a given individual and the variability in the organization of language cortex between individuals with the usual pattern of cerebral dominance, as determined by preoperative intracarotid sodium Amytal testing. Next we consider studies subdividing this global language function into different mechanisms, including rate of production of speech sounds, nonverbal oral movements, perception of speech sounds, language structure (syntax), and verbal memory. Here we identify two major cortical subdivisions of human verbal communication, a motor-discrimination system and a short-term verbal memory system; we also identify local areas concerned with syntax. Finally, we discuss subcortical language mechanisms at the level of the lateral thalamus. There an interface with attentional mechanisms important both in verbal memory and in language output has been identified. We close with a discussion of the implications of these findings for the evolutionary aspects of social communication.

The Organization of Language in Human Cortex as Indicated by Alterations in Naming

Localization

A global assessment of the cortical areas involved in language in an individual patient and the variability in this area between patients has been provided by stimulation mapping during object naming tasks in a series of 10 patients undergoing left frontal-parietal-temporal craniotomy for medically intractable epilepsy (Whitaker and Ojemann, 1977; Ojemann, 1978b; Ojemann and Whitaker, 1978b). In these patients, the left brain was shown to be dominant for language by preoperative intracarotid Amytal testing. Preoperative IQs ranged from 77 to 110 (mean, 94). Naming was measured by showing slides of common objects; these could be named preoperatively by the particular patient without hesitation. Each slide incorporated a test phrase such as "This is a" printed above the object picture, which the patient was trained to read aloud before naming the object pictured. Each slide was shown for a constant period (4 sec in some patients and 5 sec in others). When stimulation occurred, the current was maintained for the entire time one slide was exposed. The same current, the largest that did not evoke afterdischarges in the area to be sampled, was used for all stimulations at all sites in a given patient. Because of the low threshold for afterdischarge, sites within the epileptic focus (in the anterior temporal lobe in most patients) were often not stimulated. Stimulation never occurred on immediately succeeding slides, and sequential stimulations were never done at the same site. Multiple sites were sampled in each patient, varying from 6 to 28 depending on the size of cortical exposure, but always sampling peri-Sylvian cortex. In addition, multiple samples of naming were obtained at each site (a mean of 2.8 samples per site). Naming errors were rarely seen on control trials without stimulation. When control errors occurred, only those sites with multiple errors during stimulation of $\leq 5\%$ probability of chance occurrence were considered as indicating language cortex.

This study indicates that in an individual patient, naming may be disrupted by stimulation applied to a wider area of cortex than the classic language area (Figure 5.1). Within that wide area, however, there is often discrete localization of naming; naming may be disrupted on every sample at one site, while no change occurs on samples of naming at sites within 2 cm on the same gyrus using the same stimulating current, as illustrated for the inferior parietal lobe in Ojemann and Whitaker (1978b). Graded responsiveness is also seen; at a constant current, some sites show errors on all samples (100% sites) and some on only part of the samples (partial sites) (Whitaker and Ojemann, 1977).

Variability

The stimulation sites in peri-Sylvian cortex in these patients were then pooled by their relation to sensory motor cortex and the end of the Sylvian fissure. Peri-Sylvian cortex was arbitrarily divided into a series of zones, and the percentage of patients showing disturbances of naming in each zone was determined (Ojemann, 1978b; Ojemann and Whitaker, 1978b). Only one zone showed naming changes in all 10 patients sampled there: the narrow band of cortex immediately in front of the motor strip (Figure 5.1). It represents only a small portion, the posterior one-third, of the traditional

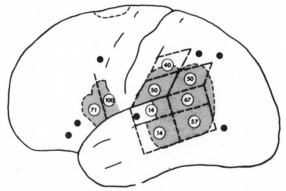

FIGURE 5.1. *The location of the classic language areas in the dominant hemisphere as adapted from Rasmussen and Milner (1977) are shown by the shaded areas. The small black dots show the most extreme locations where sites of 100% naming errors were evoked in our series of 10 patients with known left brain dominance for language. Partial naming errors were evoked from a much wider area of cortex in these patients, including all of the anterior temporal lobe. The straight dashed lines enclose the cortical zones where individual variability in localization of naming changes was assessed in these 10 patients. There are at least 5 patients with one or more samples of naming in each zone. The figures within each zone represent the percentage of patients who showed evoked naming errors in that zone, regardless of type or whether partial or 100%. The high degree of variability, except for the small area of inferior premotor cortex, is apparent.*

Broca's area. Elsewhere in the frontal cortex and throughout all of the parietal and temporal cortex, there is considerable individual variability in the localization of naming changes; in each of these zones, 29 to 86% of the patients with samples there (representing at least 2 patients in each sample) demonstrated no naming errors. This applies equally to the anterior two-thirds of Broca's area and to all of the posterior language areas. Indeed, in the classic Wernicke's area (superior temporal gyrus), not only were there patients who showed no naming errors, but all of the remaining patients demonstrated only partial naming errors at this site, in contrast to all other portions of classic language areas where at least some patients demon-strated 100% naming sites.

There seems to be a high degree of individual variability in the cortical organization of language in this patient population. Of these 10 patients, 9 had the onset of seizures at an early age; this may predispose them to some reorganization of the language cortex. Unfortunately, there is no com-parable information in terms of the detail of language organization in other patient populations. However, studies of morphological variability in "nor-mal" human cortex in both the language area (Rubens *et al.*, 1976) and the occipital visual cortex (Stensaas *et al.*, 1974) have indicated substantial variability as well. Of course, it is not known whether this morphological variability also reflects functional variability, but our data would suggest that. There is a hint that this functional variability may correlate with overall verbal performance. Patients with low verbal IQs are over-represented in the patients showing naming errors when inferior parietal sites were stimulated, as compared with patients with high IQs. At frontal and temporal sites the errors are seen about equally in patients with both high and low IQs, raising the possibility that a particular localization of naming may be related in some way to poorer overall functioning of the verbal systems.

Another possible source of variability is that the brain areas used for naming at any point in time can change, not over the course of hours, for repeated mapping during an operation usually shows the same pattern, but with brain maturation or with changing experience with a language. A sug-gestion of this comes from stimulation mapping of naming in bilingual pa-tients (Ojemann and Whitaker, 1978a). Each patient showed only a partial overlap of sites where stimulation altered naming in each language. Nam-ing in the patients' second language was altered from larger areas of cortex than the patients' primary language, even though in one case the secondary language had comparable nonstimulation error rates, was the language of the patient's surroundings and at the time of mapping the language most frequently used by the patient. One hypothesis to explain this difference in the size of the cortical areas used for naming in primary or second languages is that large numbers of neurons spread widely over the cortex are used for naming when a language is first acquired; with continued use or greater proficiency, the number of neurons or cortical areas needed for executing naming shrinks.

Stimulation mapping during naming at peri-Sylvian sites in the non-dominant hemisphere shows naming changes only from the inferior sensorimotor cortex (Ojemann and Whitaker, 1978b). Thus, stimulation-evoked naming changes from other than Rolandic cortex followed the lateralization of language as indicated by intracarotid Amytal testing.

We propose that this high degree of individual variability in language location in the dominant hemisphere (usually left hemisphere), present in all areas except the immediately premotor inferior cortex, is probably an evolutionarily significant aspect of our model of the neurological basis of human communication. This proposal is further discussed at the end of this chapter.

Subdivisions of Cortical Language Processes:

Motor Mechanisms

Localization of motor mechanisms has been assessed by stimulation effects on three different measures: naming, duration of the production of the speech sound /s/, and the ability to mimic repetitive and sequential nonverbal oral-facial movements.

INDICATIONS FROM NAMING

Naming errors were subdivided into those in which the patient said nothing at all (arrests) and those in which the patient was unable to correctly name, but demonstrated his ability to speak by either saying the test phrase "This is a" or misnaming the object (anomia). Arrests of speech were identified only in the sensorimotor cortex (bilaterally) and, in the dominant hemisphere, from cortex immediately anterior and posterior to it. In the 10 patients with the usual left dominance, and left cortical stimulation, arrests occurred in 56% of the patients with naming errors in the strip of cortex immediately anterior to sensorimotor cortex, 33% in sensorimotor cortex, and 25% in the parietal cortex immediately behind the Rolandic strip. Elsewhere, naming errors were all of the anomic type. The bilateral occurrence of arrests after Rolandic stimulation is similar to the results reported by Penfield and Roberts (1959), in which arrests of speech were frequently evoked from either sensorimotor strip. However, our data suggest that arrest responses in the dominant hemisphere are more localized to cortex immediately in front and behind Rolandic cortex than they were in Penfield and Roberts' data. Arrests during naming seem more likely with disruption of the final common pathway motor mechanism, whereas these are certainly intact in the anomic type of error. This, then, suggests that the final motor pathway mechanism is localized in para-Rolandic supra-Sylvian cortex.

INDICATIONS FROM TIMING OF SPEECH SOUNDS

Further information on the localization of motor speech mechanisms was obtained by measuring the effects of stimulation on the time taken to produce the speech sound /s/ as indicated by oscillographic analysis. These data were obtained from tape recordings of naming in seven of our patients by Dr. Bruce Smith of the Speech and Hearing Sciences Department at the University of Washington, using standard procedures and the Honeywell Visicorder 1508A run at 50 mm/sec (Smith, 1978). In three patients, the duration of the speech segment /is/ from "this" was measured; in the remaining four cases the segment /s/ was measured in a variety of naming and reading contexts, although never in the phrase final position. The average durations of /s/ segments under nonstimulation conditions and with stimulation at each cortical site for which speech samples were available were compared. Increases in the duration of /s/ of up to 90% were observed with mean increases, at affected sites, of about 30%.

Samples of the speech sound /s/ were available from frontal lobe sites in five patients undergoing dominant hemisphere stimulation. Slowing during stimulation, as indicated by at least a 15% change in phoneme duration, was observed in 61% of the 36 frontal lobe sites, many of these quite distant from what is traditionally considered to be motor or premotor cortex. Six patients had samples from stimulation of temporal and parietal lobe sites. Slowing during stimulation was evident in 18% of the 30 parietal sites and 40% of the 36 temporal sites. The relation of these duration changes to naming changes is also of interest. Only 35% of the pre-Rolandic frontal lobe sites that showed slowing also showed naming errors, indicating that there were a substantial number of frontal sites where slowing occurred even though naming was intact. A similar separation of naming and slowing occurs in the temporal lobe, where only 28% of the sites demonstrated such an association. On the other hand, 67% of the parietal lobe sites that showed slowing demonstrated naming changes, indicating a relatively close association between these functions in the parietal lobe. Only two sites, one in the frontal lobe and one in the parietal lobe in one patient, demonstrated significant decreases in the duration of the /s/ sound. Cortical sites where speech sound timing could be altered with stimulation did not tend to cluster in the immediate pre- or post-Rolandic cortex. The /s/ speech sound was measured in one patient during nondominant stimulation. No changes were noted from any of the 21 frontal, parietal, or temporal cortical sites outside the sensorimotor cortex. This study suggests that timing of speech production can be altered over wide areas of dominant cortex, including areas not identified as part of language cortex by evoked alterations in naming.

INDICATIONS FROM NONVERBAL ORAL-FACIAL MOVEMENTS

A recent study by Mateer and Kimura (1977) suggests that all aphasic speech disturbances, even when speech is fluent, are associated with some degree of impairment in acquisition and performance of complex nonverbal oral movements. We have measured the effects of cortical stimulation on nonverbal oral-facial movements in the dominant hemisphere of four patients and in the nondominant hemisphere of one patient. The patients were trained preoperatively to mimic simple oral-facial postures in response to slide pictures of the movement. Postures represented the terminal position for simple movements such as lip protrusion or tongue lateralization. Slides were each composed of three separate photographs. In one series the three photographs depicted the same posture, and the patient was instructed to repeat the same movement three times. In the other series, three different postures were depicted that required different movements to be repeated in sequence. Thus, this task measured the patients' ability to produce both the same repetitive oral-facial movement and a sequence of different oral-facial movements. Oral-facial movements were recorded by a television camera on video tape, along with markers indicating the onset and termination of stimulation. They were evaluated blindly without use of the stimulation marker channel.

Production of even a single movement was disrupted by stimulation at all five motor cortex or immediate premotor cortex sites in the three patients with samples in the dominant hemisphere. In each case, consistent speech arrest during naming was observed at the same site. Single movements and their repetition were intact at all other sites, but the ability to carry out sequential movements was disrupted by stimulation at three of eight frontal sites anterior to this premotor area, four of five parietal sites, and two of seven temporal sites in the four patients with samples in the dominant-hemisphere areas. The location of the sites showing disruption of sequential motor tasks on multiple stimulation trials is illustrated in Figure 5.2. These sites are in the frontal premotor area, parietal cortex immediately posterior to Rolandic cortex, and just beneath the Sylvian fissure in

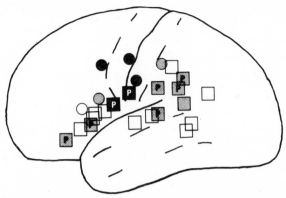

FIGURE 5.2.   *Sites where the effects of stimulation on oral-facial motor movements and phoneme identification have been assessed. Oral-facial motor movements were assessed in four patients at all sites included in this figure; phoneme identification was assessed in only three of those patients, at the sites identified by squares. The black circles and black squares indicate sites where single oral-facial movements were disrupted by stimulation; and stippled circles and squares indicate the sites where only sequential oral-facial movements were disrupted by stimulation, single movements being intact. The letter P within a square indicates the sites where phoneme identification was disrupted by stimulation. The following are apparent: the disturbance of single nonverbal oral movements in the motor and premotor cortex; sequential nonverbal oral movements in frontal, parietal, and superior temporal lobe; and the close association between the sites at which nonverbal oral-facial movements could be altered with stimulation and the sites at which phoneme identifications was altered. The sites at which stimulation alters these functions outline what we have called the motor-discrimination subdivision of language.*

posterior superior temporal gyrus. The alteration in sequential movements included completely incorrect movements, hesitation, multiple attempts at revisions, incomplete excursions, and errors in ordering. Such errors were not seen in control trials. Thus, within the classic language zone there are sites where sequential nonverbal oral-facial movements can be disrupted.

Stimulation of peri-Sylvian cortex in the nondominant hemisphere of one patient showed no sites where only sequential oral-facial movements could be altered, though stimulation of a site in the inferior motor cortex, but not a site in the premotor cortex, disturbed both single and sequential movements.

Data on stimulation effects on both the rate of speech articulation of the /s/ sound and sequential nonverbal oral-facial movements are available in three patients. Two-thirds of the sites that showed changes in the rate of articulation of speech sounds also showed changes in the ability to produce the oral-facial movements. Thus, there is considerable overlap between these two phenomena, suggesting that they measure the same system. The motor system for oral movements includes the premotor area of the dominant hemisphere, where even simple oral-facial movements are disrupted, suggesting that this is a part of the final common motor pathway, and wider areas in frontal, superior temporal, and parietal lobes, where more complex sequential movements are altered by stimulation. These latter areas are probably involved in the complex organization of motor activity leading to speech, which is then probably relayed to the final common pathway. Note that the areas where performance on a wholly nonverbal oral-facial motor task is altered occupy a substantial portion of both anterior and posterior classic language areas.

Phoneme Identification

The alterations in identification of imbedded phonemes evoked by cortical stimulation were measured in three patients. Patients were presented with taped live voice auditory stimuli consisting of the stop consonants /p/, /b/, /t/, /d/, /k/, and /g/, imbedded in the same nonsense carrier syllable, "/ae_____ma/." They were trained to identify the imbedded phoneme and report this aloud. Stimulation was applied during the presentation of the auditory stimulus and terminated before a response was to be given, to ensure that there would be no interference by potential disruption of motor mechanisms. Phoneme identification was measured at the sites illustrated in Figure 5.2. Sites where this was disrupted on more than one sample are also identified in that figure. Errors without stimulation were uncommon on this test, occurring on an average of 13% of the items (range, 3 to 28%) in these patients.

Errors in phoneme identification involve predominantly frontal and parietal sites, half of the frontal sites, three-fourths of the parietal, but only one out of seven temporal sites. The most striking finding was the association between the sites where phoneme identification was altered and the sites where oral-facial movements were altered in these patients. Oral-facial movements, either single or sequential, were disrupted at nine sites in these three patients; at eight of those nine sites, phoneme identification was disrupted as well. There was no site at which phoneme identification was disrupted that did not also show disturbance of oral-facial movements.

This association between motor movements and phoneme identification is strong and unique direct evidence for a motor model of speech perception (Liberman *et al.*, 1967). Additional evidence for this association between oral motor and speech perception mechanisms is available from the observations of Darwin, Taylor, and Milner (1975) that nonaphasic patients who had surgical face area excisions were impaired in single phoneme identification, despite good pure tone acuity and comprehension of connected speech. In addition, patients with nonfluent aphasias, though they are frequently said to have intact comprehension, often show defects on multiple commands or nonredundant speech, such as used in the Token Test (De Renzi and Vignolo (1962). The model of cortical organization of speech supported by these data, then, is that of a system involving posterior inferior frontal lobe, anterior parietal lobe, and a portion of the superior temporal lobe that coordinates sequential oral-facial movements (whether these movements are verbal or nonverbal) and is involved in speech comprehension at the level of phoneme identification. The same areas of cortex appeared to be common to both.

Syntactic Ability

Syntactic abilities were measured as part of a reading task. The patient viewed a black and white slide showing a printed sentence of eight or nine words, constructed so that the patient reads an auxillary verb form and supplies the future form of some verb appropriate to the meaning of the sentence where a blank is indicated. Stimulation at various cortical sites was applied during the entire exposure of the slide that was to be read. Data are available from the stimulation of 61 frontal, parietal, or temporal sites of the left dominant hemisphere in five patients.

This is a more difficult task than those discussed above and allows the patient more opportunity for variable responses. Indeed, errors in the form of incompletely or inaccurately read sentences occurred on 22% (range, 5 to 76%) of the nonstimulation trials in the five patients studied. Reading errors were most commonly evoked by stimulation of the 24 frontal lobe sites anterior to the motor strip. Fifty-eight percent of these sites showed errors on two or more samples of reading; errors occurred at at least one frontal site in all patients. Reading errors were least common at parietal lobe sites, being present in only 28% of these sites in only three patients. Each patient also showed at least one temporal site where reading was altered by stimulation, and this represented 37% of the temporal lobe sites tested. Reading and naming deficits with stimulation of the same site were common frontally, where 80% of the sites showing reading deficits also showed naming deficits, but less common elsewhere. Only 33% of the parietal sites and 36% of the temporal sites showed this relationship.

Reading errors fall into two general categories on the basis of this association with naming errors. One category includes total or partial arrests of reading and perseveration. These errors include no speech production during stimulation, or correct reading of only the first few words on the slide, or perseveration on one of these words. These types of errors were the common error on control trials and represented the reading error made during stimulation at all 11 sites where naming was also altered, and at 2 of the other 11 sites where naming was correct.

The second category includes jargon and grammatical reading errors. Jargon responses evoked during stimulation demonstrated fluent reading with many errors. For example, in response to the sentence, "If it is sunny next Saturday" the patient said "If it is searnest sucky." A grammatical error is, for example, responding (during stimulation) to "If my son is late for class again he _____ principal" with "If my son will getting late class today he'll see the principal." Jargon and grammatical errors were never seen on control trials. They never occurred at any of the 11 sites where naming changes were also evoked. These types of errors were evoked at 9 sites in the frontal, parietal, and temporal lobes, all sites where naming was correct.

Thus, the first category of reading errors probably represents evoked changes in language mechanisms common to both reading and naming. This is probably the motor-discrimination system, as either speech slowing or oral-facial movement disruption occurred at 80% of those sites showing both evoked reading and naming changes where the additional functions were sampled. The second category of reading errors can be further subdivided by the relation to this motor system. Jargon errors were seen only from parietal and temporal sites and were associated with speech slowing or oral-facial movement disruption at three-quarters of these sites. On the other hand, sites showing grammatical errors have little relation to the motor system. None of these sites showed motor disruption where tested and only 40% showed speech slowing. Rather, the sites of grammatical errors seem to identify specific areas concerned with syntax. One type of grammatical error involved agreement of verbs. These errors were seen at three sites in two patients, in each at a mid–frontal lobe site and in one at a superior temporal site. A second type of grammatical error evoked at two sites in the parietal-temporal junction of one patient consisted of omissions or incorrect use of small words: pronouns, prepositions, and modifiers.

Short-Term Verbal Memory

Short-term verbal memory has been measured using an adaptation of the single-item paradigm of Peterson and Peterson (1959). Input to memory is object naming, a standard verbal distraction is then provided, in previously reported data, by counting backward by 3's from a pictured two-digit number greater than 30, in new data by the reading task. Following this 6-sec distraction, retrieval from short-term verbal memory is cued by the word "recall" appearing on the screen. The patient has been trained to read back the name of the previously pictured object on this trial. Stimulation occurs only during the input, storage, or output phases of the test on selected trials and during both the input and output phases on other trials. These are pseudo-randomly interspersed with trials without stimulation, which measure control performance.

Data from the dominant left hemisphere of six adult patients have been reported by Ojemann (1978a) and are shown in Figure 5.3. The sites at which stimulation altered naming and those at which short-term memory was altered were largely separate. Indeed, currents that had reliably produced anomia in these patients on no occasion disrupted short-term memory, regardless of whether they were applied during input, storage, or output. In addition, the effects of stimulation during different parts of the test varied with different cortical sites. Stimulation during input and particularly during storage disrupted short-term memory when applied to inferior parietal and posterior temporal cortex, surrounding but not touching the sites where naming was altered. Stimulation applied during the output part of the test altered short-term memory largely at the posterior frontal lobe sites, immediately anterior to those disrupting naming. These findings suggested that episodic and generalized word memories are largely separate at the cortical level, and that the frontal lobe sites are particularly involved in retrieval from short-term memory, whereas the parietal and temporal lobes, which are involved in the input and (especially) storage aspects, might be the site of the active storage process of short-term memory. These

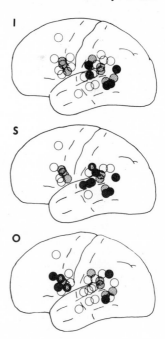

FIGURE 5.3. *Sites at which naming and short-term verbal memory were assessed in the six patients previously reported by Ojemann (1978a). Sites with naming changes are shown in each schematic brain by shading. Sites with evoked short-term verbal memory changes under each of the different test conditions are shown by the filled circles. Short-term verbal memory test conditions: I, stimulation during input only; S, stimulation during the storage phase only; O, stimulation during output only. Storage stimulation data are available for only five patients. The letter B indicates those few sites where both short-term verbal memory and naming were altered at the same location in the same patient. The general separation between sites where naming is altered and sites where short-term verbal memory is altered by stimulation is apparent, as is the difference in the sites where stimulation during output and stimulation during input or storage alter short-term verbal memory.*

findings are complemented by those of Fedio and Van-Buren (1974), where no changes were evoked from stimulation of the nondominant cortex during a somewhat similar short-term verbal memory task.

Short-term verbal memory measurements have also been made in three additional adult patients, with naming, reading, speech, and nonverbal oral motor and phonemic discrimination measurements taken as well. The only other difference from the earlier series is that reading was used as the distractor task instead of counting. The results are generally similar. Input and storage errors represented 8 of 12 short-term memory errors at parietal and temporal sites, but only 1 out of 7 frontal sites, the remainder being output errors. Sites at which short-term verbal memory can be altered by stimulation are generally separate from those sites at which the other functions are altered. Two-thirds of the sites where all of these functions were measured that showed short-term verbal memory changes showed no evoked changes in other functions. Sites showing input and storage short-term memory errors are particularly likely not to show any other evoked changes. Seventy-five percent of these sites showed no change in the other functions with stimulation.

Output short-term memory errors showed two different relationships to other functions. Two-thirds of these occurred at frontal (and one temporal) sites, where naming and reading were also altered, usually arrested, often also with alteration in both single and sequential oral motor movements. These evoked memory errors probably reflect only a disruption of the final common motor pathway, rather than any alteration of memory per se. However, the remaining output errors, usually from frontal sites, but occasionally from temporal and parietal sites, occurred without alteration in naming, reading, speech rate, or motor movements. At these sites, stimulation probably alters short-term verbal memory retrieval mechanisms.

Subcortical Language Function

Attentional Mechanisms

Up to this point, we have considered the cortical organization of the human language communication system. We turn now to similar evidence on the role of lateral thalamic mechanisms in human language. Access to the thalamus in neurosurgical operations under local anesthesia occurs during the course of an operation called thalamotomy, used for the treatment of a group of motor disabilities (dyskinesias): Parkinson's disease, dystonia, and intention tremor. These operations are performed under local anesthesia so that physiological landmarks can be used to place the lesions, including the effects on the dyskinesia, which disappears under anesthesia. Electrical stimulation is one of the techniques for determining physiological placement of the lesions. We have adapted this to our standard measures of naming and short-term verbal memory.

These findings have been thoroughly discussed elsewhere (Ojemann *et al.*, 1971; Ojemann, 1975, 1977). Human lateral thalamic stimulation in the dominant hemisphere appears to alter a mechanism that we have called the *specific alerting response*, which has the following characteristics. Stimulation during input to short-term verbal memory decreases later retrieval errors. This effect can be followed in long-term memory as well (Ojemann, 1975, 1976, 1978c). The same current applied to the same sites blocks retrieval of material already in short-term memory. Lower currents at these same sites accelerate memory processes, but not speech (Ojemann, 1974).

Naming errors of the anomic type have also been evoked from the lateral thalamus, in discrete areas of the posterior medial central parts of the ventral lateral thalamus, extending into the anterior superior pulvinar. Anomia is similar to the naming changes generally evoked only from language and not motor cortex (Penfield and Roberts, 1959). Evoking this type of naming error from this discrete portion of lateral thalamus suggests that it, too, is involved in language processes (Ojemann *et al.*, 1968; Ojemann and Ward, 1971). Study of language changes after spontaneous thalamic lesion, usually hemorrhage, also suggests a language role for the thalamus in the dominant hemisphere (Fisher, 1958; Mohr *et al.*, 1975; Ojemann, 1976; Reynolds, Harris, Ojemann, and Turner, 1978). We now interpret these naming changes as most likely representing defects in retrieval of the object name from long-term memory, based on a high negative correlation between the severity of the language disturbance after a thalamic lesion and a measure of the intensity of the stimulation-evoked short-term memory changes that make up the specific alerting response in that same patient (Ojemann, 1975, 1976). This correlation suggests that there is a common substrate to both short-term verbal memory and language at the thalamic level, the "specific alerting response."

This response in the ventral lateral thalamus acts as a gate determining access to memory at any point in time and modulating the likelihood of later retrieval. It is strongly lateralized in man, in that nondominant thalamic stimulation shows none of these effects for verbal material, though they are present for visual spatial information; the converse is true for the dominant thalamus (Ojemann, 1977). When this response is activated, attention is focused on material in the external environment, which then enters short- and long-term memory in such a way that its later retrieval is enhanced. Material already internalized is less available when this response is activated. The relative intensities of this response in the right or left brain, then, may well determine the ease with which verbal or visual spatial features of the environment enter memory or are available from memory.

This altering response is evoked from parts of the lateral thalamus (especially the ventrolateral nucleus) that are usually considered to be part of the motor system, based on major anatomical connections and other known functions. The ventrolateral thalamus seems to coordinate altering and motor processes in a mechanism phylogenetically important for motor learning in animals, which has been adapted in man for verbal processing in the dominant hemisphere and for visuospatial processing in the non-dominant hemisphere. This, then, would be another site in addition to the cortical motor-discrimination system where functions related to human communication develop in motor portions of the brain.

Motor Mechanisms

In addition to this attentional mechanism, the lateral thalamus most likely also interfaces with other evolutionarily older aspects of motor function. Slowing of motor speech mechanisms occurs during left, not right, ventral lateral thalamic stimulation (Mateer, 1978), and medial portions of the lateral thalamus may well interface with the respiratory substrate for speech. Stimulation there evokes an inhibition of respiration in expiration, the appropriate respiratory change for speech, with a substantially lower threshhold from the left thalamus than from the right (Ojemann and Van-Buren, 1967). The thalamic sites at which stimulation evoked respiratory inhibition and those at which anomia was evoked in a separate patient series showed considerable correspondence (Ojemann, 1977).

The Evolutionary Significance of the Cortical and Subcortical Patterns of Language and Memory Organization as Derived from Stimulation Studies

The major features to be derived from these studies, then, are as follows.

1. There is a high degree of individual variability in the cortical organization of language, at least as measured by naming. Only from a narrow portion of the inferior premotor cortex, probably a final motor pathway for speech, was naming uniformly altered in all of our left hemisphere-dominant patients with samples in that hemisphere.

2. There are two major subdivisions located in the language-dominant hemisphere. One is a motor-discrimination subdivision, controlling the production of sequences of discrete movement. In man, this seems to be concerned with oral-facial movements, though there is considerable evidence that it is also involved in the control of complex movements, gestures, and manual communication systems, including those of the deaf (Kimura, 1976). The same subdivision seems to be involved in the discrimination of speech sounds. It is centered principally in frontal and parietal cortex immediately anterior and posterior to the motor strip. A second, separate subdivision serves short-term verbal memory. It seems to have two portions: a retrieval portion most often located anterior to the frontal lobe part of the motor-discrimination system, and a portion concerned with input and especially storage aspects of short-term memory, which is located in the parietotemporal lobe. Short-term verbal memory is rarely altered at the same brain sites as are the traditional language functions, naming and reading. Rather, naming and reading largely but not wholly overlap with the motor-discrimination system, but there are also areas that subserve only one of these functions. One of these areas subserves syntactic aspects of reading. There is considerable individual variability in the exact cortical location of these subdivisions, but the general relationship between subdivisions has been relatively constant in each dominant hemisphere we have studied to date. Such subdivisions of cortical functions are not entirely new or original. Wernicke, and, before him, Meynert, considered frontal lobe function to be related to movement and temporal lobe function to be related to memory (Boller *et al.*, 1977).

3. At a subcortical level there is an attentional mechanism that also involves a traditionally motor portion of the brain, the ventral lateral thalamus. In man this attentional mechanism in the dominant hemisphere modulates language and verbal memory.

One would expect to find each of these systems represented in lower forms, particularly in various primates. Thus, a lateralized sequential control system should be identifiable, presumably arising in relation to motor cortex. Lateralized cortical representation of short-term memory should develop at the periphery of this sequential motor system, and thalamic attentional mechanisms should be identifiable in motor learning. There is some evidence of the appearance of these systems in animals. For example, Dewson (1977) has identified a lateralized auditory short-term memory system in the temporal lobe of monkey. Chimpanzees can be taught a manual communication system, though whether this is based on lateralized brain mechanisms is presently unclear. The stimulation studies of Mahut (1964) and of Wilburn and Kesner (1972) suggest that a system like the specific alerting response may be present in the thalamus of various animals and may be involved with motor memory.

The finding of a high degree of individual variability in the overall organization of language and its subdivisions at the cortical level has major evolutionary implications. Substantial variability is often a characteristic of rapidly evolving species. This may be true of man in relation to language mechanisms. If some patterns of individual functional cortical organization correlate with high levels of verbal ability, as suggested by our observation that language representation in the parietal operculum seems to be associated with lower verbal abilities, then an environment generating substantial selective pressures of a cultural nature for facile verbal processes is likely to lead to a comparatively rapid further biological evolution of human language systems.

Acknowledgments

Some of the patients included in these studies were under the care of Dr. A. A. Ward, Jr., and Dr. A. R. Wyler. Their cooperation in our testing of these patients is appreciated. Dr. C. Dodrill provided IQ and intracarotid Amytal data.

References

Boller, F., Kim, Y. and Mack, J. (1977). *Studies in Neurolinguistics 3*, 1–64.
Darwin, C., Taylor, L., and Milner, B. (1975). *Neuropsychologia 13*, 132.
DeRenzi, E. and Vignolo, L. A. (1962). *Brain 85*, 665–678.
Dewson, J. (1977). *In* "Lateralization in the Nervous System" (S. Harnael, R. Doty, L. Goldstein, J. Saynes and G. Krauthamer, eds.), pp. 63–74. Academic Press, New York.
Fedio, P. and VanBuren, J. (1974). *Brain and Lang. 1*, 29–42.
Fedio, P. and VanBuren, J. (1975). *Brain and Lang. 2*, 78–100.
Fisher, C. M. (1958). *In* "Pathogenesis and Treatment of Cerebrovascular Disease" (W. S. Fields, eds.) pp. 318–342. Charles C. Thomas, Springfield, Ill.
Kimura, D. (1976). *Studies in Neurolinguistics 2*, 145–156.
Liberman, A. M., Cooper, F. S., Shankweiler, D. P., and Studdert-Kennedy, M. (1967). *Psychol. Review 74*, 431–461.
Mahut, H. (1964). *J. Comp. Physiol. Psychol. 58*, 390–395.
Mateer, C. (1978). *Neuropsychologia 16*, 497–499.
Mateer, C. and Kimura, D. (1977). *Brain and Lang. 4*, 262–276.
Mohr, J. P., Watters, W. C., and Duncan, G. W. (1975). *Brain and Lang. 2*, 3–17.
Ojemann, G. (1974). *Neuropsychologia 12*, 1–10.
Ojemann, G. (1975). *Brain and Lang. 2*, 101–120.
Ojemann, G. (1976). *Studies in Neurolinguistics 1*, 103–138.
Ojemann, G. (1977). *Ann. N.Y. Acad. of Sci. 299*, 380–396.
Ojemann, G. (1978a). *Brain and Lang. 5*, 331–340.
Ojemann, G. (1978b). *J. Neurosurg. 50*, 164–169.

Ojemann, G. (1979b). *In* "Modern Concepts in Psychiatric Surgery" (E. Hitchcock, H. Ballantine and B. Meyerson, eds.) pp. 103–110. Elsevier, Amsterdam.

Ojemann, G., Blick, K., and Ward, A. A., Jr. (1971). *Brain 94*, 225–240.

Ojemann, G. and Fedio, P. (1968). *J. of Neurosurg. 29*, 51–59.

Ojemann, G., Fedio, P., and VanBuren, J. (1968). *Brain 91*, 99–116.

Ojemann, G., Hoyenga, K., and Ward, A. A., Jr. (1971). *J. of Neurosurg. 35*, 203–210.

Ojemann, G. and VanBuren, J. (1967). *Arch. Neurol. 16*, 74–88.

Ojemann, G. and Ward, A. A., Jr. (1971). *Brain 94*, 669–680.

Ojemann, G. and Whitaker, H. (1978a). *Arch. Neurol. 35*, 409–412.

Ojemann, G. and Whitaker, H. (1978b). *Brain and Lang. 6*, 239–260.

Penfield, W. and Jasper, H. H. (1954). "Epilepsy and the Functional Anatomy of the Human Brain." Little, Brown, Boston.

Penfield, W. and Perot, P. (1963). *Brain 86*, 595–696.

Penfield, W. and Roberts, L. (1959). "Speech and Brain Mechanisms." Princeton Univ. Press, Princeton.

Peterson, L. and Peterson, M. (1959). *J. Exp. Psychol. 58*, 193–198.

Rasmussen, T. and Milner, B. (1977). *Ann. N.Y. Acad. of Sci. 299*, 355–369.

Reynolds, A., Harris, A., Ojemann, G., and Turner, P. (1978). *J. of Neurosurg. 48*, 570–574.

Rubens, A., Mahowald, M., and Hutton, J. (1976). *Neurology 26*, 620–624.

Schaltenbrand, G. (1975). *Brain and Lang. 2*, 70–77.

Smith, B. (1978). *J. Phonetics 6*, 37–67.

Stensass, S., Eddington, D., and Dobelle, W. (1974). *J. Neurosurg. 40*, 747–755.

VanBuren, J., Fedio, P., and Frederick, G. (1978). *Neurosurgery 2*, 233–239.

Whitaker, H. and Ojemann, G. (1977). *Nature 270*, (5632), 50–51.

Wilburn, M. and Kesner, R. (1972). *Exp. Neurol. 34*, 45–50.

6

Language Representation in the Brain[1]

JASON W. BROWN

Ce qui nous manque aujourd' hui, ce ne sont pas
les faits, ils surabondent, mais plutôt un esprit
synthétique qui saura les comprendre et les
analyser tout en les rapprochant et qui saura en
faire ressorter la profonde unite.

C. v. Monakow, 1920

In spite of more than a century of research on the organization of
language in the human brain, a period that has produced a wealth of
knowledge concerning the various forms of language pathology with focal
brain lesions and the effects of brain stimulation in various sites on the surface and in the depths, we still do not have a coherent model of the neural
bases of language. In the classical accounts of the early neurologists, a
series of interconnected cortical areas or centers, as implied by damage in
the aphasias, were assumed to be the repositories of the physiological processes of language function. The various schemas, both old and new, that
resulted from this approach have by now been largely abandoned. That
they still to some extent survive is a reflection of the lack of alternative concepts of brain–language organization. Clearly there is a need for a fresh
look at the whole problem.

[1] The preparation of this chapter was aided by NIH research grant NS 13740

NEUROBIOLOGY OF
SOCIAL COMMUNICATION IN PRIMATES

Localization: Some Problems

What Is the Effect of a Brain Lesion?

The simplest response to this question is that the lesion damages or destroys something—a function, a process, a strategy—located in or mediated through the damaged area. But what can be said of this "something"? A lesion in the left Wernicke area (posterior T1) may lead to paraphasia. What does paraphasia signify? Is it a product of a damaged Wernicke area? Is it, as Wernicke himself thought, a product of a disinhibited Broca area? Does it arise from neighboring intact brain, or is it a manifestation of an aphasic right hemisphere? Cases have been described of a left Wernicke area lesion where the paraphasia has either disappeared or remained unchanged after a second lesion in the right hemisphere (Kuttner, 1930; Nielsen, 1944; Schilder, 1953). Symptoms in unilateral cases with recovery may reappear following a second lesion in the same hemisphere, while repeated partial lesions in the left-hemispheric speech zone, for example, multiple surgical excisions, eventually destroying that zone, may not necessarily lead to permanent aphasia. Nonfluent aphasics may show "recovery" during psychotic states (Robinson, 1977), and I have observed a fluent aphasic with word finding difficulty and some paraphasia to improve during a psychotic regression. I have had under my care a jargon aphasic whose speech became more intelligible during bouts of sleeptalking. I have also observed nonfluent aphasics to show transient improvement in speech during ictal or focal seizure episodes. Cases of (generally) posterior lesions may have relatively normal speech and deteriorate to jargon after several minutes of conversation. In such cases we speak of "fatigue" or "blocking." There is also the observation of improved speech under emotion, or while singing. In all of these cases, we have a static brain lesion but a changing clinical picture. How are we to understand such phenomena?

Recovery and Localization

The problem of recovery of function is the same problem as that of symptom formation. The global symptoms of an acute lesion and the more restricted effects of a chronic lesion reflect a state of pathology no less directly than do the symptoms of an intermediate phase. There is a tendency to attribute the acute symptoms to diaschisis, the chronic symptoms to right-hemisphere compensation, and the subacute aphasic syndrome to the effects of the lesion itself. These ad hoc interpretations point out the fundamental inadequacies in our understanding of this problem. Clearly, if we could explain symptom formation at any stage in the life history of the lesion, we could explain it at every stage. The problem that is avoided in such roundabout interpretations is not just that of lesion and symptom, but of structure and function; in other words, it is the central issue of the mind–brain relationship. Seen in this light, such theories as diaschisis or "remote effects," right-hemisphere compensation, reorganization, and so on do not attack the problem of the symptom and the lesion head-on, but simply transfer the problem to a different region of the brain or to a new level of discourse.

Diaschisis

Von Monakow (1914) proposed a continuum to exist between patients without aphasia, those with recovery over several days, weeks, or months, and the occasional aphasic with restitution over a period of a year or more. This continuum was taken to reflect the gradual fading away of a state of inhibition (elevated discharge threshold) induced in neighboring and distant (homologous) cortex by brain damage. This inhibitory state was termed diachisis:

> A group of functional symptoms of "inhibition" at a distance (analogous but not identical with "shock") resulting from the sudden interruption of a permanent excitation which acts on more or less distant regions which are in direct associative or subordinate relationship with the destroyed area. When the influence of this cortical area is suddenly suppressed, the function of those distant regions connected to it is profoundly disturbed, inhibited, paralyzed, and able to be re-established only by striking a new path over neuronal complexes [Ladame and von Monakow, 1908.]

According to von Monakow, two types of symptoms could be discerned: those due to diaschisis and those due to the underlying brain lesion. These were not distinguished on a qualitative basis. When diaschisis wore off, the symptoms of the lesion itself were those that remained.

The neuronal concept of diaschisis is supported by some experimental work (Keminsky, 1958; see also Teuber, 1974) and has been explained on a vascular basis by Meyer and colleagues (1971). The view that transcallosal inhibition of the opposite (intact) speech area plays a role of aphasia symptomatology (Monakow, 1914) preceded by many years current notions of dominance establishment through callosal-mediated suppression of the right hemisphere.

A variant of the diaschisis theory has been proposed by Luria (1963), who described two types of function loss: permanent symptoms resulting from cell destruction and temporary symptoms due to interference with neuronal excitability and conductivity. Similarly, Zaimov (1965) argued that aphasic symptoms reflect varying degrees of cellular impairment in areas surrounding the lesion center, occurring in a manner comparable to

Wedensky inhibition, an alteration in the conductivity of a peripheral nerve following acute injury.

There seems little question but that something akin to diaschisis does occur, particularly in the sense of a "suppression" by an acute lesion of the opposite hemisphere. The effect may be more pronounced with lesions of the dominant side and may also occur with subcortical lesions. An example of this would be a sudden lesion of the left thalamus producing a state resembling the end stage of slowly progressive bilateral pathology (Brown, 1979).

If we accept that inhibition of the right hemisphere by an acute left-sided lesion accounts for the initial state of the patient, say a global aphasia, then a relaxation of this effect would permit the evolution of the disorder to a partial, for example, Broca's aphasia. This is another way of describing right-hemisphere compensation (see below). The real question is: What is the relation of global to Broca's aphasia, and the intervening stages between these forms, and how are we to understand each symptom complex, that is, that of global aphasia and that of Broca's aphasia, in relation to some neural substrate? The concept of inhibition at a distance is a subterfuge explanation. What is it that is inhibited, and how does this inhibition lead to change from one symptom to another? This is the central issue. We see that the problem of a *lesion* of an area, and the resultant symptom, is the same problem as that of *inhibition* of an area and the resultant symptom. In one case there is anatomical destruction and in the other physiological inactivation. However, from the point of view of an understanding of symptom formation, the fact that the latter is reversible is irrelevant. The problem is in our conceptualization of the overall bihemispheric structure that underlies language production both in the normal state and in pathology. A brain model that can account for linguistic change in aphasia, that is, the transition from one aphasic symptom to another, and the correspondence of these transitions to brain structure, should also be able to accommodate the effects of "diaschisis"—whether these effects are local, involving neighboring cortex, or at a distance, involving the opposite hemisphere.

Right-Hemisphere Compensation

The possible role of the right hemisphere in aphasia has been discussed since Hughlings Jackson first attributed to it the residual utterances and automatisms of the motor aphasic. Henschen (1920) repeatedly argued for right-hemispheric function in a variety of aphasic and related disorders. His argument was based on cases where large or strategically located left-hemispheric lesions obviated performance through that hemisphere, or on the effects of a second lesion on the right side.

These early observations were supported by findings of some language following complete surgical removal of the left hemisphere for glioma (Hillier, 1954; Smith & Burklund, 1966) and speech loss on injection of intracarotid Amytal to the right hemisphere in left-damaged aphasics (Kinsbourne, 1971). Further evidence comes from studies in commissurotomized subjects (e.g., Zaidel, 1977) and a well-studied case of early language deprivation (Curtiss, 1977). Dichotic listening studies in aphasia (Johnson *et al.*, 1977; Yeni-Komshian, 1977) provide inferential support for the probable role of the right hemisphere in aphasic symtomatology. How can we characterize the right-hemisphere contribution to aphasic (and normal) language? Does the right hemisphere account for the various degrees of insufficiency in aphasia, or is the right hemisphere truly *aphasic*? In other words, does aphasic language emanate from the right hemisphere or does the right hemisphere support whatever level of adequacy the aphasic achieves? And, with regard to a presumptive brain model, is the right-hemisphere effect a compensatory one, in which a mirror system either limited in capacity or differing in design "takes over" for the damaged left side, or are structures in the right hemisphere part of a unitary bilateral organization mediating language production? We may also ask: What are these brain structures and how are they organized?

Studies of right-hemisphere language in callosal patients indicate that comprehension is determined by the semantic and lexical, rather than phonological, features of the test material. On the other hand, "right-hemisphere speech" is characterized by its agrammatic nature. This was the most striking feature of the Genie case (Curtiss, 1977), as well as of Smith and Burckland's (1966) left-hemispherectomy patient. It is of interest in this respect that agrammatism is the predominant aphasic manifestation of right-hemisphere lesions in dextrals (Brown and Hecaen, 1976), whereas right-hemisphere lesions in nonaphasic dextrals have been noted to produce semantic impairments in comprehension (Lesser, 1976). In other words, from the point of view of pathology, right-hemisphere language comprehension is organized about a semantic base, and right-hemisphere language expression is characterized by agrammatism. These findings have structural implications to which we will return later.

Regarding Amytal studies, these have to be interpreted with great care. Goodglass (1971) has raised many salient criticisms of Kinsbourne's (1971) paper, which have gone unanswered. In my own limited experience with this technique, the effects of administering Amytal to the right hemisphere appear to have depended on the type of aphasia. In one phonemic (conduction) aphasic left injection had little effect on aphasic performance, while right injection produced marked phonemic paraphasia approaching jargon. Yet, in another aphasic with neologistic jargon, right injection had no effect on the jargon, which was abolished by injection on the side of the lesion.

One explanation for this difference is that the aphasia type itself reflects the degree of participation of the right hemisphere. Thus, in conduction aphasia, there are phonemic errors in speech with good comprehension and, in neologistic jargon, there are severe phonemic (and possibly semantic) errors in speech with poor comprehension. Keeping in mind the inner relationship that exists between comprehension and expression in the fluent aphasias (Brown, 1972), it is likely that the milder phonemic errors and good comprehension in the first case pointed to right-hemisphere participation, and thus there was deterioration with *right*-side injection, whereas the severe jargon and lack of comprehension in the second case pointed to an absence of right-hemisphere participation, and so there was deterioration with *left*-side injection. Evidence in support of this view has been described by Hamanaka *et al.* (1976), in an Amytal study of a conduction aphasic. In Amytal studies of left-handers (Milner *et al.*, 1966), a dissociation in language errors was noted in some patients according to the side of injection. Thus, word finding errors were present with left-side injection, whereas errors in series speech and counting were noted in association with injection of the right side. While the error analysis of such cases is not sufficiently detailed to be certain of aphasia *type* following left- or right-side injection, this finding, together with the above observations, suggests that an aphasia represents the *level* to which the utterance can be processed by both hemispheres working together as a single system; that is, that an aphasia secondary to a brain lesion or an Amytal injection reflects the degree to which the lesion or injection attenuates language processing in a bilaterally organized system.

Plasticity

Brain damage early in life may have less serious consequences on language than similar lesions later on. Studies of childhood aphasia (Hecaen, 1976) and early hemiplegia and hemispherectomy (Krynauw, 1950) demonstrate relatively good recovery (though see Dennis and Whitaker, 1976). Similarly, brain lesions in young animals tend to be less disruptive than similar lesions in older animals (Kennard, 1936; Stein, 1974). However, the situation is more complex than this. The effect of an early lesion also depends on the maturational stage of the system in question (Goldman, 1976). The less severe effect of serial or multistage operations as opposed to acute large lesions (Goltz, 1881; Stein, 1974) also depends on some version of the plasticity hypothesis.

What is meant by plasticity? Again, the real issue is not one of substitution, compensation, or reorganization, but of how brain substance (e.g., a damaged area) is to be conceived in relation to psychological function (e.g., a deficit). Generally we are faced with a choice between several competing physiological theories, such as disinhibition, adaptation or new learning, or, in the reorganization version, a new but lowered functional level achieved by the brain as a whole. Regardless of which one or combination of these possibilities happens to be correct, it does not take us any closer to an understanding of the nature of psychological deficits in relation to brain structure. The one conclusion that can be made of this work is that theories of plasticity are at least consistent with the idea that restitution, by which is meant *a change from one symptom to another*, is related to the degree of functional specification within (and between) the cerebral hemispheres, putting aside for the moment a discussion of the way in which this specification may be achieved.

Regeneration and Recovery

There is some evidence for central nervous system regeneration following a destructive lesion. Cotman *et al.* (1973) have demonstrated an increase in acetylcholinesterase-staining terminals in the dentate gyrus following lesion of the entorhinal cortex in rat. Lynch (1974) cites evidence that axonal growth establishes functional contacts and that it is more prominent in younger animals. Axonal sprouting has been described in the visual system of hamster (Schneider, 1974) and in cat spinal cord (Goldberger and Murray 1972). A related mechanism is that of denervation hypersensitivity (Raisman 1969). This has been described by Ungerstedt and Arbuthnott (1970) in rat striatum following lesion in the substantia nigra and by Glick and Greenstein (1973) after lesion in the frontal cortex. While regeneration, sprouting, and synaptic hypersensitivity are intriguing phenomena, it is doubtful that they can account for recovery from aphasia. Certainly, they do not explain, among many problems, the variable time course of restitution, the pattern of linguistic change, or the role of the right hemisphere. Moreover, even given that a recovery level is associated with one or more of these mechanisms, how do we explain that same functional level or symptom complex when it occurs as the initial phase of an acute lesion or as a phase in deterioration? We see how fruitless it is to pursue morphological theories of this type for an explanation of structure–function relationships.

Disconnection

Since Wernicke (1874), it has been customary to attribute some aphasic symptoms to interruption of pathways between various cortical regions. The essentials of the theory were best formulated by Bastian in 1896

(Brown, 1980) and have changed little since. This theory, which is an outgrowth of nineteenth century sensory psychology, at least of the version that contaminated the early days of neurological study, is open to all of the criticisms of areal localization previously discussed, since it presupposes a localization of function in precise brain areas. However, rather than emphasizing the consequences of damage to these areas, the focus is on the interruption of information flow between them. The symptom is a result of this structural break, or at least of degraded flow between the separated regions.

In the disconnection theory, we again have an anatomical interpretation that says nothing about the organization of the disturbed psychological process. What is disconnected from what? In the case where a perceived object is assumed to be disconnected from the name of that object, the theory does not detail the nature of the perceptual or naming process that goes on in the "disconnected" regions, nor does it specify the nature of the information presumed to flow between these regions. It accounts neither for the change in symptom expression from one moment to another, nor for recovery in the face of a persistent anatomical rupture, nor for the matrix of other symptoms within which the "disconnection symptom" appears. Even in the instance of *callosal disconnection*, where all of the dynamics of the clinical picture are attributed to hemispheric isolation, we have learned very little about the organization of psychological function *within* the hemisphere, and, not knowing that, we have learned still less of the organization of psychological function in the brain as a whole. These objections are not to be dismissed as merely theoretical in the face of anatomical fact. The anatomical fact, for example, that of a callosal lesion, is not a demonstration of a disconnection. The disconnection is a theoretical interpretation of the effects of the lesion; the lesion itself is not proof that the theory is correct. In aphasia study, the disconnection concept is utterly without foundation [see Brown (1972, 1980) for specifics]. The split-brain work is also open to reinterpretation and, in my judgment, has not yet received the scrutiny it deserves.

What Is a Symptom?

We begin with the question, what is a symptom? In the case of a language error we must also ask, what is an utterance? An utterance is a type of act in which language is displayed in the context of a motor performance. The utterance does not exist in isolation; it is embedded in a matrix of perception and motility. Postural tone, gesture, and facial expression are all part of the language act. And, even if we look only at the utterance, there is a background of meaning and inference that is part of its propositional content. The vocal elements of the utterance are only the exteriorized surface of its structure. In this structure, all levels actualize together, and this total presentation, constituting the utterance in its entirety, is itself a constituent of a cognition also hierarchically organized.

The neuropsychological study of language suggests that language is an emergent or microgenetic process that develops over a series of psychological levels. These levels correspond to or map stages in the evolutionary and maturational history of the brain. An aphasic symptom represents a segment of this emergent process. If we attempt to isolate an aphasic symptom, in order to study it more closely, we find that we lose sight of this prehistory and the position of the symptom in relation to the utterance as a whole. On the other hand, when we attend to these aspects of symptom formation, the symptom is seen as an orderly and predictable event, appearing in a context and changing together with that context.

Accordingly, we may say that, in pathology, levels in language production appear as symptoms. A symptom reveals a stage in language production that is traversed in the realization of the normal utterance. Brain damage has the effect of allowing symptoms—contents from more preliminary levels—to come to the fore. There may also be a regression to a more preliminary level. Accordingly, a brain lesion does not disrupt a mechanism, or a center where that mechanism is situated. Rather, it involves that structural level through which the (pathological) content is normally elaborated. Thus, we can say that every aphasia points in two directions: to a level in normal language and to a level in brain structure.

A Brain Model of Language

Though there have been various psychological accounts of aphasia over the years, only one—Wernicke's—anatomical theory has ever been proposed. According to this theory, speech perception occurs over a series of stages in the posterior-superior temporal region (Wernicke's area), which both transmits to (for repetition) and modulates Broca's area. Language produced in the posterior brain is conveyed by an uncertain path to Broca's area for articulation. In more modern accounts, mechanisms are assumed to be located within the anterior and posterior regions, which can account for the diversity of clinical symptomatology: for example, mechanisms for sound, word, and meaning perception; for discrimination and sequencing; for various short-term memory (STM) and long-term memory (LTM) processes; a parietal dictionary attached to Wernicke's area accounting for anomia; impaired verbal or auditory feedback. Luria's classification is derived from Wernicke's approach, since similar mechanisms are inferred ad hoc from the clinical picture. In such theories, every new observation implies a new mechanism and results in a chaotic patchwork of areas, mechanisms, and interconnecting paths.

The anatomical model of aphasia presented in this chapter differs fundamentally from this classical account. Limbic and neocortical zones identified with aphasic disturbances are viewed as strata or levels in the evolutionary and maturational structure of the brain. In this model, there is no rostral conveyance of language content, but rather a simultaneous realization of the entire hierarchical system within its anterior and posterior sectors. Nor are there separate processes or strategies operating on language content; instead, there is a resubmission of emerging abstract content at each hierarchical level to the same reiterated process—in other words, one process at multiple levels, rather than multiple processes at the same level. These differences from the traditional approach will become more apparent as we proceed into the actual model.

Evolution of the Language Area

The neural substrate of language consists of a complex hierarchical system of levels corresponding to stages in neocortical evolution. The system has an anterior (frontal) and posterior (temporo-parieto-occipital) component. The two main classes of aphasia, the nonfluent and the fluent aphasias, refer to these components, while the various aphasic syndromes within each class point to different levels within the anterior or posterior sector. The structure as a whole develops out of medial and paraventricular formations through several growth planes of limbic and paralimbic (transitional) cortex to a stage of generalized ("association," "integration") cortex. Within the latter, developing through a process of core differentiation, the primary motor (gigantopyramidalis) and sensory (koniocortical) zones appear. For example, with respect to motor cortex, the evolutionary wave that had been assumed to lead from precentral to premotor area (Campbell, 1905; Bailey and Bonin, 1951) may actually be the reverse of the true direction of neocortical growth. The premotor area may in fact be the older of the two areas.

Evidence for this reinterpretation of neocortical morphology has been steadily accumulating in recent years. An important element of this new approach is the view that fiber size is related to evolutionary stage. This notion derives largely from the work of Bishop (1959), who noted a tendency for small fibers to project to phylogenetically older sites in brainstem, and larger fibers to project directly to newer targets in the dorsal thalamus, suggesting an evolutionary trend toward increasing fiber diameter. Diamond and Hall (1969) reported studies of the optic system which were in support of this idea. Drawing on work in comparative anatomy and perceptual physiology, they proposed that the geniculo-striate system develops in evolution subsequent to a system relating tectum and pulvinar to the association cortex; that is, the primary visual area is the more recent zone in neocortical phylogenesis. They noted that the cytoarchitecture of the secondary visual area is more primitive than the primary visual area, and its neurons have larger (more primitive) receptive fields.

A similar view has been expressed by Sanides (1975): "What has been recognized as a main feature of neocortex evolution in primates, namely the enormous development of the supposedly secondary integration cortices was actually not understood . . . the most *generalized* neocortical structure is bound to become the most predominant one with the widest scope for further differentiation [Sanides, 1975]." The myelinogenetic studies of Flechsig (1920), which showed that the classical motor and sensory areas were among the first to begin myelination in the perinatal period, while the secondary or "association" areas showed a more drawn-out pattern of myelination, were misinterpreted to indicate the evolutionary priority of the primary sensorimotor areas. In fact, these areas have the *heaviest* definitive myelin content (Vogt and Vogt, 1919) and for this reason show early myelination. The more heavily myelinated systems are more recent in phylogenesis. Thus, the area gigantopyramidalis, which forms the heaviest myelinated core of the motor region, represents the most recent stage in motor cortex evolution, while the hypergranular koniocortex, forming the heaviest myelinated core of the sensory region, represents the most recent stage in sensory cortex evolution. According to Pandya and Sanides (1973), the "development of koniocortex cores as the last wave of sensory neocortex differentiation occurred during evolution only in the *visual, auditory* and *somatic* sensory systems mediating the sharpest objectifying and localizing representations of the periphery."

This is consistent with observations in cat of a two-stage development of visual cortex, an early primitive stage characterized by the activity of a system of fine fibers projecting to a wider extent of cortex, and a later more differentiated stage of large fibers projecting to the primary visual cortex only (Marty, 1962). In the cat there is evidence for a more diffuse geniculocortical projection on both anatomical (Niimi and Sprague, 1970) and electrophysiological (Bignall *et al.*, 1966) grounds. In contrast, in monkey the geniculate projection is exclusively to the striate cortex. These findings are in agreement with the concept of an evolutionary and maturational progression from a diffuse to a focal organization, and with the evolutionary concept that in neocortical phylogenesis the primary "sensory" cortex differentiates *out of* the generalized "association" cortex, as at the thalamic level the geniculate bodies develop out of the lateral-posterior: pulvinar complex. Thus, we might consider the frontal sector to be organized into several evolutionary tiers or planes, leading from limbic to paralimbic (e.g., parts of cingulate gyrus) cortex, to generalized ("association") neocortex, and finally to the primary precentral motor area (gigantopyramidalis) (Figure 6.1). Moreover, levels in brain structure correspond to levels in behavior. The paralimbic region supports a preliminary stage in the development of the motor act. Here the motor pattern is global and organized about the axial and proximal musculature in relation to drive or motivational systems. This stage is more bilaterally organized. At the subsequent level of generalized neocortex, the act develops outward toward the distal muscles. Finally, in primary motor cortex, the most highly differentiated phase (the fine paw or digital movements of the contralateral limb) is achieved.

Accordingly, studies of motor cortex demonstrating a rostrocaudad topography (Woolsey, 1952) may indicate that the proximal movements associated with more rostral stimulation represent phylogenetically and ontogenetically older (i.e., earlier) phases in the elaboration of the act than the fine distal movements elicited on precentral stimulation. In general, it appears that, as one goes from limbic cortex to premotor to precentral area, stimulation tends to elicit a progressively more distal and more contralateral response. These findings are consistent with the idea of a microgenetic (moment-to-moment) elaboration of the motor act up through this sequence of evolutionary levels.

The temporo-parieto-occipital sector is organized in a manner parallel to that of the frontal zone (Figure 6.1). With regard to language, limbic-derived insular and medial temporal cortex form a preliminary stage leading to generalized posterior "association" cortex. The primary sensory (e.g., auditory) cortex differentiates—together with its thalamic [medial

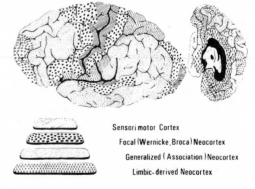

FIGURE 6.1. *Levels in brain evolution and cognitive microgenesis and corresponding regions on an early cytoarchitectural map. The map is quite general and serves to illustrate a conceptual organization.* [Modified after Campbell, 1905.]

Sensori motor Cortex
Focal (Wernicke, Broca) Neocortex
Generalized (Association)Neocortex
Limbic-derived Neocortex

geniculate body] nucleus—out of this posterior isocortical field. The organization of perceptual systems is identical to that of motility. Perception does not occur through a constructive phase of in-processing that begins in koniocortex. Rather, *it develops in the reverse direction over a series of levels to a koniocortical end phase.* In other words, the process of object formation unfolds in a cognitive sequence leading from a brainstem preobject through a limbic and generalized neocortical phase, to a final modeling achieved through "primary" visual cortex (Brown, 1977). In a fashion similar to that of the frontal sector, the perception leads from a global pre-object in a unitary field at a preliminary level to the final more or less contralateral hemifield representation at a koniocortical end stage.

A series of recent studies (Jones and Powell, 1970; Petras, 1971; Pandya and Sanides, 1973; Jacobsen and Trojanowski, 1977) has helped to clarify the relationship between various levels in the described infrastructure. Connections, direct and reciprocal, have been shown to exist in a "vertical" direction from one level to the next and across levels; in a "horizontal" direction, excepting sensorimotor cortex, there are connections between homologous levels in the anterior and posterior systems of the same hemisphere, as well as commissural connections to the corresponding anterior or posterior level of the opposite hemisphere. In other words, long intrahemispheric fibers run between the anterior and posterior sectors joining the first three levels in each system. Commissural fibers connect each of these levels, excepting sensorimotor cortices, to the same level in the other hemisphere, and, within each of the four sectors (right/left, anterior/posterior), there are reciprocal connections between each level. It is maintained that these various intra- and interhemispheric connections do not serve to transfer contents, percepts or verbal commands, from one point on the surface to another. Rather, it is proposed that they link up temporally, that is, maintain *in-phase*, homologous levels of different brain regions (Figure 6.2).

Maturation and Theory of Lateralization

The nature of cerebral dominance is the central problem for an anatomical theory of language organization. Studies that demonstrate a fixed or native hemispheric asymmetry, whether structural (Teszner *et al.*, 1972) or functional (Molfese *et al.*, 1975; Kinsbourne and Hiscock, 1977), lend support to a "prewiring" concept of language mechanisms. On the other hand, there is considerable evidence (Brown and Jaffe, 1975) that language is gradually biased to one (the left) hemisphere. This argues for a more dynamic model of dominance establishment. However, a theory of continuing lateralization is not necessarily inconsistent with a theory of

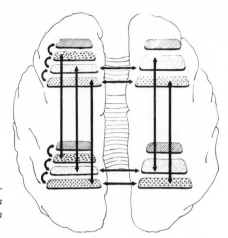

FIGURE 6.2. *Cortico-cortical fibers connect homologous levels within and between the hemispheres. The areas are defined on Figure 6.1.*

early structural or functional specialization. Indeed, some combination of the two seems to be required. According to the "prewiring" model, it is difficult to explain findings of considerable right-hemispheric language capacity in commissurotomized adults (Zaidel, 1976), not to mention the longstanding problem of right-hemispheric compensation following aphasia and/or left hemispherectomy. Similarly, the genetic model depends to some extent on the nativist theory to account for the consistent left-side tendency.

In previous articles, an ontogenetic model of the differentiation of the language area has been proposed based on studies of age specificity in aphasia (Brown, 1975). In brief, the model sought to account for age-dependent qualitative differences in aphasia type on the basis of degree of dominance establishment. Confirmation of the model was obtained through a study of aphasia in states of atypical dominance, that is, crossed aphasic dextrals and aphasic left-handers (Brown and Hecaen, 1976).

The evidence was consistent with the following hypothesis: that the degree of dominance reflects the degree of completion of a process of "core differentiation" of focal zones within generalized neocortex of the left hemisphere. These focal zones are conceived as levels that, in the course of development, build up both language and neural structure. Further consideration of this ontogenetic "building up" of the language zone, and its relationship to the described stages in neocortical evolution, has led to a more explicit model of the lateralization process.

According to this model, the differentiation of asymmetric focal neocortex repeats in ontogenesis the pattern established in the evolution of the sensorimotor areas, namely, a core differentiation of a specialized zone within a more generalized field. This phase, which may continue into late life, is characterized by the gradual appearance, in generalized neocortex of both anterior and posterior sectors, of an increasingly more focal and asymmetric (left-lateralized) zone.

One implication of this model is that the evolutionary sequence from transitional to generalized to sensorimotor cortex, which is characteristic of the development of the primate brain, would become, in man, one from transitional to generalized *to focal* to sensorimotor cortex. The emergence of this interposed level (of focal neocortex) out of a penultimate (generalized neocortex) and not a terminal (sensorimotor cortex) phase of neocortical evolution is consistent with the general pattern of evolutionary branching from earlier, less individuated forms rather than from end stages of specialization.

The progression of encephalized neocortex from a bilateral to a unilateral (contralateral) hemispheric organization is shown schematically in Figures 6.3–6.6. The essential feature of this model is that cerebral dominance does not come about as a (higher) stage beyond that of the contralateral representation of sensorimotor cortex. Rather, cerebral dominance—or lateral representation—occurs as an intermediate step between bilateral and contralateral representation. The bilateral organization that is characteristic of limbic and transitional cortex develops, in man, through a stage of (left) lateral representation to one of contralateral or crossed representation. Thus, while cerebral dominance constitutes a final phase in neocortical evolution, it is the stage of crossed representation and not that of cerebral dominance that is the end point of motor and perceptual realization (microgenesis).

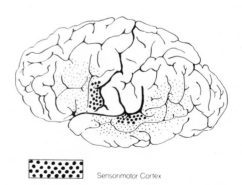

FIGURE 6.3. *A generalized mammalian brain showing sensory (auditory) and motor cortex and surrounding generalized neocortex.*

Sensorimotor Cortex

Generalized Neocortex

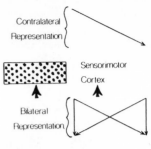

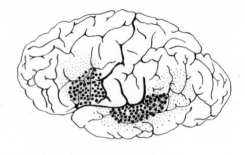

FIGURE 6.4. *Levels of representation in the mammalian brain. Cognition develops from a bilateral to a contralateral representation.*

FIGURE 6.5. *The human brain with asymmetric neocortex (Broca and Wernicke areas) differentiating within generalized neocortex.*

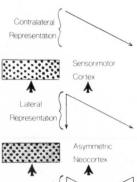

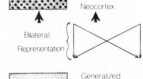

FIGURE 6.6. *The differentiation of asymmetric neocortex establishes a level of lateral representation between bilateral and contralateral representation. The degree of differentiation of this zone determines the degree to which language is "lateralized" to the left hemisphere.*

The Fate of the Right Hemisphere

What is the nature of right-hemispheric organization in the face of continued specification of the left-hemispheric language zone? According to the proposed model, the differentiation of asymmetric or focal neocortex out of generalized neocortex would occur primarily in the left hemisphere, at least in the average right-hander. In dextrals, this implies that in the right hemisphere a direct path exists from generalized to sensorimotor cortex without, or with minimal development of, the intervening stage of focal neocortex (see Figure 6.2). The evolutionary level of generalized neocortex would then typify the organization of the right hemisphere. This is consistent with evidence concerning the possibility of a more diffuse organization in the right hemisphere (Hecaen *et al.*, 1956; Semmes, 1968), with a demonstration of a semantic but not phonological processing in the "isolated" right hemisphere (Zaidel, 1976), and with the demonstration of "right-hemispheric language" in a language-deprived adolescent with presumed "disuse" of the focally differentiating left-hemispheric language zone (Curtiss, 1977). Moreover, the nature of "right-hemispheric language" demonstrated in the above studies corresponds in certain respects with that mediated by generalized neocortex in the left hemisphere as inferred from the aphasia material (see below, p. 181).

It is probable that the degree of focal differentiation within generalized neocortex of the right hemisphere is variable from one individual to another with some relation to handedness. It seems likely that the degree of focal differentiation in the right hemisphere is linked to that in the left hemisphere, such that some "balance" between the hemispheres is achieved. The nature of this balance would determine right (and consequently left)-hemispheric language capacity, as well as the potential for recovery in aphasia.

The Neural Structure of Language

Through the described process of neocortical phylogenesis, and its ontogenetic continuation in the formation of focal asymmetric neocortex (lateralization), a dynamic system is constructed that mediates the process—the microgenesis—of language production. The anterior and posterior components of this system develop in parallel, symmetrically, supporting complementary constituents of the unfolding *language act*.

Each level in the anterior component is in relation to the corresponding level in the posterior system, as well as to the homologous anterior or posterior level of the opposite hemisphere, through long intra- and interhemispheric fibers. These connections maintain a simultaneity across levels as the content unfolds, in both the anterior and the posterior sector, and between the hemispheres. Both the anterior and posterior components of this system show a progression from bilateral through lateral to contralateral organization. This is inferred from the effects of pathology. In general, symptoms referable to an early (e.g., limbic) level require bilateral lesions; symptoms referable to an intermediate level (e.g., generalized neocortex) result from lesions that are relatively left-lateralized but not strongly focal. This proceeds to an association with focal left-side lesions at the level of asymmetric neocortex and finally to focal lesions of sensorimotor cortex with a still more restricted predominantly contralateral expression.[2]

Language is elaborated through a process of development over this sequence of anatomical levels. This process can be dissected through a close study of pathological cases. These reveal the infrastructure of normal language, its preliminary levels, and the transitions between them. The aphasias reflect a destructuration within the anterior or posterior component, or sometimes within both. Or there is disruption at a specific level within one component. There is a constant relationship between symptom and level of pathological involvement. The symptom points to the level; the level mediates the language content to which the symptom refers. Thus, we can say that the described sequence of anatomical levels corresponds—in pathology—to a transitional series of symptoms, and these in turn reflect stages in the production of normal language.

Clinical Pathology of Language

Language is not composed of elements, but unfolds in a direction toward those elements. This process of unfolding transforms abstract content to a new state at each successive level. These levels also correspond to stages in the phyletic history of the brain. At each stage, abstract content presents itself in a cognitive mode expressive of that evolutionary level. To the extent that language represents that cognitive mode, it can be said, according to its level, to have a different existence in the world.

This microgenetic sequence leads from a semantic to a phonological level. These levels are really states of the emerging abstract form. In pathology, they are the expressions of this form as a continuous process is

---

[2] The lesser degree of contralateral representation in auditory cortex, in comparison with somaesthetic, visual, and motor cortices, raises the possibility of selective pressures acting on the evolution of this region. The contralateral advantage, however, points in the direction of the major trend in sensorimotor cortex evolution, namely, toward contralateral representation.

"sliced" at successive moments. The dynamic nature of this process will become clearer when we focus on its infratemporal or transitional characteristics.

A description of syndromes of the posterior component follows. First to be considered is the possibility of a limbic disorder of language. *Confabulation* is taken to represent the most prominent *posterior limbic* symptom of language change. There is a transition from confabulation to semantic jargon. This transition is illustrated through a consideration of the expositional (conversational, contextual) speech of patients.

Subsequently, the focus shifts to referential (naming) defects. The sequence of language disorders, in relation to structural brain levels, is then displayed through the varieties of naming errors. A central point is that the language form, the utterance, or the naming error always refers to a level in the described structure. At early levels, bilateral lesions are essential for symptom formation. At later levels, symptoms occur with an increasingly more unilateral (and focal) lesion.

Language Disorders of the Posterior Sector

LIMBIC–LEVEL DISORDERS

A limbic core underlies and provides a foundation for the anterior and posterior systems. It serves as a common base out of which these components will arise. Structures of the limbic core support an early, perhaps initial, stage in language production. This limbic role in language, though suggested by stimulation studies in subhuman primates (Myers, 1976), has not been adequately demonstrated in man. It is known that bilateral lesions of the human limbic system produce a severe amnesic syndrome (Scoville and Milner, 1957), but there are also important effects on speech and language. In the posterior system, bilateral lesions of inferomedial and lateral temporal cortex lead to confabulatory states and/or semantic jargon. The failure to find such changes in the "surgical" amnesic probably reflects the anterior location and more limited extent of the lesion. The effect of lesions of medial transitional cortex (e.g., insular cortex) is not fully established (see p. 174).

One of the most striking clinical symptoms of bilateral lesion of posterior limbic-derived cortex is *confabulation*. The confabulation that accompanies an amnesic (Korsakoff) state is not something "added on" to the memory deficit. Rather, it represents a disruption at a stage in cognition where the intimate relationship between the linguistic and mnemic aspects of the deficit is most apparent. Confabulation may be thought of as a "deep-level" aphasia.

Amnesia and Confabulation

Confabulation, or false recollection, is characteristic of the amnesic (Korsakoff) syndrome, but occurs in other pathological states as well. Confabulation is usually explained as an attempt by the patient to compensate for or "fill in" an unrecollected gap in memory. It is true that confabulation may signal the restitution of an amnesic syndrome. However, the occurrence of spontaneous confabulation, confabulation for *future* as well as past events, and the dynamic features of the confabulatory content all argue against a compensatory mechanism. In the course of recovery, a patient may progress from an encapsulated amnesia through a confabulatory stage to partial and then more or less complete recall. In such cases, the confabulation points to a stage in the retrieval of a content. The following case is instructive in this regard.

> *Case 1:* An 80-year-old woman had sudden dizziness and collapse, with a brief period of unconsciousness. Neurological examination was negative except for amnesia surrounding the spell; otherwise behavior and language were normal. The following day, marked confusion and confabulation were noted. She described an attack at home by a band of robbers and said her home had been vandalized and that she had been hit over the head and beaten. She suspected the nurses to be in collusion with her assailants, and felt they were trying to kill her. The following day, this picture completely disappeared. She was able to correctly recall events just up to, and immediately following, the period of unconsciousness.

In this patient, a short period of retrograde amnesia led to a state of agitated confusion and confabulation in the course of recovery. This was interpreted by some physicians as a sign of deterioration, when in fact it represented a phase in the resolution of the amnesic segment. Cases of this type indicate that confabulation refers to a level between complete irreminiscence and full recall.

Banal Confabulation

Confabulation is something that happens to a content in the course of its retrieval. In the amnesic without confabulation, the content simply is not evoked. The patient may state that he cannot remember the test material. However, some confabulation can almost always be induced by a strong insistence on full recall during the "decay" period. This is briefly illustrated in Case 2.

> *Case 2:* This 53-year-old man had the subacute onset of a profound amnesic syndrome and spinal fluid findings suggestive of herpetic encephalitis. Past memory, digit span, and intelligence were preserved in the presence of severe anterograde (learning) deficit. Except for two or three events there was no recollection for any experiences subsequent to the onset of the amnesia.

The patient was given the following story to read and immediately after relate it back to the examiner:

> *The Hen and the Golden Eggs:* A man had a hen that laid golden eggs. Wishing to obtain more gold without having to wait for it he killed the hen but he found nothing inside of it for it was just like any other hen.
>
> P: It's about a golden hen. . . . The owner of the hen wanted the hen to lay a golden egg . . . so that he could get prosperous I guess. . . . Somehow the hen refused to lay the egg.

This story was repeatedly read to and by the patient, and initially there was no ostensible improvement in immediate recall, though there was a diminution in the (weak) confabulatory response. When the story was shown to him again after several minutes of distraction, there was evidently a type of déjà vu feeling for the story, though he could not say definitely whether he had seen or heard it before. Over a longer time this receded into a complete amnesia.

Semantic Errors in Confabulation

The link from a banal confabulatory amnesia to semantic jargon occurs through a semantic elaboration of the confabulatory content. The content of the confabulation in the spontaneous speech of Korsakoff patients does not always show the disarray that characterizes jargon.[3] However, this can often be brought out by tests of story recall. Such tests provide a standard against which the recollection (spontaneous speech) can be judged.

The following patient had relatively normal conversational speech and an obvious amnesic syndrome with banal confabulation. On story recall, there was a marked deterioration approaching a jargon pattern.

> P: [Reads story of *Hen and Golden Eggs* correctly.]
> E: Can you tell me that story?
> P: It's an old hunk of mythology. You have four or five of the same thing don't you. . . . It was about a horse that bred one year or something like that but then didn't breed anything because there was no contact . . . ya, it comes from a thing out of modern politics too. . . . It's sort of an esophagus deal they're trying to sell today. [E: What's an esophagus deal?] . . . . Do you remember the old esophagus [Aesop?] stories? If you've read some, they're very much like that.

Later that day, the patient was again tested with the same story. As before, he read it aloud correctly.

> P: It might be, it sounds like . . . I don't know. . . . Well, the old missions about all sorts of things . . . about the old hen, what do you mean, about Colorado, the hen

---

[3] Schizophrenic paramnesia also appears to be linked to semantic errors in retrieval, differing from confabulation, at least in part, by its more subjective (i.e., less manipulable) character (Brown, 1977).

out there, what was the name of that hen? [E: The one with the golden eggs?] No, it wasn't and the thing was written by Charles Dickens, but he didn't write full verse and all that. It wasn't accepted as a literary piece.

In this sample, there is relative preservation at the sentence level, but the lack of intersentential relatedness and the propagatory nature of the confabulation lead to rather striking productions that, from the point of view of the material to be recalled, would have to be judged as semantic jargon. There are also occasional word substitutions, for example, "esophagus deal," which foreshadow the lexical errors characteristic of aphasic jargon.

SEMANTIC JARGON

The patient with semantic jargon produces utterances with fairly good or correct syntax but aberrant meaning. There are three major error types: noun substitution, derailments in the speech flow, and circumlocution. The disorder is present in referential speech (naming) and in conversation. As with all of the jargonaphasias, the condition is rare in young patients and when it does occur is generally associated with bilateral lesions of the temporal lobe, possibly involving underlying limbic structures (Weinstein *et al.*, 1966). In older aphasics, as in the present case, the disorder can result from a lesion about Wernicke's area or (deep?) lesion of posterior T2, though precise anatomical studies are lacking. Semantic jargon also occurs in acute confusional states and resembles the word-salad of schizophrenia.

Case 3: This 72-year-old man had a left temporoparietal infarction (Figure 6.7). Although neurological examination was normal, except for the aphasia, there was evidence for bilateral involvement. An electroencephalogram demonstrated some degree of right posterior slowing, and a history was obtained of several left-sided seizures post-onset. Initially there was severe semantic jargon with occasional neologism. The following is an example of the conversational speech of this patient.

And I say, this is wrong. I'm going out and doing things and getting ukeleles taken every time and I think I'm doing wrong because I'm supposed to take everything from the top so that we do four flashes of four volumes before we get down low. . . . Face of everything. This guy has got to this thing—this thing made out in order to slash immediately to all of the windpails . . . This is going right over me from there—That's up to is 5 station stuff from manatime—and with that put it all in and build it all up so it will all be spent with him conversing his condessing [condessing?]. Condessing his treatment of this for he has got to spend this thing. [E holds up handkerchief: What is that?] Well—this is a lady's line—and this is no longer what he wants. He is now leaving their mellonpush [''mellonpush''?]. Which is spelled "U" something or other which also commence the fact that they're gonna finish the end of that letter which is spelled in their stalegame and opens up here and runs across what "M"—it wasn't "M" it's "A" and "M" is the interval title and it is spelled out with all of this.

On tests of naming, the patient named some objects correctly, but many

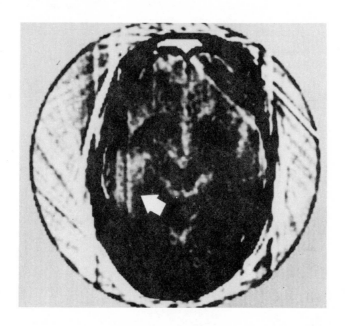

FIGURE 6.7. *Vascular lesion in the left temporal lobe [computerized tomography (CT)].*

objects were misnamed. These misnamings, or paraphasias, often showed a semantic link to the target object:

earring	"haircone"
pipe	"smokin mob"

At other times no clear relationship to the target object could be ascertained:

ashtray	"mouse looker-atter"
wallet	"lover lob"

In cases such as this, there is generally little or no recognition by the patient of the erroneous nature of the verbal substitution. One has the impression that the closer to target the substitution, the greater the awareness of the error, and the greater the tendency toward self-correction. Immediately following a misnaming, presenting the erroneous name to the patient, for example, asking the patient if the object is a "haircone," will usually elicit a positive response. If the examiner invents a word or a compound such as the above, it is more likely to be rejected by the patient than his own substitution. The ability of patients to reject the examiner's substitution may also increase as the substitution approaches in meaning the target item. Patients also seem better able to reject "words" that do not closely approximate those in their own language than words that are, or could reasonably be expected to be, items or possible items in their own lexicon. Patients are generally able to reject "jabberwocky," as well as sound errors in otherwise correct words. If one speaks in semantic jargon to such a patient there is no clear indication that the jargon is perceived as such by the patient. The patient may recognize when he is addressed in a foreign language but may respond as if he has just heard a meaningful utterance. Bilingual patients, both semantic and neologistic jargonaphasics, may switch their jargon to the language in which they are addressed. From such observations, one forms the impression that patients more readily reject phonological than semantic errors and, within the latter, better reject those closest in meaning to the target item, and that this is true for both speech production and the perceived speech of others.

In addition, patients with semantic jargon are usually euphoric, or at least have a heightened mood without frustration over a poor performance. It seems that the affective tonality is specific to the utterance, as is the awareness experience, so that one sees a change in these parameters moment to moment depending on the predominating error type.

With regard to comprehension, this is usually only moderately impaired, or at least not so severely involved as in neologistic cases, and repetition may be fairly good at times. There also appears to be a Korsakoff-like state with defective verbal (and visual) memory. Cases of this type have been described by Clarke *et al.* (1958) and Kinsbourne and Warrington (1963). The present case was also an example of this and in the course of recovery evolved into a fairly characteristic Korsakoff state. During the initial stage, reading showed semantic jargon as severe as in expositional speech, but gradually reading improved to the point where it was relatively intact, and speech became more intelligible with some word-finding difficulty and semantic paraphasias. At this time, the patient was given the story *The Hen and the Golden Eggs* to read, which he did so correctly and was then asked immediately to recall the story. His performance was as follows.

Story: A man had a hen that laid golden eggs. Wishing to obtain more gold without having to wait for it he killed the hen but he found nothing inside of it for it was just like any other hen.

P: It seems to be a lot about something . . . it was a command on the hen and he told her to do certain things and she didn't do anything . . . and he tried to get everybody to do things and nobody did. He turned everything over to the marine base. He couldn't see out of this eye [patient had been discussing his visual problem before] doesn't see anything the way it should [E: What about that story?] That's what I'm trying to come back to. He gave it to other people and something took charge and he liked the pieces. He had to put this thing like this because he had to keep this not looked at . . . and then he began to hear things from the outside, which he liked very much, and he let the thing go through because horn [?] this horse he was able to tell exactly who did what with her and who did the wrong with her. [Reads story again, correctly.]

E: Can you tell the story now?

P: I'll try to though I have to keep things . . . centered. There was a man who had a hen, and she gave golden eggs and they collected the water and sent it into the water and brought it out, and he had to keep it this way or he couldn't see it and when he sent her into the water other things took charge and he had to hide this from them all the time.

E: Do you think he did the right thing in killing the hen?

P: No, I do not . . . . Well, because it distributed itself and brought up everything out of there, not only brought every other hen but it brought any other thing that might be invested by it, and he did that despite my keeping this thing normal. I don't know whether I'm saying it to you or not.

This type of language is similar to the confabulation of Korsakoff patients, especially that seen during the acute confusional prelude to the disorder. It is probable that there is a close link between semantic jargon and confabulation. In the latter, the memory defect is in the foreground of the clinical picture, and confabulation might be considered a mild or attenuated form of semantic jargon. Semantic jargon aphasia, on the other hand, may be thought of as a deteriorated confabulation where the language defect becomes more prominent. One can say that in semantic jargon there is an amnestic syndrome embedded in the language disorder, whereas in the amnestic syndrome there is an aphasia embedded in the confabulation. Put another way, in banal confabulation the intrasentential organization is relatively intact. The defect is at the intersentential level. This gives the impression of a cognitive or memory defect. The flow of discourse is more comprehensible; the derailment is ideational, not linguistic. However, as the disturbance approaches in the direction of word selection, substitution *within* the sentence becomes more pronounced and the goal of the utterance becomes uncertain to the listener. Now the defect is intrasentential and seems to be linguistic rather than conceptual. However, these disorders lie along a continuum. The Korsakoff patient in the acute stage may have typical semantic jargon, which may be indistinguishable from the semantic jargonaphasic. Ordinary amnesics commonly show aphasic misnaming during this stage (Victor, 1974), and errors tend to be of the semantic type, that is, verbal paraphasias. Such patients have been shown to have a deficit in semantic encoding (Cermak and Butters, 1973). On the other hand, semantic jargonaphasics have an amnesic disturbance. This is apparent on clinical study of such patients but has not yet been the subject of careful investigation. It is of interest, however, that patients generally do not recall the period of their aphasia after recovery.

LEVELS OF SEMANTIC REALIZATION

The transition from confabulation to semantic jargon takes us into aphasia proper. In order to simplify the account of transitions from one aphasic pattern to another, we will set aside our descriptions of conversational speech and turn to an examination of naming.

The errors that patients make on naming tests reflect very closely the nature of their aphasic disorder. When asked to name an object, an aphasic may make one of several types of paraphasic (substitutive) errors. His response may vary greatly from moment to moment. At one time the patient will produce a semantic or verbal (word) substitution, at another a phonological error. However, the response pattern tends to be clustered around predominantly one type of error, and the nature of this error determines what "syndrome" the patient is said to "fit." Syndromes, therefore, are not stable entities but rather refer to the qualitative mean of performance.

To my mind it is preferable to consider separately each utterance or performance of an aphasic patient. These represent a kind of momentary slice through the infrastructure of the patient's cognition. In other words, a paraphasia reveals the level at which—for the instant of that paraphasia—the cognitive structure has actualized. In a sense, therefore, the utterance—the paraphasia—is only the focal point in a cross section of cognition. The paraphasia is seen in relation to other aspects of cognition that, though more subtle in their presentation, accompany and are simultaneous with the paraphasia at the same cognitive level.

The repertoire of paraphasic errors can be viewed along a continuum that leads from an asemantic (or nonmeaningful) substitution, through a substitution in which there are unusual "associative" bonds to the correct item, to categorical or "in-class" errors, and finally to correct selection of the word but failure in its evocation (Figures 6.8 and 6.9). This progression can be conceived as an emergence of an abstract representation of the utterance-to-be up through a series of semantic fields, from those of wide "psychological distance" (asemantic) through a level of (?) experiential and/or affective bonding ("associative") to a categorical selection and

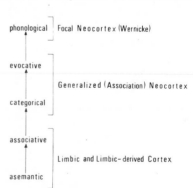

FIGURE 6.8.   *Levels in language production in the posterior sector correspond to phylogenetic and maturational levels in brain structure.*

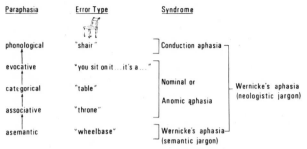

FIGURE 6.9.   *Aphasic syndromes correspond to levels in semantic or phonological realization.*

finally correct word choice. At this final stage, the lexical item, the word, or an abstract representation of that word has been correctly selected but cannot be evoked. At the next level, the stage of evocation has been adequately traversed but the phase of phonemic encoding, the realization of that word in phonological form, is deficient. This results in phonemic errors in otherwise correctly selected words.

Asemantic

In semantic jargon, paraphasic errors are characteristically *asemantic.* This means that the link between the substitution and the object to be named is obscure. Sometimes there are shared elements or attributes ("smokin mob" for pipe), while at other times there is no clear relationship to the required object name (e.g., "wheelbase" for chair). The morphological features of the object do not appear to play a prominent role in determining the paraphasic response. Asemantic errors may occur as "clang" or rhyming responses (e.g., "hair" for chair), but this usually occurs in the context of phonological rather than verbal substitution.

In younger patients, semantic jargon (asemantic paraphasia) may be due to a *bilateral* lesion of limbic transitional cortex. The condition may also occur with a left temporal lesion in the presence of a "physiological" depression of the opposite hemisphere. In older aphasics the disorder may result from a unilateral left temporal lesion, though anatomical studies of such cases are lacking. The disorder also occurs in acute confusional states. There is some resemblance to the word-salad of schizophrenia, where limbic system dysfunction seems likely (Torey and Peterson, 1974).[4]

[4] Recent papers that sharply distinguish schizophasic language from aphasia make the error, in my judgment, of considering only relatively mild schizophrenics, and not the more deeply regressed (and nowadays rare) cases in which word-salad occurs. Schizophasia is not be be compared to aphasia in general, but only to semantic jargon and "associative" paraphasias. Goldstein (1943) has described patients of this type. Whatever differences exist between such forms of schizophrenic and aphasic language can be attributed to the chronicity

"Associative"

Paraphasia in which there is a special type of "associative" relationship to the target item occurs in semantic jargon and in confusional states. The misnaming may reflect situational, experiential, and affective factors. The substitution shares an attribute with the correct object name and commonly has a pedantic or facetious quality, as in the following examples:

bedpan   "piano stool"
doctor   "butcher"

At times the misnaming is correct or vaguely acceptable but of an unexpected nature. In such instances, the paraphasia may be of a considerably lower word frequency than the required name, as in:

glasses   "spectacles"
matches   "ignition system"
red   "fuchsia"

In spite of the low frequency response, the errors may tend to involve lower frequency object names (Rochford, 1974). Accompanying such errors is usually a euphoric mood with little or no insight into the aberrant or tangential nature of the misnaming. Having (mis)named an object, the patient rarely accepts the correct name from the examiner and steadfastly refuses to alter his performance. However, it does seem that such patients are less willing than those of the previous group to accept the asemantic productions of the examiner.

As is the case in semantic jargon, there is generally a bilateral lesion, probably involving medial or inferolateral temporal or limbic structures. The disorder frequently occurs in the acute stage of the Korsakoff syndrome and is common in postconcussive or postanesthetic states.

Categorical

The progression from asemantic substitutions to those with an experiential and/or "associative" relationship to the intended or demanded object name leads to substitutions within the object class or category.[5] At this

---

of the former, the co-occurrence of delusional and/or paranoid trends in schizophrenia, and its fluctuation over time. In addition, in contrast to the schizophrenic, the aphasic brings a normal personality to the (abrupt) pathological change.

[5] Paraphasias of this type represent the most common responses on word association tests in normal subjects. One can say that the word association technique usually penetrates only to a limited depth in the described system, while aphasic responses (e.g., "associative," asemantic) reflect lexical organization at a deeper strata.

stage the utterance has achieved (or been selected to) the category of the object name, as in:

comb	"brush"
table	"chair"
red	"green"

Categorical substitutions do not show the ease and lability of asemantic and associative paraphasias. There is an impression of word search and, at times, a hesitancy or a tentative quality in the naming response that, though not approaching the next stage, that of evocative difficulty, does signal a change from the more fluent and effortless paraphasias of the preceding level. With utterances of this type, there is also a mitigation of the euphoric mood in the direction of a more appropriate affective stage. There is also increased awareness of error. The patient tends to be dissatisfied with his performance and makes attempts at self-correction.

The pathological correlation is with a lesion of left generalized neocortex, especially middle temporal gyrus (T2) and its (usual) posterior continuation into angular gyrus. The fact that a naming disturbance of this type is also seen in diffuse pathological states and with lesions throughout and beyond the classical language zone is consistent with an association with the more widely distributed level of generalized neocortex.

Evocative

This refers to the anomic who can "block" a paraphasia and gives descriptive or tip-of-the-tongue responses, but is unable to produce the correct object name. This is not a result of the capacity to suppress a potentially erroneous response. Rather, the abstract representation of the word has achieved its correct lexical form; that is, it has been selected beyond the point of a category error, but the item, the word, cannot be evoked. One can say that such patients have intact word meaning in the presence of word finding difficulty, whereas those patients at a more preliminary level (asemantic, associative) have facile word production with impairment of word meaning.

The affective state accompanying this disturbance is one of frustration, with relief on success and at times catastrophic reaction on failure. Awareness of the difficulty is more acute and self-critical, with corrections and efforts to achieve an adequate response. The anatomical correlation is, as in the preceding form, a lesion of generalized neocortex somewhat peripheral to, or less severely involving, the central language zone. This form of anomia is also the initial aphasic pattern in the atrophic (and other) dementias, where the degeneration preferentially affects "association" (generalized) neocortex.

PHONOLOGICAL REALIZATION

Phonemic (Conduction) Aphasia

The final stage in the semantic series corresponds to the realization of an abstract representation of the correct lexical item (Figure 6.9). At this point, prior to the process of phonemic selection, the word exists incipiently, its evocation into consciousness taking place as it is realized phonologically. This process leads through a phonological selectional operation analogous to that in the semantic component. Disorders at this level are characterized by a disturbance in the phonological realization of otherwise appropriate target items. Among the many errors that may occur are those of substitution, deletion, and transposition (metathesis):

> president   "predisent"
> kite   "dite"
> green   "reen"

The disturbance is typically present in conversational speech as well as in naming and repetition and is often referred to as conduction (phonemic) aphasia. This designation emphasizes the repetition defect as the primary impairment, though in fact the disorder of repetition is only one instance of the central defect in phonemic encoding (Brown, 1975). There is evidence that in such cases substitution errors respect distinctive feature distance (Blumstein, 1970). Clinically, this does seem to be true for the mild case with a restricted phonological disorder. However, it may not be true of the phonological errors in neologistic jargon, where phoneme substitutions often violate the expectations of distinctive feature theory. Such cases suggest that there may be a narrowing or targetting down in the phonological sphere, from distant to close phonemic paraphasias, comparable to that occurring in the process of lexical selection.

The anatomical correlation of phonemic aphasia is a lesion of the posterior superior temporal gyrus (posterior T1) and especially its continuation into supramarginal gyrus (Ajuriaguerra and Hecaen, 1956). There is no evidence for involvement of underlying white matter in this disorder. Occasionally, phonemic aphasia occurs with lesion of the angular gyrus (Brown, 1972; Green and Howes, 1977).

With regard to neologism, this may represent either a deteriorated (i.e., unintelligible) phonemic paraphasia or a phonemic paraphasia in combination with semantic paraphasia. The former is a type of phonemic or undifferentiated jargon (see below); the latter is neologistic jargon.

Neologism and Neologistic Jargon

Isolated neologisms are not uncommon in a variety of aphasic forms. Lecours and Rouillon (1976) have argued that the occasional neologism observed in the Broca aphasic is structurally identical to that of jargon aphasia. This is open to question. Perhaps it may be the case in resolving global aphasics, but it is unlikely that this is so in documented anterior aphasics. In such patients, neologism is more likely determined by distortion or misarticulation. Phonological errors in the presence of dysarthria may result in words judged to be unintelligible.

The isolated neologism preferentially attacks the content word, often with preservation of syntactic agreement over a range of neologistic forms. Thus, a patient of Caplan *et al.* (1972) said: "They will have to *presite* me . . . [and] yes, because I'm just *persessing* to one." A personal case remarked: "It was my job as a *convince*, a *confoser* not *confoler* but almost the same as the man who was *commerced*." Another said, "Your *patebelin* like the mother . . . and his mothers got to go in his *stanchen*." Another patient, asked to name eyeglasses, said, "Those are *waggots*, they have to be *fribbed* in."

Buckingham and Kertesz (1976) give a number of such examples. As can be seen even in these brief excerpts, the preferential attack on the content word often leaves the initial part of the utterance unaffected.

As the incidence of the neologism increases, there is a corresponding decline in the intelligibility of the utterance. At some uncertain point, say, when neologisms constitute 20–30% of the words in an utterance, the patient is said to have neologistic jargon. It is rare that this incidence goes beyond about 80%. Cases where speech is virtually 100% neologistic should then appear as "undifferentiated" jargonaphasics. In such cases, either the patient is no longer producing wordlike segments or the listener can no longer segment the utterance into wordlike sequences. In fact, undifferentiated jargon is probably *not* a deteriorated neologistic jargon, for reasons discussed on p. 169. In addition, we should keep in mind that, as the incidence of neologism increases, the utterance is not simply being filled with more and more nonsense words. The entire fabric of the aphasia is changing, as well as the patient's affective and cognitive state.

In typical cases of neologistic jargon, the jargon invades all language performances more or less equally. Lhermitte and Derousné (1974) have described a case with good writing, but this appears quite exceptional. There is no selective sparing of sung or automatic speech as in the Broca aphasic; that is, lyrics are jargonized even though the melody may be preserved. Comprehension is (invariably?) severely impaired, and the patient may almost seem deaf, though this can usually be ruled out on careful testing. Generally, even in such cases a response of bewilderment occurs when the patient is addressed in jargon. In a number of personal cases this has been the only language-related response that could be obtained. In one patient even a galvanic skin reflex (GSR) could not be elicited to abusive comments, though surprise was registered to the examiner's jargon. In spite of the severe disorganization of language, some patients can function fairly well in the world. Performances on tests such as Ravens Matrices may be quite good. A personal case was able to continue playing chess, albeit at a rather unsophisticated level. Affect is generally heightened; the patient may be euphoric, even manic. This may change with the utterance. A patient may be euphoric during jargon and depressed when speaking *correctly* about his illness or family.

Frequently during the acute stage patients produce sequences of numbers instead of words. In time, this will develop into a more typical jargon pattern. At this point there is often a relative preservation of some thematic material. For example, a patient with completely unintelligible jargon might produce a (usually short) series of correct though often stereotyped utterances about his illness or his work. A personal case spoke incessant jargon except when describing the accident that brought about his condition. He was run over by a car driven by his (jilted) girlfriend and was ordinarily quite lucid in his repetitive accounts of this incident. Commonly, patients lapse into these islands of speech preservation when asked to produce utterances that would otherwise result in jargon. The affective element in such phenomena should not be overlooked.

Lecours and Rouillon (1976) have described predilection units and predilection themes in jargon patients. This means frequently recurring neologistic segments and topics and is confirmed by my own experience. Conceivably there is a relationship between the islands of preserved speech and those of preferential impairment. One may inhibit the other since each seems to occur at widely separated linguistic moments or at different stages in resolution. In spite of the occurrence of predilection units, there is little or no consistency in the jargon on a word-by-word basis. There is no evidence that jargon constitutes a private language. However, it is of interest that neologistic jargonaphasics may engage each other in "conversation" though rejecting a jargon produced by the examiner. Cases have been cited where the jargonaphasic has seemed to accept his own recorded jargon, but not the same jargon after it has been transcribed and rerecorded by another person. Other well-known features of this type of jargon include the relative preservation of intonation and probably word stress even when it falls on a neologism. Augumentation is common in such patients; that is, they produce more syllables than in the target words. This may also affect intact as well as jargonized utterances, though possibly not to the same degree, and may account in part for the logorrhea.

Within the group of neologistic jargonaphasics several subtypes can be distinguished. There are fluent productive forms with only a slight tendency for alliteration or sound association, while in other patients, or in the same patients at other moments, reiteration on the basis of sound similarity can become quite prominent (e.g., "much tereen, will I could talk, sorlip, gorrip, grip, grip, stick, eye, grin, grim, greeda"). This has been discussed by Kreindler *et al.* (1971) and Lecours and Rouillon (1976). In cases of this type, fluency is impeded and this limits the jargon to a restricted repertoire of sounds. These patients will approximate the jargon stereotypy of the global or motor aphasic. In such patients Lecours and Rouillon (1976) have argued, correctly I believe, that this error points to a phonological element. In the neologistic jargon case it also reflects the involvement of the phonological system.

The anatomical lesion of jargonaphasia was discussed by Henschen (1922). It was found that a *focal* lesion of the left superior temporal gyrus (T1), regardless of whether it involved a more anterior or posterior section, might cause word deafness in the presence of normal speech. This was also true for combined lesions of T1 and T2, even if subcortex was involved. However, diffuse softening of left T1 and T2 was invariably accompanied by jargonaphasia. More recently, the problem was studied by Kertesz and Benson (1970) in light of several anatomical cases. These authors found involvement of the classical Wernicke zone. In view of the fact that jargon is associated with the late-life aphasic, it is conceivable that in younger patients the disorder may require bilateral pathology. This anatomical correlation can be understood if we interpret neologistic jargon as a combined phonological and semantic defect. For the reason of this constituent phonological disorder, the lesion correlation of neologistic jargon is similar to that of the more restricted phonological disturbance in phonemic aphasia. This corresponds to the level of focal asymmetric neocortex.

Phonemic Jargon

We owe to Alajouanine *et al.* (1952) the first description of undifferentiated (phonemic[6]) jargon, a form of aphasia in which speech consists of a series of phonemes without meaning or grammatical organization, possibly similar in structure to the jargon stereotypy of the motor aphasic. Unlike neologistic jargon, there are few if any recognizable words or wordlike segments, the jargon consisting of phoneme strings of variable length and diversity. Their example was the following brief excerpt: *sanénequéduac-quitescapi.* Subsequently, Alajouanine and Lhermitte (1963) emphasized

---

[6] The term phonemic jargon is applied by the French school to cases of conduction aphasia. Such cases, however, can hardly be said to have jargon, in that speech is generally quite intelligible. It is more appropriate to use phonemic jargon for cases such as the present one rather than for conduction aphasics. The term undifferentiated jargon, which is applied to such cases, is unsatisfactory since the jargon is, arguably, not undifferentiated, and the term does not capture the essential feature of this disorder, namely, that it is a jargon (at least) at the phonological level.

the tendency to stereotypy and perseveration and the involvement of repetition and reading. However, oral and written comprehensions were said to be preserved. A buccofacial apraxia was uniformly present.

This description appears to be based on the clinical impressions of one or several cases of this type without a more detailed study of the jargon production. Indeed, the term *undifferentiated jargon* has been applied both to phonemic jargon [glossolalia of Lecours and Rouillon (1976)] and to cases of aphasic mumbling. It is usually considered that phonemic jargon is a form of jargonaphasia with posterior lesion and that it occurs in quite elderly patients. In contrast to mumbling, such patients have good articulation and speech is readily transcribable. The anatomical lesion is not known. The following case, reported in detail elsewhere (Perecman & Brown, 1980), is an example of this disorder.

Case: A 74-year-old German–English bilingual carpenter was evaluated 2 months following a stroke. There was a transient left-side weakness. Subsequent neurological examination was normal except for the aphasia. A computerized tomogram demonstrated two large areas of softening, two confluent lesions involving the left temporoparietal area, and a lesion involving the right temporoparietal region (Figure 6.10). After reviewing the scan findings, an effort was made to obtain information on a prior stroke, but no

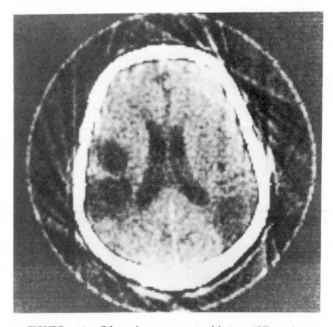

FIGURE 6.10.   *Bilateral temporoparietal lesions (CT scan).*

such history was obtained. Evidently, he was in good health and had normal language up to the present hospitalization. The following is a transcription of a reading sample.[7]

Text: The Hare and the Tortoise
A hare made fun of a tortoise because it was so slow.
Then the tortoise said to the hare: "I will challenge you to
a race." The hare thought this was a silly idea but agreed
to the competition. They started off but the hare
ran so quickly that soon he was far ahead. Then he thought:
"I'll take a nap until that stupid tortoise catches up
with me." The tortoise crawled along very, very slowly,
passed the hare who was fast asleep, and had nearly reached
the finishing line when the hare woke up. The hare
ran as fast as he could but he didn't manage to get there first.
The tortoise had won the race.

Reading aloud:

der hera unt der heriterstudraytsisus
awoyts vwayts ders of a densderayisterzes pirasolrivu inays orvayri
dirdiruf derbaynliesuses odederaybiurefye abaif vridimirspuristu dirs
avwaseve ther amebatpiristedvayservwaysfuzen apervuvites erigam
ibilis thi ers obadamber distkomeris mans enteneversofer ayisirs
fer aynbeverspferisipus dzer ayvayvwosder dmosbeze dzomdzerinidzani
of adtheryisovwayis omimifor odhaimitzonimenitzertzatzamambiv
nurbaysbobriz der gamnesersf fomays fromasforiboravidpandrimemzirs
barofder vaynirsveyirso asberyiunzes svandafareindzem
ther vursoenstermorelspisasferay viemertzipif
a aybesesder kamerm arfderfofts rememi berere inder menye firt
ther kamder tsinerits faynedivaytsder ruberkamisbos

A detailed analysis of the frequency of speech sounds in this patient was carried out (Perecman & Brown, 1980). Vowel and consonant frequencies were compared across two readings of the same English test, and one reading of a German text, and reading was compared to spontaneous speech in both languages. The phoneme distribution of the jargon was found to be essentially identical between English readings, and very similar across languages. However, in terms of sequences of phonemes, there was little or no consistency between two readings of the same English text. There was a reduction of the consonant: vowel ratio, though not to the extent noted by Mihăilescu *et al.* (1967) in a population of mixed Rumanian aphasics. In spontaneous speech, the most common English consonants were /s/, /b/, /r/, and /m/ and, in German, /r/, /m/, /p/, and /b/, in decreasing order of occurrence. Of these, only /r/ in standard English, and /r/ and /s/ in German, have a relatively high frequency. In the German reading, /s/, /R/, /t/, and /n/ were most common. This differed from a control reading (/n/, /t/, /R/, /s/) that corresponded to the rank order of

---

[7] There is a line-by-line correspondence of the text to the jargon reading.

frequency in German. In English reading, /t/, /s/, /R/, and /d/ were most frequent in a control and /R/, /s/, /d/, and /n/ in the patient. These findings were interpreted as showing that spontaneous speech does not reflect the normal relative frequencies of consonants in either language, primarily because of excessive production of labials, and that the written texts may have an influence on the reading production. If one consults the sample above, it is clear that some correspondence in the reading is present at the very beginning of the text in the title, but this disappears quite rapidly. There is no correspondence on a line-by-line basis across texts. However, relative phoneme frequency is altered in the direction of the control reading. Moreover, the patient produces many more sounds than are present in the text. In this regard, it is of interest that, during the reading, the patient followed each word with his finger, seemingly unaware that his reading was unintelligible, and stopped only when his finger reached the last word of the text. It was in this way that we could evaluate performance on a line-by-line basis. With regard to vowels, the frequency distribution is quite similar across readings and in spontaneous speech with a preference noted for /a/. An additional finding was the great predominance of /s/ in the "word"-final position.

In spontaneous speech there was an increase in volume and he could inflect questions and commands. The intonation pattern seemed relatively normal, although it was difficult to determine this precisely. The determination of "word" boundaries was possible only through a reliance on pause durations and release of plosive consonants. The utterance did not have the wordlike segmentation found even in severe neologistic jargon. One can suppose that it is the relative preservation of functors in the latter that accounts for this, but it is my impression that such patients *use* neologisms as real words. In neologistic jargon, there appears to be preservation of "word" stress, whereas in phonemic jargon either the stress pattern is lost—though the intonation pattern may be preserved—or it is distributed more generally over the utterance so that stress is not helpful in segmentation. The problem may be that, in phonemic jargon, abstract word frames are no longer available.

Studies of neologistic jargon (see above) have demonstrated that relative phoneme frequency corresponds to that in the patient's language. Our preliminary analysis of a corpus in Buckingham and Kertesz (1976) has confirmed this observation. However, this was not completely true in our case of phonemic jargon and is evidence that this disorder may not represent a deteriorated form of neologistic jargon.

There are some clinical features of phonemic jargon that should be emphasized. In the diagnosis of this condition, confusion may occur with cases of jargon stereotypy showing some degree of phonemic variability in the jargon production. Indeed, this question of phonemic variability is often a determining factor in the clinical diagnosis of phonemic jargon, the assumption being that in jargon stereotypy only a limited number of reiterated phonemes are produced, whereas in true phonemic jargon all or almost all of the phonemes in the language are employed. This matter is of some theoretical and practical import in that jargon stereotypy is generally considered to occur in the context of a severe motor or global aphasia, that is, presumably with an anterior or central lesion, whereas phonemic jargon appears to relate to a posterior lesion, even though anatomical studies of such patients are lacking. For this reason, the position of phonemic jargon among the aphasias is unclear. Is there a link between phonemic jargon and jargon stereotypy in relation to the anterior aphasias? And, in relation to posterior aphasia, is phonemic jargon severe phonemic paraphasia or deteriorated neologistic jargon, or something else again?

With regard to jargon stereotypy, this is a reiterated phonological string that, as with other automatisms and residual utterances, tends to intrude on all attempts at volitional or propositional speech. Cases of this type are readily distinguished from phonemic jargon. The difficulty occurs when the phoneme repertoire expands coincidently with a mitigation of the stereotypic pattern. Continued expansion within the sphere of the stereotypy leads to increasing phoneme variability. There is a corresponding reduction in the explosive nature of the stereotypy, though apparently the lack of awareness by the patient for his erroneous productions is similar in both stereotypic and phonemic jargon. It may be that the tendency for reiteration and predilection units in phonemic as in other types of jargon represents the transitional element to the pattern of jargon stereotypy.

In addition to this evidence of the present case report, there are several clinical features that serve to distinguish phonemic jargon from a deteriorated neologistic jargon. The patient with phonemic jargon is more attentive to the speech of others and follows the conventions of the speaker–listener situation. The patient does not have the logorrhea and pressured speech of the neologistic jargonaphasic and can engage the examiner in a "conversation," even though appearing to understand nothing of what is said (by either party), and himself unintelligible to others. Moreover, the presence of some reading comprehension and writing, with, at least in the present case, a self-critical behavior in writing though not in speech, is in contrast to the usual case of neologistic jargon where these behaviors are impaired to the same degree (compare Lhermitte and Derouesné, 1974), the lack of awareness of deficit being distributed over all language (and, presumably, other) performances.

SOME PROBLEMS CONCERNING THE
POSTERIOR APHASIAS

Classification

There are various classifications of the posterior aphasias (Leischner and
Peuser, 1975). In some laboratories, there is a subdivision into several independent types (e.g., anomia, conduction aphasia, Wernicke's aphasia),
whereas in other laboratories these forms are treated as varieties of a single
Wernicke aphasia, differing along a gradient of severity. However, there is
no agreement as to what constitutes Wernicke's aphasia. For some, this
refers only to the comprehension impairment, which may or may not be
associated with an expressive deficit. According to this view, word
deafness would represent the "pure" form of Wernicke's aphasia and the
nucleus of the comprehension impairment in the posterior aphasias; that is,
there would be either a severe or a mitigated word deafness in association
with one or more other component disturbances. For others, however,
Wernicke's aphasia is as much characterized by the pattern of expressive
speech as by the comprehension deficit. In some classifications
jargonaphasia is considered to be the chief expression of Wernicke's
aphasia or at least a severe form of Wernicke's aphasia, with conduction
aphasia and anomia representing milder or recovering forms. It is recognized that there are several different types of jargon—semantic, neologistic,
and undifferentiated—though it is unclear whether all of these belong to
the group of Wernicke's aphasias.

The difficulty in the classification of individual cases and the complexity
of the literature on the subject have led some workers to revive the simpler
distinction of fluent (posterior) and nonfluent (anterior) forms. This is an
expeditious way of segregating experimental populations, but it has the effect of lumping together quite different clinical syndromes under a generic
label of little or no theoretical value. To say that a patient is fluent is to say
that with respect to fluency he is normal, fluency being the normal state
and the goal of recovery. Patients evolve toward fluency in recovery and
only exceptionally in the reverse direction. Moreover, we occasionally see
fluent anteriors and nonfluent posteriors, the latter especially during the
acute stage (Karis & Horenstein, 1976); patients may be fluent for some
performances and nonfluent for others, while in many patients fluency is
not clearly determinable and may fluctuate greatly. Certain cases classified
as fluent (e.g., conduction aphasics with phonological errors) may resemble more closely—both clinically and on experimental testing—the
behavior of cases classified as nonfluent (e.g., the Broca aphasic with articulation errors) than either resembles other aphasic types within the same

category. Fluency is always performance specific and is always a question of degree. For these various reasons, it seems preferable to develop a classification based on predominant error type, for example, semantic, phonological, whether expressed in expositional or referential speech or both, and so on, rather than on the traditional basis of syndromes or on a meaningless fluent–nonfluent dichotomy. Were agreement to be reached on such a classification, patients could then be grouped in such a way as to satisfy both the clinical valuation of individual differences and single case study and the experimental demand for analysis by groups.

Aphasic Mumbling

Mumbled words are common in aphasic patients, and mumbling also occurs in severe dementia, Parkinsonism, and other neurological disorders, but "fluent" inarticulate and essentially untranscribable mumbling with infrequent if any audible words or even discrete speech sounds in a patient with a focal brain lesion is quite unusual. It is not a rare disorder, and most aphasiologists encounter at least a few cases of this type. However, to my knowledge, there is only one case report in the literature. This is the 42-year-old woman of Kähler and Pick (1879) who mumbled constantly all day long, the only articulate sounds being Tschen and Tscho. There was no evidence of comprehension and no paralysis. The autopsy showed bilateral superior temporal lobe softening.

The place of mumbling in the aphasias is not known. It has been linked to phonemic jargon, perhaps as a more severe form of the latter. Lecours (1978) indicates that, though distinct, it is often grouped with the glossolalias. In view of the scarcity of case material, one can only speculate on the pathology. In the present case (see below), a left temporal lesion was demonstrated. However, the severity of the comprehension impairment suggests the possibility of a right temporal lesion as well. I have also seen another case in which a bitemporal lesion was suspected. However, the disorder can probably result from a large left central or posterior lesion in which the mumbling is a kind of vestige of fluency in the context of a global aphasia. It may be that in younger patients [e.g., the case of Kähler and Pick (1879)] bilateral temporal lesions are necessary, whereas in older patients (e.g., the present case) a large unilateral (temporal?) lesion will suffice. The following case is an example of this disorder. Because of its unique nature, it is reported in somewhat greater detail than the others.

Case: A 69-year-old man with right-side weakness and aphasia for 5 months following a stroke. Since this time vocalization has consisted of unintelligible mumbling sounds at low intensity. Initially, he was said to be able to write only the letters "AUDD," and he could not copy these or other letters or shapes. An EEG revealed left posterior temporal

FIGURE 6.11. *Large Sylvian lesion demonstrated by nuclear medicine brain scan.*

slowing, and a brain scan showed a large area of uptake in the left temporoparietal region (Figure 6.11). There was gradual recovery of function in his right side but speech remained unchanged over a year of observation. He was right-handed without family history of left-handedness. During the first 6 years of his life he spoke only French, but was said to have had equal fluency in French and English. There was no motor or sensory deficit. Spontaneous speech consisted of a continuous stream of mumbling sounds. No neologisms or paraphasias were heard. There was a tendency for a decrescendo pattern. The mumbling would decrease in volume and trail off, at which time a new burst would initiate another decrescendo. There was marked speech pressure and it was difficult to interrupt the patient. Speech consisted chiefly of prolonged vowel sounds, introduced by a consonant, occasionally the first syllable of a word spoken to him. There was little or no intonation, but gesture was quite active, though noncommunicative. He did not appear fully aware that his "speech" was incoherent. When addressed in French, a definite French accent and an occasional French word were noted. For example, when he was told, *Fermez les yeux,* the word *yeux* was distinctly heard in the mumbled response. Motor series were impossible. The patient did attempt to sing with assistance; the melody was vaguely approximated, but the lyrics were unimproved over spontaneous speech. There was no dysarthria. On repetition and naming, the first sound of the target word was occasionally produced as the initial sound of the response. Otherwise, there were no correct repetitions or namings. Comprehension was markedly impaired. During moments of spontaneous silence one could occasionally elicit a correct pointing response to a single object prior to the intrusion of the mumbling. Once this occurred, comprehension could no longer be tested. He was able to carry out commands such as "lift your arm" and "point to the table," given in French or English, during moments of spontaneous—but not enforced—silence. There was no echolalia. Reading aloud was identical to spontaneous speech. The patient could not follow simple written commands. There was some ability to copy and transliterate simple words. Writing to dictation was impossible, but when given a sentence in French or English he did attempt to write in the appropriate language (Figure 6.12). The patient could copy a simple square but had difficulty with three-dimensional figures. Simple calculations were usually impossible; he once summed 12 and 29 correctly, but usually could not add single digits. Praxis could not be evaluated.

There are several features of this case that deserve emphasis. There was an assimilation into the mumbling of some perceived speech sounds, as well as the initial syllable of target nouns on naming tests. This behavior was also noted in the case of phonemic jargon. In my experience, it is also common in neologistic cases. The ability to read and write at the single-word level is also similar to that noted in the case of phonemic jargon, while the pressured, logorrheic quality of the mumbling and the loss of the speaker–listener constraints resemble more closely the situation in neologistic jargon. It is of interest that a change in accent could be perceived according to the language employed by the examiner, even though

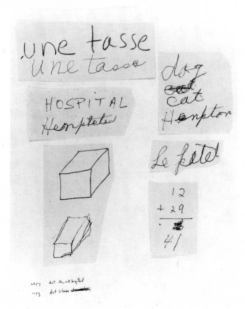

FIGURE 6.12. *Writing and drawing in case of aphasic mumbling. On left side, copy and transliteration; on right, to dictation (dog, cat, hospital, le livre) and calculation.*

this change was carried by a continuous stream of mumbled sounds almost completely devoid of phonemic or morphemic properties. The nature of aphasic mumbling, however, remains to be determined. It appears to be (related to) a form of *posterior* aphasia, but further case studies are needed before even this can be accepted.

The Role of the Insula

An important but still unsettled question in aphasia study is that of the relationship of insula to language pathology. Wernicke postulated on a theoretical basis only that a lesion of insula would produce an aphasia. Dejerine (1885) argued that an aphasia secondary to insula lesion would result not from damage to insula cortex but from the interruption of subcortical association fibers traversing this region. Among the early case reports, a widely quoted case was that of a left-handed boy with a type of pure motor aphasia and destruction of right insula, described by Wadham (1869). Saundby (1911) described a questionable case with a similar aphasia and softening of the left insula. Additional cases were reviewed by Dufour (1881) and then by Monakow (1905, 1914), who could not come to a definite conclusion. The problem was not greatly clarified by Henschen (1922), and there have been few if any cases reported since his comprehensive review.

The frequent involvement of insula in cases of aphasia and the continuity of this area with the cortical language zones led some workers to include it as part of these latter regions. Thus, Niessl von Mayendorff (1911) considered the anterior insula to be part of Broca's area, and Pick (1931) included the posterior insula with Wernicke's area. In more recent times the strongest advocate of a role for the insula in language has been Goldstein (1948), who claimed that it represented a central zone in the anatomical organization of language and that lesions of insula interrupted the language process at a (central) stage in the "thought–speech transition." He even reported (Riese and Goldstein, 1950) a case of a skilled orator with asymmetric enlargement of the insula.

Subsequent work, however, has not confirmed Goldstein's thesis. Penfield and Faulk (1955) did not obtain speech arrest on stimulation in insula, nor did they find aphasia after unilateral—chiefly right-sided—excision in a few cases. Evidently, there are no cases of bilateral surgical lesion of insula on record. Rasmussen (1978) notes that transient aphasia may follow surgical ablation of the left insula. This is similar to that after anterior lobectomy and is attributed to postoperative edema. However, these observations may have to be reconsidered in view of a recent report of dysphasia in a patient during stimulation in the anterior left insula (Ojemann and Whitaker, 1977).

The following case is of interest for several reasons, in addition to the presence of a lesion in the left insula. It is presented in the hope of renewing a dialogue on this basic, and in recent years neglected, problem in the pathological anatomy of aphasia.

Case: A 21-year-old right-handed man, deaf since age 8 following a high fever, developed sudden right-side paralysis. Previously, he used sign language with some speech accompaniment. According to hospital notes, and friends who visited him daily during his hospitalization, he was unable to express himself vocally or in sign with the left hand for the first few days. He was unable to form words with his mouth. He was unable to spell with the left hand. This lasted for about 10 days, at which time he was said to have shown some recovery in signing with the left hand. At 2 weeks post-onset, spelling and signing with the left hand were said to be normal. The right arm and leg showed gradual improvement in strength over 4–6 weeks. Evidently, there was a point where strength had returned but signing was poor with that hand. By 6 weeks post-onset, he was normal, with respect to both signing and vocalization and the neurological examination. The patient was strongly right-handed with no family history of left-handedness. The CT scan showed a lesion in the left insula (Figure 6.13). Subsequent angiography was normal.

This case indicates that a lesion of the left insula can produce at least a transient aphasia. It also demonstrates that, in deafness acquired at least as early as age 8, the temporal region may persist as a language zone.

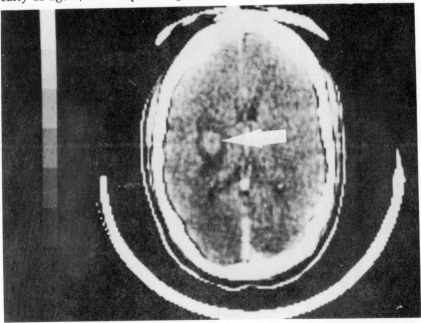

FIGURE 6.13.   *Demonstration of vascular lesion in the left insula with surrounding edema (CT scan).*

Some studies of forebrain development (e.g., Yakovlev, 1972) have shown that the insula is a preliminary stage toward (i.e., in evolutionary continuity with) the generalized neocortex of the posterior hemisphere. In view of this, it would be surprising if the insula were completely unassociated with language or cognitive function. Part of the insula is transitional or *meso*cortex and corresponds to cingulate gyrus on the medial surface of the hemisphere. Unilateral lesion of anterior cingulate gyrus is not, so far as I know, firmly identified with any clinical symptomatology, whereas a bilateral lesion leads to profound mutism and akinesia. Since bilateral lesions are required for symptom formation in anterior mesocortex (cingulate gyrus), it is likely that a similar requirement obtains for posterior mesocortex (insula). If so, this would explain the mild or transient nature or even the lack of symptoms with left insula lesion and emphasizes the need for well-studied bilateral cases.

Language Perception and Production: A Comment and a Speculation

Clinical study suggests that in the posterior aphasias there is an inner bond between the production error and the disturbance in comprehension. One has the impression that a change in one determines a change in the other. If this is so, then we can say that, even when we look at production only, we are indirectly assaying the status of comprehension. This observation raises the first and perhaps most central problem regarding these disorders, namely, the relationship between language comprehension and the language *corpus* that is so prominently affected in posterior aphasia.

Language is organized in the brain largely within a perceptual field. Wernicke's area and its wider extent develop in relation to perceptual areas. The anterior (Broca) zone is in relation to motor systems. Speech errors resulting from pathology in this region are generally those of misarticulation or omission. One has always the question of a (probably systematic) dissolution or impoverishment of motor structure rather than a specific disruption of language. Certainly, experimental studies of syntax in the anterior aphasic have been inconclusive with regard to the specificity of syntactic deficits in these patients (e.g., Kellar, 1978). This is not the case with involvement of the posterior sector. Here the rich inventory of error types points to a selective language disorder. It is of interest, therefore, that morphological studies of the hemispheres have demonstrated asymmetry primarily in the posterior zone [see Rubens (1977) for a review]. This is the main *language* area.

What is the nature of the relationship between perception and production? Clearly the perception of language involves more than listening to the speech of others. The speaker can only have knowledge of his own utterance through his perception of that utterance. For the speaker, the mental representation of the utterance is largely if not completely perceptual. There is little or no motor experience. More than this, the utterance itself appears to be generated or realized through perception. The motor region seems to contribute the initial kinetic and its derived intonational pattern, the final programmation of sound sequences, and perhaps also the feeling of intentionality that (except in some pathological cases) accompanies the utterance, but this motor aspect unfolds too rapidly to persist in awareness as a stable content. The problem here is the same as that posed in relation to the perception of limb or eye movements and in past discussions of feelings of effort (Innervationsgefühle). In more recent times this has been treated as a feedback or corollary discharge phenomenon [see Roland (1978), and discussion]. However, at least in the case of language and insofar as we can infer from the aphasia material, the perception of an utterance is not a secondary or following effect of the motor discharge; it is, for the speaker, the primary event. Thus, it would seem that every language act involves a simultaneous discharge over levels in both perception and motility (Brown, 1979). I believe that the content of an utterance is represented as an end stage in a perceptual development, its articulation being a manifestation of the transition leading to this end stage. In other words, in perception there is an awareness of content without an awareness of the development leading to that content; in action, there is a development—the act itself—without a content. These represent two phases of a common process.

We may recall Bastian's (1890) idea that the motor speech center is really a kinaesthetic area, and that motor aphasia is really a kind of sensory disorder. This is related to Luria's (1966) syndrome of kinaesthetic aphasia. The present argument is in some ways the reverse, that the sensory language area is concerned with language production, and that the comprehension disturbance in the sensory aphasias presents itself in the form of a production deficit. This is offered only as a speculation for the reader without a fuller discussion, as this would take us far from the objective of this chapter. The nature of acts and objects (perceptions) and their derivations in language will be the subject of a monograph that is currently in preparation.

Language Disorders of the Anterior Sector

LIMBIC-LEVEL DISORDERS

As in the posterior system, the anterior sector also develops out of a common limbic core. The anterior division of this core mediates the initial stage in motor speech differentiation and so underlies the anterior aphasias, just as limbic amnesia and confabulation were shown to be transitional to the posterior aphasias. Moreover, as with the posterior limbic disorders, the anterior limbic structures require bilateral lesions for symptom expression, which accounts for the relative scarcity of cases of anterior limbic pathology. The most typical example of an anterior limbic syndrome is the condition of anterior cingulate gyrus mutism.

The most characteristic effect of *stimulation* in anterior cingulate gyrus in man is a state of arousal or vigilance not directed to the external world. Motor responses of the limbs tend to be oriented toward the body, and there are changes in mood—fear, anxiety, or sadness—which are usually not verbalized. The impression is of an affective and instinctual state organized about an archaic level in behavior (Bancaud *et al.*, 1976). These effects of stimulation are consistent with the clinical description of bilateral pathological lesions of the anterior cingulate gyrus. In such cases there is a state of mutism, motor akinesis, apathy, and lack of affective responses. In the first case report of this type (Nielsen and Jacobs, 1951), there was no spontaneous speech or movement, but the patient could be aroused for brief periods of time, during which she was said not to be aphasic. Subsequent reports have confirmed this picture, though a case has been described in which the mutism was broken by excited outbursts (Farris, 1969). This problem was reviewed by Buge (1975) in light of three further case reports, and the complete absence of spontaneous speech was emphasized, as well as the rarity with which communication could be established. With strong arousal, an occasional, often fortuitous, command could be elicited. In contrast to these profound effects, bilateral stereotactic lesions of human anterior cingulate gyrus lead mainly to a state of placidity. This is probably a partial expression of the more severe condition seen with anterior cingulate destruction.

These findings are consistent with the view that anterior medial limbic cortex mediates an early or "deep" stage in the anterior component of the cognitive process. This stage is characterized by a consciousness level of arousal or vigilance, while behavior is organized about the proximal and/or axial musculature and is not directed to extrapersonal space. The emotional pattern also suggests a "deep" (drive) level in affect differentiation. This stage is referred to as that of the "motor envelope," in which the various preparatory elements of the final utterance—the utterance itself, its accompanying gestures, the postural tonus within which it emerges—are all present together, incipiently, along with an affective component pointing to a drive level in affect derivation. This (pathological) level prefigures and gives rise to a subsequent cognitive stage that is mediated by (left) generalized frontal neocortex. The most striking defect in the syndrome of the anterior cingulate gyrus, loss of spontaneous vocalization with occasional brief emotional utterances and short repetitions, forms the nucleus of the next syndrome (level) to be discussed, transcortical motor aphasia.

LEVELS IN THE REALIZATION OF THE MOTOR COMPONENT

The developing utterance issues out of a preliminary (limbic) cognition. Simultaneously, the action of the limbs proceeds outward toward objects in extrapersonal space. The diffuse, labile affect of the limbic level is derived into more differentiated partial expressions, while consciousness of speech and action becomes increasingly more critical and acute. The anterior aphasias can be thought of as moments or segments of this process at sequential stages in its emergence.

From the point of view of the speech act,[8] there is a progression from akinetic mutism, an inability to evoke or further realize the motor envelope of the speech act, through an attenuated form in transcortical motor aphasia, to agrammatism and finally Broca's aphasia proper. In this progression, the cognitive stage, as it is revealed in pathology, becomes increasingly more focused as it develops toward the final articulatory units (Figure 6.14).

Differentiation of the Motor Envelope

The motor envelope contains the embryonic speech act together with its accompanying gestural and somatic motor elements. At this stage, the elements are only prefigurative, bearing little resemblance to the perfor-

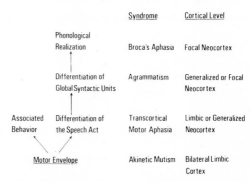

FIGURE 6.14. Syndromes of the anterior sector, with corresponding brain structural level, and level of speech act differentiation.

	Syndrome	Cortical Level
Phonological Realization	Broca's Aphasia	Focal Neocortex
Differentiation of Global Syntactic Units	Agrammatism	Generalized or Focal Neocortex
Associated Behavior / Differentiation of the Speech Act	Transcortical Motor Aphasia	Limbic or Generalized Neocortex
Motor Envelope	Akinetic Mutism	Bilateral Limbic Cortex

[8] This refers to the utterance as an action that unfolds, emphasizing the fact that the anterior component of language is bound up with motility as a type of specialized motor performance.

mances (speech, gesture) to which they give rise. In the microgenesis of this
motor configuration, certain elements precipitate out early, for example,
postural elements, while others undergo a separate microgenetic fate.

At the point where the speech act differentiates out of this background
organization, pathological disruption is characterized by a lack of spontaneous or conversational speech with good repetition. At times, naming
and reading aloud are also spared. This constellation of findings has been
termed *transcortical motor aphasia*; it is similar to the dynamic aphasia of
Luria. In this syndrome there is more than just a lack of spontaneous
speech; nonspeech vocalizations and gestures are also reduced and there is
often an inertia of behavior generally, which may suggest Parkinsonism.
These "associated" symptoms all point to a common level or origin in the
unfolding motor act; they are also a sign of its proximity to akinetic
mutism.

This disorder reflects an involvement at the level of generation of an utterance, a speech act, without a disruption of the (potential) constituents of
that act. This may concern the initiation of the utterance, its organization,
and/or its differentiation from other motor elements at that level. Repetition aids in achieving this differentiation and in providing a configuration
through which syntactic differentiation can occur. Naming is frequently
preserved. The presence of an object may help to provide a structure
through which this differentiation may occur. The presence of good naming in this disorder marks a transition to the next level, that of agrammatism, where there is also superior noun production.

The affective state of such patients is usually one of apathy or indifference. The apparent lack of emotional responsiveness is another feature
indicating a link with cingulate gyrus disorders. However, the pathological
anatomy is uncertain. At least it can be said that the preservation of repetition is not to be understood by the sparing of a repetition pathway; that is,
it is not a transcortical defect (Brown, 1975). In many cases there is partial
involvement of the left Broca area (Goldstein, 1915). I have described a
case with subtotal destruction of Broca's area (Brown, 1975). On the other
hand, the most frequent cause is probably occlusion of the anterior cerebral
artery, which entails damage to the anterior cingulate gyrus, supplementary motor area and contiguous structures on the medial surface of the
frontal lobe. Conceivably, the disorder may follow a left cingulate lesion as
a partial form of the severe mutism with bilateral cingulate destruction.
The association with supplementary motor area is now well-established.
These contain limbic-derived neocortical zones. In any event, the correlation is either with limbic-transitional or *meso*cortex or with the next level in
neocortical phylogenesis, generalized neocortex of the left frontal region.

Syntactic Realization

As the speech act differentiates, the simpler and more global units of the utterance-to-be are the first to emerge. Along with the appearance of nouns and uninflected verbs, there is an unfolding into the forming utterance of the small grammatical or function words. The appearance of the functors is thus delayed to a level of individuation subsequent to that of holophrastic noun and verb production. Disorders at this stage are characterized by incomplete differentiation of emerging syntactic units (agrammatism).

Conceivably, this disturbance can be conceptualized as an incomplete elaboration of a phrase structure tree, with a premature appearance of representations at the noun and verb phrase level. If so, the nouns and verbs of agrammatic patients should not have precisely the same value as those same lexical units in normal speech. Certainly, agrammatics use nouns in a more diffuse, more propositional (holophrastic) way. Their difficulty in classifying nouns (Lhermitte *et al.*, 1971) may reflect an attenuation of noun phrase differentiation. However, this is only a way of characterizing agrammatic language and may not reflect real psychological events. The preferential sparing of nouns and simple verbs in agrammatism may well have more to do with the initial use of these items as activity concepts, in relation to the cognitive mode of their acquisition. This is not to suggest that the agrammatic has a child's grammar; there is considerable evidence against this point of view. The evidence suggests, rather, that the young child's use of language in relation to activity may play a determining role in the psychological representation of these lexical items.

There is yet another way of looking at agrammatism. In this disorder there is a disruption at an intermediate stage in the unfolding of the kinetic melody of the utterance. We may consider this "melody," at the level of the motor envelope, as a prefigurative rhythmic organization that is derived, at the level of agrammatism, into the intonational pattern, and then into the temporal programming of sound sequences. Involvement at an intermediate stage in this series might produce a flattened intonational contour, inability to inflect (tones as well as words), and loss of prosodic values. This might be accompanied by an increased salience for stressed content words, reflecting the posterior development of the content words and the lexical basis of word stress. This would give the appearance of an attenuated syntactic differentiation. This might explain the infrequency of agrammatism in French aphasics. In French, the stress pattern is more evenly distributed over the sentence, allowing the production of functors to survive disruption of the intonational pattern. I believe that the correct interpretation of this disorder, as well as the other "motor" aphasias, is to be found in the concept of successive levels in the derivation of a rhythmic series, though this does not necessarily exclude an account on the basis of a syntactic impairment.

The link between the level of agrammatism and the preceding level can be seen from individual case studies. Patients with transcortical motor aphasia may have agrammatic repetitions. As spontaneous speech returns, it often goes through an agrammatic phase, or there may be agrammatism in writing. Similarly, the improved performance on repetition in this disorder carries over into agrammatism proper where performance is generally better on repetition. In fact, cases of agrammatism in conversational speech with normal repetition have led to the view that agrammatism is a type of economical speech. While there does not seem to be any basis for this belief, the frequent dissociation between good repetition and conversational agrammatism establishes a link with the more preliminary disorder.

The anatomical correlation of agrammatism is typically with generalized neocortex of the dominant frontal lobe, that is, a partial or peripheral lesion of Broca's area. It corresponds, therefore, with anomia (evocational stage) in the posterior series, which is also linked to generalized neocortex. Like anomia, agrammatism is imprecisely localized; the possibility of a frontal anomia is as open as the possibility of a posterior agrammatism. Another similarity is that, in anomia, the abstract representation of the lexical item also fails to achieve phonological realization. In anomia this is more pronounced for nouns than functors, whereas in agrammatism it is the reverse. Both levels converge toward a final stage of phonemic encoding.

An intriguing aspect of agrammatism is that it is the chief manifestation of aphasia in young children, and in dextrals with right-hemisphere lesions (crossed aphasics). In such patients, agrammatism occurs with posterior as well as anterior lesions, presumably because their incomplete lateralization goes along with more diffuse intrahemispheric language organization; in other words, there is less focal differentiation (of a "Broca" area) within generalized neocortex of either hemisphere. The relation of agrammatism to generalized neocortex accounts for its occurrence in cases of "right hemispheric speech; for example, in cases of left hemispherectomy in adults, and in the case of feral child where left hemispheric regional specification apparently did not develop (see p. 149).

Phonological Realization

The anterior development terminates as the emerging lexical frames actualize into phonological sequence. This transition corresponds to that of the posterior component, as both systems converge toward a final phonological stage. Disruption at this level gives rise to phonemic-articulatory disturbances in otherwise well-formed utterances. The disorder is taken to represent Broca's aphasia proper[9] and is distinct, though transitional, from agrammatism. Within this final segment, several sublevels have been described (Lecours & Rouillon, 1976; Alajouanine *et al.*, 1939).

The transitional nature of the phonological level out of the agrammatic complex is evident in the overlap between the two disorders. Agrammatics commonly show some phonemic-articulatory deficits, while the latter group commonly have some agrammatic features. Patients with articulation errors in speech may show agrammatism in writing. In both groups, performance is generally improved for repetition. Moreover, in cases where the impairment is mainly phonological, the deficit is more prominent for functors than for content words.

As the utterance achieves a stage of phonemic encoding, there is an improved affective state; mood and awareness of disability are comparable to that in the posterior phonological disorder (phonemic aphasia). One can say that the affective tone has proceeded to the same derivation as the utterance that accompanies it. Patients also show active, at times labile, gesture, though there is little capacity for sign language or pantomime. The prominent gestural activity may be understood by a microgenetic completion which differentiates from the final speech act development.

The anatomical correlation is with Broca's area (mainly postero-inferior F3; see p. 184) of the dominant hemisphere, in other words, with focal asymmetric neocortex. From this point, the further elaboration of the utterance occurs through inferior precentral (motor) cortex, which mediates the terminal (phonetic) phase of the anterior microgenetic sequence. Disruption of this segment gives rise to the "syndrome of phonetic disintegration" (Alajouanine et al, 1939).

SOME PROBLEMS CONCERNING THE ANTERIOR APHASIAS

Comprehension in Anterior Aphasia

The inner bond between comprehension and expression, which seems to exist in the posterior aphasias, is not so readily apparent in the anterior disorders. One has the impression that patients with very similar production deficits vary greatly in the severity of oral comprehension. There is also some evidence (Hecaen and Consoli, 1973) that the severity of the comprehension deficit is related to the depth of the pre-Rolandic lesion. However, in the more restricted cases there is a suggestion of a relationship

---

[9] Probably the same as "verbal apraxia," a term that emphasizes the basis in motility but does not otherwise have any explanatory value.

between the comprehension deficit and the *pattern* of expressive speech. Patients with mutism generally do not show comprehension ability, or at least it is difficult to demonstrate. Patients with intermediate-level disorders, such as "transcortical motor aphasia," have a moderate comprehension impairment, especially when this is in association with echolalic repetition. In agrammatism there is evidence that the mild comprehension disorder is specific to the production pattern (Zurif and Caramazza, 1976), whereas Broca aphasics with primarily phonemic-articulatory errors show relatively good comprehension ability. Clearly, further work is needed in this area; yet there is the suggestion that the disruption in aphasia affects a common level in a system elaborating both language production and perception.

Broca's Area: A Brief Review and Conclusions

Historically, Broca (1863) maintained that the frontal speech zone included the posterior portion of the inferior or third frontal convolution (F3), that is, the pars opercularis and possibly the pars triangularis, and perhaps also part of the middle frontal gyrus (F2). Various authors have included the inferior precentral convolution (i.e., the "face area"), although this zone is generally distinguished from Broca's area proper. Some writers (e.g., Goldstein, 1948; Mills and Spiller, 1907) included the anterior insula in an extended frontal speech zone. Niessl von Meyendorff (1911) maintained that the frontal operculum was the motor speech center.

For Monakow (1914), Broca's area included F3, anterior insula, the connecting gyrus between F3 and the precentral convolution, and the Rolandic operculum. This is similar to the description given by Dejerine (1914): the posterior part of left F3, frontal operculum, and neighboring cortex ("cap" of F3 and "foot" of F2), extending to the anterior insula but excluding Rolandic operculum. Henschen (1922) argued that the cap of F3 (pars triangularis) was not a part of the motor speech zone on the basis of a negative case with bilateral lesion reported by Bonvicini (1926).

In sum, there has traditionally been general agreement on the importance of the posterior part of F3 (pars opercularis) and neighboring frontal operculum and uncertainty as to pars triangularis, posterior F2, and anterior insula.

However, shortly after the observation by Broca that a lesion of left posterior F2 produced motor aphasia, exceptional cases were described. In fact, Broca (1865) took up this problem himself in relation to a case of Moreau (1864), a 74-year-old nonaphasic epileptic woman found to have congenital atrophy of the left Sylvian area, particularly involving the posterior inferior frontal region.[10] Recently there has been renewed interest

[10] The absence of aphasia in this case with extensive lesion of the motor speech area led Broca to remark that many times before, in studying the brains of aphemics, "a lesion of the

in this problem, in light of cases of destruction or removal of Broca's area without or with only transient aphasia.

Cases of destruction of the left Broca area without persistent aphasia are well known from the early literature (Moutier, 1908; Monakow, 1914; Nielsen, 1946), though in the majority handedness was not specified. Such patients tended to be younger, though several cases were recorded with recovery at an advanced age [e.g., a 70-year-old patient of Bramwell (1898)]. Moreover, there was no simple relationship between duration of aphasia and lesion size. Persistent aphasia occurred in several patients with small lesions restricted to the foot of F3, while recovery occurred in some patients with extensive destructions. Involvement of the anterior insula in addition to pars opercularis appeared to be a factor in many cases. Other cases are described in Mohr (1973) and Hecaen and Consoli (1973).

Of greater interest are patients with bilateral lesions of Broca's area with and without aphasia. Negative cases were described by Moutier (1908). Cases of bifrontal lesion with aphasia are not comparable in that an initial left-side lesion would tend to produce a persistent aphasia so the effect of a second right-side lesion cannot be clearly determined. Thus, only the patient with an initial left-side lesion without aphasia tends to be reported. Such a case was described by Barlow (1877), a 10-year-old boy with speech loss and right hemiparesis following a fall. After 10 days, speech returned except for some paraphasia. Subsequently, he developed a left hemiplegia and permanent motor aphasia. Postmortem examination showed symmetrical lesions in F2 and F3.

Another case was reported by Charcot and Dutil (in Monakow, 1914). This patient developed, at age 44, a right hemiparesis and Broca's aphasia, which resolved to a partial agraphia. Eleven years later he had another attack with right hemiparesis and complete motor aphasia. After 2 years speech returned; then 5 years later there was a third attack with mild articulation difficulty. Shortly after this, he developed a motor aphasia that persisted to death. At postmortem examination, a small focus of softening was found in the left Broca's area and, on the right side, softening of the Rolandic operculum and pars opercularis. In this case, the terminal aphasia was related to the more recent right-side focus.

Tonkonogi (1968) described a 58-year-old woman who developed weakness of the right side and mild motor aphasia. Speech was described as

---

left third frontal convolution was not always directly related in intensity with the alteration of language; . . . (having) seen speech completely abolished by a lesion from 8–10 mm in extent, while in other cases lesions ten times larger altered only partly the faculty of articulate language." Broca commented that "in all probability, the two hemispheres contribute to language . . . [and that in this case] it was perfectly evident that the right third convolution had substituted for the absence of the left." He went on to say that he "wondered why this was not the case in all aphasic patients."

slow with reduction of words. There was complete recovery in 1 month. One year later, she developed a mild left hemiplegia and dysarthria. Speech was unclear but nonaphasic. Autopsy revealed, in the left hemisphere, softening in the triangular and opercular portions of F3 and a 2.5 × 1.5 cm cyst in the underlying white matter involving insula. In the right hemisphere, the lesion involved internal capsule, insula, the inferior portion of the frontal lobe, and the upper temporal lobe.

There are obvious problems in the interpretation of the clinical material, and for this reason cases of surgical excision of Broca's area are of special import. In 1891, Burckhardt reported a series of patients with psychiatric disorders undergoing cortical topectomy. In two of these cases, there was a unilateral surgical excision of Broca's area.

> The patient was a 26-year-old right-handed male psychotic. Prior to surgery, speech fluctuated between logorrhea and mutism, with frequent neologism. Initially, left posterior T1 and T2 were removed, following which the patient spoke little and mumbled, but utterances were well articulated. Auditory comprehension was said to be intact; auditory hallucinations persisted but were diminished in intensity. Six days after surgery, a transient aphasia was noted. Two years later, because of the return of verbigeration, Burckhardt resected the pars triangularis and foot of the left F3. Following this second operation, the patient spoke with less fluency, with a reduction of words.

> This case was a 51-year-old right-handed woman with long-standing maniacal agitation. In the first operation, part of the left superior parietal and supramarginal gyri was resected. Three months later part of the left T1 was removed, and 2 months after this a third operation was done with excision of posterior-superior temporal cortex and angular gyrus on the left side. While no change was noted in speech, auditory hallucinations were diminished. The final operation was a resection of left pars triangularis, with no postoperative aphasia.

With regard to these cases, the first patient, who was relatively young, had a more extensive excision of Broca's area, with a change in fluency and some vocabulary loss. In the second case, only pars triangularis was resected as the fourth in a series of operations. These factors might account for the lack of language change following surgery. It is of interest that Burckhardt decided to carry out only a partial removal in the second patient because of the patient's age, noting that compensation after complete removal might be less pronounced than in a younger patient.

Subsequently, two patients were described with *bilateral* topectomy of Broca's area (Mettler, 1949):

> The patient was a 27-year-old woman with regressed catatonic schizophrenia, described as seclusive and nearly mute. At operation, the inferior frontal gyri (Area 45 and parts of Areas 10, 44, and 46) were ablated bilaterally. Following operation the patient was unresponsive for at least 6 days. By the 12th day, she still did not speak spontaneously and a motor aphasia was suspected. Though on psychological examination it was recorded that 1 month following surgery the patient "spoke clearly and distinctly," even

on preoperative neurological testing the patient was said to have spoken in monosyllables. At 4 weeks postsurgery, the neurologist noted that she spoke in thick monosyllables with prominent dysarthria and bilateral facial weakness as well as weakness of tongue movements and dysphagia. There was no evidence of aphasia but rather a "severe degree of dysarthria." Reading and writing were intact within educational limits; naming was also preserved.

The patient was a 31-year-old female with hebephrenic schizophrenia. Prior to surgery speech was confused with "word-salad." At operation, Area 44 (pars opercularis) was removed bilaterally. After operation, speech was said to be "greatly improved with little or no babbling." Questions were answered directly and accurately. This improvement lasted only a short time and within 3 months the patient was speaking as before, that is, word-salad. There was no dysarthria or dysphagia.

In the first case, with a somewhat larger topectomy, some degree of postoperative articulation difficulty appears to have been present, though the recorded notes do show discrepancies as to the extent and duration of this difficulty. In the second case, the cortical removal was more restricted, which may account for the lack of postoperative aphasia. Another and somewhat speculative possibility that has not been considered, which could explain the Burckhardt and Mettler cases, concerns the fact that these were all severely regressed schizophrenics. If the cognitive level or state is important in determining symptomatology—and if Broca's area mediates a cognitive as well as a language level—then the more preliminary cognition of the psychotic may permit a relative escape following Broca's area lesion.

These are the most dramatic reports; however, other surgical patients have been described by Chavany and Rougerie (1958), Robb (1948), and Zangwill (1975). In addition, Jefferson (1949) observed that excision of Broca's area, avoiding deep undercuts in the white matter, produced no more than a transient speech disorder. Burcklund (1975, personal communication) has stated that "bilateral removal of the frontal lobes, including the frontal opercular cortex, results in temporary mutism which persists for only a few days" and that "unilateral excision of the 'dominant' frontal opercular area produces mutism, lasting from several days to one month, following which a slight dysarthria can be detected in most cases." According to Burcklund, dysarthria does *not* occur after bilateral frontal opercular excisions. Finally, Mettler (1972) has emphasized that motor aphasia has not been described following frontal lobotomy in spite of the insertion of the leucotome in many cases through pars opercularis.

From this review we may at least conclude that there are sufficient cases on record to allow us to distinguish a central zone in the posterior frontal operculum, lesions of which cause some motor speech impairment. The question that we are left with, however, and one that we have attempted to deal with in the preceding sections, is how we are to interpret the functional organization of this anterior zone in relation to both aphasic symptoms and normal speech. Certainly, the evidence cited is consistent with the claim that size of lesion and age at time of pathology are of crucial importance. The surgical topectomies producing minimal or no aphasia were limited cortical resections and were carried out in relatively young individuals. There is improved recovery in younger subjects, as well as in left-handers regardless of side of lesion. We have argued that this can be explained by the more diffuse organization of language in younger subjects and in non–right-handers. But what is meant by "more diffuse"? Diffuseness of representation implies an incomplete differentiation of focal neocortex out of background *generalized neocortex.*

Recent studies in pigment architectonics indicate that the Broca area has relatively fixed boundaries in the operculum. This is in contrast with the Wernicke area, which, in agreement with the thesis of this chapter, shows a greater individual and interhemispheric variability (Braak, 1979). However, the differentiation to which we refer is *functional.* It is a conclusion from the lesion data and not purely morphological. Moreover, the postulation of a progression from generalized neocortical to focal neocortical representation applies not only to language but probably to other perceptual and motor systems. Studies of evoked potentials might be expected to show a gradual narrowing down of the zone from which such phenomena are elicited in the course of ontogenetic development.

Rather than a circumscribed zone in a geographic map of the hemispheres, Broca's area undergoes a *development* out of homotypical (generalized) isocortex. This development occurs to a variable extent and determines the degree of lateralization in a given individual. This maturational transition is similar to that which occurs in evolution as homotypical isocortex develops out of mesocortical limbic structures. These structures appear to be represented in the anterior forebrain by portions of supplementary motor area and cingulate gyrus. Evidence that these latter zones participate in language production has been discussed above. There is also the recent demonstration of increased cerebral blood flow to supplementary motor area during vocalization (Lassen *et al.,* 1978). The result of these lines of evolutionary and maturational development is that a hierarchical structure is built up in the anterior and posterior brain that mediates or supports the process of language formulation. Broca's area does not represent a language center but rather a *phase* in this dynamic stratification.

"Thalamic" Aphasia

There is some evidence (see Brown, 1979) that lesions of the (left) thalamus can produce an aphasia. Historically, thalamic lesions have been associated with a nonfluent aphasia, actually a type of mutism, but recently a fluent jargon has been described. Both occur with acute left thalamic lesion, but only mutism has been described in progressive bilateral cases. It is unclear whether these disorders reflect the disruption of intrathalamic mechanisms or are referred effects on overlying cortex; we also do not know what mechanisms (nuclei) are involved.

Thalamic differentiation appears to follow an evolutionary course parallel to that of neocortex. For this reason the symptomatology of thalamic lesions should occur as a destructuring of this phyletic organization; that is, thalamic symptoms should show a level-by-level correspondence with those of cortical lesions. Thus, the picture of a lesion of the pulvinar should be of the same general type as that of a temporoparietal lesion, the picture of a lesion of limbic thalamus should resemble that of limbic cortex, and so on.

Studies in primate have demonstrated prominent connections between the inferior parietal lobule and the pulvinar (Walker, 1938). Using the horseradish peroxidase technique, Kasdon and Jacobson (1978) have shown a more heterogeneous input to this area in monkey but have confirmed the presence of major projections to the inferior parietal lobule from the pulvinar. This arrangement is similar to that in man, where strong connections have been shown to exist between the pulvinar and the inferior parietal and posterior temporal region (Van Buren and Borke, 1969). In contrast, the anterior speech area receives the major projection from the dorsomedial nucleus. This has been demonstrated in monkey by Tobias (1975) and appears also to be the case in man. In sum, the pulvinar is in relation to the generalized neocortex of the posterior sector, and the dorsomedial nucleus is in relation to the neocortex of the anterior sector.

Accordingly, we might suppose aphasic jargon to be associated with a pulvinar lesion and the dysnomia that occurs on pulvinar stimulation to correspond to the semantic disorders that occur with lesions of posterior (transitional and generalized) neocortex. Similarly, mutism may be associated with a lesion of the dorsormedial nucleus. This would correspond to the mutism that occurs as a symptom of anterior (transitional and generalized) neocortical lesion. The lack of phonological or articulatory deficit in the thalamic syndromes would indicate that these nuclei are in functional relationship to the generalized neocortex surrounding the Broca and Wernicke areas, *sensu stricto,* since lesions of these latter zones give rise to deficits at the phonological stages in language processing.

General Conclusions

The structure of the anterior and posterior sectors can be inferred from the pathological material. In the *anterior* system the progression is over a series of levels from:

1. A deep bilaterally represented limbic state incorporating the as-yet-undifferentiated speech act in a matrix of instinctuo-motor activity in relation to a rhythmic kinetic pattern about the body midline (akinetic mutism)
2. Separation of the incipient vocal configuration from its simultaneously emerging nonvocal motor and affective accompaniments, represented in transitional and/or dominant generalized neocortex (transcortical motor aphasia)
3. The possible appearance within this configuration of the early-differentiating global (i.e., holophrastic) syntactic units through dominant, that is, *laterally represented*, generalized neocortex; alternatively, the derivation out of the kinetic melody of an intonational pattern, disruption of which leads to a prominence of content words and the (lexically based) stress pattern (agrammatism)
4. Phonological encoding of the terminal units through the further derivation of the intonational pattern into the temporal programmation of sound sequences, represented in dominant *focal neocortex* (Broca's aphasia)
5. The final phonetic realization achieved through *contralaterally represented* motor cortex (phonetic disintegration)

Simultaneous with this progression, there is an unfolding in the posterior system from:

1. A *bilaterally represented* limbic stage at which the utterance is aroused in memory *in statu nascendi* (amnestic syndrome) leading to the separation and entry into a semantic operation of a verbal component (confabulation)
2. The selection of the verbal component through a series of progressively narrowing semantic fields, by way of limbic transitional and dominant, that is, *laterally represented*, generalized neocortex (semantic aphasic disorders)
3. Adequate selection of an abstract representation of the constituent lexical items, possibly with more facile emergence through the semantic layer of the less meaning-laden functors through dominant, that is, *laterally represented*, generalized neocortex; abstract items emerge in anticipation of a stage of phonemic encoding (anomic aphasia)
4. Phonological encoding of the more or less fully realized utterance through dominant *focal neocortex* (phonemic or conduction aphasia)
5. The final perceptual (and expressive?) exteriorization of the "physicalized" percept through (more or less) *contralaterally represented* sensory (konio)cortex ("word-deafness" and related disorders)

Inspection of the complementary sequences of levels in the anterior and posterior sectors strongly suggests a correspondence between homologous levels in each system. This correspondence is presumably maintained by level-specific inter- and intrahemispheric fibers (Figure 6.2). Interuption of these fibers has not been shown to produce aphasia, though there are some speculations to the contrary. The proposed model, though in disagreement with the concept that language or perceptual information is conveyed over these pathways, does predict some effect of pathway interruption, for example, anterior–posterior asynchrony or elevation of discharge thresholds in the interconnected areas. The central point is that the cortico-cortical fibers relate to timing or to phase relationships in separate, conically organized systems, rather than serving as conduits for the transfer of cognitive packets.

According to this theory, aphasia is the result of disruption at some level in either of two distinct systems. The disruption displays that level and does not destroy a specific language mechanism situated in the damaged site. An aphasic symptom is a fragment of a disturbed level that survives into the end stage, or the development of the level is attenuated. This "regression" effect explains why similar aphasic symptoms occur in patients with small or large lesions, since the extent and nature of the regression are determined by individual differences in the degree of regional hemispheric specialization, that is, the degree of the asymmetry of focal neocortex. This is also related to potential right-hemispheric language processing, in other words, to the development of focal neocortex in right-hemispheric generalized neocortex as well.

Thus, according to the described model, the gradual differentiation that takes place in the generalized neocortex of both anterior and posterior sectors biases the bilaterally emerging utterance toward a phonological realization mediated by these newly differentiating zones. As this process continues, it tends to "asymmetrize" earlier levels, for example, generalized neocortex, which are not ordinarily asymmetric. It is in this way that language becomes lateralized, through the emergence of left focal neocortex, and not through a transfer or migration of function to the left hemisphere.

References

Ajuriaguerra J. and Hecaen, H. (1956). *Rev. Neurol.* *94*, 434–435.
Alajouanine, T. and Lhermitte, F. (1963). *In* "Problems in Dynamic Neurology" (L. Halpern, ed) Jerusalem, p. 201–216.
Alajouanine, T., Ombredane, A., and Durand, M. (1939)." Le Syndrome de Desintegration Phonetique dans L'Aphasie." Masson, Paris.

Alajouanine T, Sabouraud, O. and DeRibaucourt, B. (1952). *J. de Psychol. 45*, 158–180, 293–329.

Bailey, P. and Bonin, G. von. (1951) "The Isocortex of Man." Urbana: Univ. of Illinois.

Bancaud, J., Talairach, J., Geier, S., Bonis, A., Trottier, S., and Manrique, M. (1976). *Rev. Neurol. 132*, 705–724.

Barlow, T. (1877). *Brit. Med. J. 2*, 103.

Bastian, C. (1890). "Aphasia and other speech defects," Lewis, London.

Bignall, K., Imbert, K., and Buser, P. (1966). *J. Neurophysiol. 29*, 396–409.

Bishop, G. (1959). *J. Nerv. Ment. Dis. 128*, 89.

Blumstein, S. (1970). *In* "Studies Presented to Professor R. Jakobson by His Students" (C. Gribble, ed.) Cambridge, Slavica, pp. 39-43.

Bonvicini, G. (1926). *Wien. klin. Wochenschr. 44*, 47.

Braak. H. V. (1979). Personal communication.

Bramwell, B. (1898). *Brain, 21*, 343–373.

Broca, P. (1863). *Bull. Soc. Anthrop. 4*, 200–204.

Broca, P. (1865). *Bull Soc. Anthrop. 6*, 377–393.

Brown, J. (1972). "Aphasia, Apraxia and Agnosia." Charles C Thomas, Springfield, Ill.

Brown, J. (1975). *Brain Lang. 2*, 18–30.

Brown, J. (1977). "Mind, Brain and Consciousness: The Neuropsychology of Cognition." Academic Press, New York.

Brown, J. (1979). *In Handbook of Neuropsychology* (M. Gazzaniga ed.). Plenum, New York.

Brown, J. (1979). *In* "Biology of Language" (D. Caplan ed.). M.I.T. Press, Cambridge, Mass.

Brown, J. *Introduction to C. Bastian, Aphasia and Other Speech Defects (1890).* In preparation.

Brown, J. and Jaffe, J. (1975). *Neuropsychologia, 13*, 107–110.

Brown, J. and Hecaen, H. (1976). *Neurology 26*, 183–189.

Brown, J. and Wilson, F. (1973). *Neurol. 23*, 907–911.

Buckingham, H. and Kertesz, A. (1976). "Neologistic Jargon Aphasia." Swets and Zeitlinger, Amsterdam.

Buge, A., Escourolle, R., Rancurel, G., and Poisson, M. (1975). *Rev. neurol. 131*, 121–137.

Burckhardt, G. (1891). *Allg. Z. Psychiat. 57*, 463–548.

Burcklund, C. personal communication, 1975.

Campbell, A. (1905). "Histological Studies on the Localization of Cerebral Function." Cambridge.

Caplan, D., Kellar, L., and Locke, S. (1972). *Brain 95*, 169–172.

Cermak, L. and Butters, N. (1973). *Quart. J. Studies Alcohol 34*, 1110–1132.

Chavany, J. and Rougerie, J. (1978). *Presse Méd. 66*, 1191–1192.

Clarke, P., Wyke, M., and Zangwill, O. (1958). *J. Neurol. Neurosurg. Psychiat. 21*, 190–194.

Cotman, C., Matthews, D., Taylor, D., and Lynch, G. (1973). *Proc. Nat. Acad. Sci. 70*, 3473–3477.

Curtiss, S. (1977). "Genie: A Psycholinguistic Study of a Modern-Day 'Wild Child'." Academic Press, New York.

Dejerine, J. (1885). *Rev. de Med. 5*, 174–191.

Dejerine, J. (1914). "Semiologie des Affections du Système Nerveux." Masson, Paris.

Dennis, M. and Whitaker, H. (1976). *Brain and Language 3*, 404–433.

Diamond, I. and Hall, W. (1969). *Science 164*, 251–262.

Dufour, L. (1881). "De l'Aphasie Liée a la Lésion du Lobule de l'Insula de Reil." These, Nancy

Ey, H. (1950–1954). "Etudes Psychiatriques." vol. 1–3, Desclée de Brower, Paris.

Farris, A. (1969). *Neurology 19*, 91–96.

Flechsig, P. (1920). "Anatomies des menschlichen Gehirns und Ruckenmarks." Leipzig, Thieme.

Glick, S. and Greenstein, S. (1973). *Br. J. Pharmacol 49*, 316–321.

Goldberger, M. and Murray, M. (1972). Recovery of function after partial denervation of the spinal cord: A behavioral and anatomical study. *Meeting of the Program and Abstracts Soc. for Neurosc.*, Houston, Texas.

Goldman, P. (1976). *Advances in the Study of Behavior 7*, 1–90, Academic Press, New York.

Goldstein, K. (1915) "Die transkortikalen Aphasien." Jena, Fischer.

Goldstein, K. (1943). *J. Nerv. Ment. Dis. 97*, 261–279.

Goldstein, K. (1948). "Language and Language Disturbances." Grune & Stratton, New York.

Goltz, F. (1881). "Uber die Verrichtungen des Grosshirns." Bonn, Strauss.

Goodglass, H. (1971). *Trans. Amer Neurol Assoc., 96*, 144–145.

Green, E. and Howes, D. (1977). *In* "Studies in Neurolinguistics" (H. Whitaker, ed.) vol. 3. Academic Press, New York.

Hamanaka, T., Kato, N., Ohashi, H., Ohigashi, Y., and Hadano, K. (1976). *Studia Phonologica 10*, 28–45.

Hecaen, H. (1976). *Brain and Lang. 3*, 114–134.

Hecaen, H. and Consoli, S. (1973). *Neuropsychologia. 11*, 377–388.

Hecaen, H., Penfield, W., and Bertrand, C. (1956). *Arch. Neurol & Psychiat. 75*, 400–434.

Henschen, S. (1920). "Klinische u. anat. Beiträge z. Pathol. des Gehirns." Nordiska, Stockholm.

Hillier, W. (1954). *Neurology 4*, 718–721.

Jacobsen, S. and Trojanowski, J. (1977). *Brain Res. 132*, 209–246.

Jefferson, G. (1949). *Brit. Med. Bull. 6*, 333–340.

Johnson, J., Sommers, R., and Weidner, W. (1977). *J. Speech Hear. Res. 20*, 116–129.

Jones, E. and Powell, T. (1970). *Brain 93*, 795–820.

Kähler, O. and Pick, A. (1879). *Vierteljahresschrift für praktische Heilkunde, 1:6*.

Karis, R. and Horenstein, S. (1976). *Neurology 26*, 226–230.

Kasdon, D. and Jacobson, S. (1978). *J. Comp. Neurol. 177*, 685–705.

Kellar, L. (1978). Presentation: Academy of Aphasia, Chicago.

Keminsky, W. (1958). *Arch. Neurol. and Psychiat. 79*, 376–389.

Kennard, M. (1936). *Am. J. Physiol. 115*, 138–146.

Kinsbourne, M. (1971). *Arch. Neurol. 25*, 302–306.

Kinsbourne, M. and Hiscock, M. (1977). *In* "Language Development & Neurological Theory" (S. Segalowitz and F. Gruber, eds.). Academic Press, New York.

Kinsbourne, M. and Warrington, E. (1963). *Neuropsychologia 1*, 27–37.

Krynauw, R. (1950). *J. Neurol. Neurosurg. Psychiat., 13*, 243–267.

Kuttner, H. (1930). *Archiv. f. Psychiat. m. Nervenkr 91*, 691–693.

Kertesz, A. and Benson, D. (1970). *Cortex 6*, 362–368.

Kreindler, A., Calavrezo, C., and Mihǎilescu, L. (1971). *Rev. Roum. de Neurol. 8*, 209–228.

Ladame, P. and Monakow C. von. (1908). *Encephale, 3*, 193–228.

Lassen, N., Larsen, B., and Orgogozo, J. (1978). *Encephale 4*, 233–249.

Lecours, A. (1978). personal communication.

Lecours, A. and Rouillon, F. (1976). *In* "Studies in Neurolinguistics (H. Whitaker, ed.) vol. 2. Academic Press, New York.

Leischner, A. and Peuser, G. (1975). *Presentation*. Aachen, November.

Lhermitte, F. and Derouesné, J. (1974). *Rev. Neurol. 130*, 21–38.

Lhermitte, F., Derouesné, J., and Lecours, A. (1971). *Rev Neurol. 125*, 81–101.

Lesser, R. (1976). Presentation at European Brain-Behavior Conference, Oxford.

Luria, A. (1963). "Restoration of Function after Brain Injury." Pergamon, Oxford.

Luria, A. (1966). "Higher Cortical Functions in Man." Basic Books, New York.

Lynch, G. (1974). In "Functional Recovery After Lesions of the Nervous System." (E. Eidelberg and D. Stein, eds.) Bulletin 12, No. 2.

Marty, R. (1962). Arch. Anat. Micr. 51, 129–264.

Mettler, F. (1949). "Selective Partial Ablation of the Frontal Cortex." Hoeber, New York.

Mettler, F. (1972). J. Speech Hear Dis. 37, 278–279.

Meyer, J., Teraura, T., Sakamoto, K., and Kondo, A. (1971). Neurology 21, 247–262.

Mihailescu, L., Voinescu, I., and Fradis, A. (1967). Rev. Roum. de Neurol. 4, 81–99.

Milner, B., Branch, C., and Rasmussen, T. (1966). Trans. Amer. Neurol. Assoc. 91, 306–308.

Mills, C., and Spiller, W. (1907). J. Nerv. Ment. Dis. 34, 624–650.

Molfese, D., Freeman, R., and Palermo, D. (1975). Brain and Lang. 2, 356–368.

Mohr, J. (1973). Arch. Neurol. 28, 77–82.

Monakow, C. von (1914). "Die Lokalisation im Grosshirn, Wiesbaden." Bergmann.

Monakow, C. von (1905). "Gehirnpathologie." Halder, Vienna.

Moreau, (1864). Gaz. des Hop. 70, 70–71.

Moutier, F. (1908). "L'Aphasie de Broca." Steinheil, Paris.

Myers, R. (1976). In "Origin and Evolution of Language and Speech." (S. Harnad, et al., eds.). Ann. N.Y. Acad. Sci. 280, 745–747.

Nielsen, J. (1965). "Agnosia, Apraxia, Aphasia." (Reprint 1946 Ed.), Hafner, New York.

Nielsen, J. and Jacobs, L. (1951). Bull. Los Angeles Neurol. Soc. 16, 231–234.

Nielsen, J. (1944). Bull. Los Angeles Neurol. Soc. 3, 67–75.

Niessl von Mayendorf, E. (1911). "Die aphasischen Symptome und ihre kortikale Lokalisation." Barth, Leipzig.

Niessl von Mayendorf, E. (1911). "Die aphaschen Symptome." Engelmann, Leipzig.

Niimi, K. and Sprague, J. (1970). J. Comp. Neur. 138, 219–250.

Ojemann, G. and Whitaker, H. (1977). Presentation: Academy of Aphasia, Montreal.

Pandya, D. and Sanides, F. (1973). Z. Anat. Entwickl. Gesch. 139, 127–161.

Perecman, E. and Brown, J. W. Phonemic jargon: a case study. In preparation. In: J. W. Brown (Ed.). Jargonaphasia. Academic Press, 1980 in preparation.

Penfield, W. and Faulk, M. (1955). Brain 78, 445–470.

Petras, J. (1971). J. Psychiatr. Res. 8, 189–201.

Pick, A. (1931). Handb. d. norm. u. path. Physiol. 15, 1416–1524. Springer, Berlin.

Raisman, G. (1969). Brain Res., 14, 25–48.

Rasmussen, T. (1978). personal communication.

Riese, W. and Goldstein, K. (1950). J. Comp. Neurol. 92, 133–168.

Robb, J. (1948). Assoc. Res. Nerv. Ment. Dis. 27, 587–609.

Robinson, B. (1977). In Ann N. Y. Acad. Sci.

Rochford, G. (1974). Br. J. Dis. Comm. 9, 35–44.

Roland, P. (1978). The Behav. Brain Sci. 1, 129–171.

Rubens, A. (1977). In "Lateralization in the Nervous System" (Harnad et al., eds.). Academic Press, New York.

Sanides, F. (1975). Brain and Language, 2, 396–419.

Saundby, R. (1911). Brit. Med. Bull., 605–608.

Saundby, I. (1911). Brit. Med. J. 1, 605 (cited in Henschen, 1920).

Schilder, P. (1953). "Medical Psychology." International Univ. Press, New York.

Schneider, G. (1974). In "Functional Recovery After Lesions of the Nervous System." (E. Eidelberg and D. Stein, eds.) NRP Bulletin 12, No. 2.

Scoville, W. and Milner, B. (1957). J. Neurol. Neurosurg. Psychiat. 20, 11–21.

Semmes, J. (1968). Neuropsychologia 6, 11–26.

Smith, A. and Burklund, C. (1966). Science 153, 1280–1282.

Stein, D. (1974). *In* "Plasticity and Recovery of Function in the Central Nervous System." (D. Stein, J. Rosen, and N. Butters, eds.). Academic Press, New York.

Tes/ner, D., Tzavaras, A., Gruner, J., and Hecaen, H. (1972). *Rev. Neurol. 126*, 444–449.

Teuber, H. L. (1974). *In* "Functional Recovery After Lesions of the Nervous System." (E. Eidelberg and D. Stein, eds.) *NRP Bulletin 12*, No. 2.

Tobias, T. (1975). *Brain Res. 83*, 191–212.

Tonkonogi, J. (1968). "Insult and Aphasia" (Russian) Meditsina, Leningrad.

Torrey, E. and Peterson, M. (1974). *Lancet*, 942–946.

Ungerstedt, U. and Arbuthnott, G. (1970). *Brain Res. 24*, 485–493.

Van ' ren, J. and Borke, R. (1969). *Brain 92*, 255.

Victor, M. (1974). personal communication.

Vogt, C. and Vogt, O. (1919). *J. Psychol u. Neurol. 25*, 279–462.

Wadham, W. (1869). *St. George's Hospital Reports 4*, 245–250.

Walker, A. (1938). "The Primate Thalamus." Univ. of Chicago Press, Chicago.

Wernicke, C. (1874). "Der aphasische Symptomenkomplex." Breslau, Cohn, and Weigart.

Weinstein, E., Lyerly, O., Cole, M., and Ozer, M. (1966). *Cortex 2*, 165–187.

Whitaker, H. and Ojemann, G. (1977). Stimulation in human speech area. Presentation at Academy of Aphasia, Montreal.

Witelson, S. and Pallie, W. (1973). *Brain 96*, 641–647.

Woolsey, C. et al.: (1952). *Res. Publ. Assoc. Nerv. Ment. Dis. 30*, 238–264.

Yakovlev, P. (1972). *In* "Limbic System Mechanisms and Autonomic Function." (C. Hockman, ed.). Thomas, Springfield.

Yeni-Komshian, G. (1977). Presentation, Conference on Aphasia, C.U.N.Y.

Zaidel, E. (1976). *Cortex 12*, 191–211.

Zaidel, E. (1977). *Neuropsychologia 15*, 1–18.

Zaimov, K. (1965). "Uber die Pathophysiologie der Agnosien, Aphasien, Apraxien. Gustav, Fischer, Jena.

Zangwill, O. (1975). *In* "Cerebral Localization" (K. Zulch *et al.*, eds.). Springer, Berlin.

Zurif, E., and Caramazza, A. (1976). *In* "Studies in Neurolinguistics." (H. Whitaker, ed.) vol. 1. Academic Press, New York.

7

Neuromotor Mechanisms in the Evolution of Human Communication[1]

DOREEN KIMURA

Introduction

Communication is a very broad term. This chapter will deal with only a portion of the behaviors typically labeled communicative. Specifically, it will deal with those interactions between individuals in which a given behavior is used to represent some object or event, present or absent. There are two major forms of communication in this sense in man: speaking, and the manual signing of the deaf. In each case, the behavior, speaking or signing, consists of a complex series of self-generated movements, composed of definable units, which can "stand for" some apparently arbitrary object or action. This narrowing of the frame of reference of the term "communication" is not intended as a comprehensive definition, but rather as a rough index of the behaviors to be considered. A more exact definition would be premature at this stage since it is precisely our task to characterize communicative behaviors, to discover how they functionally resemble other behaviors, and to define the neural mechanisms that mediate them. Indeed, this chapter will argue that the past emphasis on the representational function of communication in man has given rise to two major fallacies: (*a*) that somehow the behavior involved in representational communication is mysteriously different from all other behaviors and (*b*) that the pur-

[1] This research was supported by grants from the National Research Council, the Medical Research Council, Ottawa, Canada, and the Ontario Mental Health Foundation.

197

posefulness of communication is a fundamental characteristic of communicative behaviors. It will be suggested, in contrast to these views, that communication evolved out of behaviors probably developed for other reasons and that important characteristics of these behaviors can be specified without reference to either representational or purposive function.

The emphasis will therefore be on the characteristics of the behavior emitted during communication, that is, on the movements employed. There are strong reasons for thinking that one of the most critical determinants of the capacity for communication is the degree to which appropriate movements can be employed in its service, these preadapting behaviors being to some extent freed from their prior function in order to serve as communicators. This approach suggests that the *vehicle* for communication is critical and contrasts with the approach of some linguists, for example, Chomsky (1967), who consider that language may have emerged suddenly in the course of man's development, without known precursors, and could thus be considered a unique phenomenon, with no parallels among nonhuman animals. In the latter case, the behavioral vehicle for communication would be considered relatively unimportant, since what was bestowed was some complex linguistic–grammatic capacity presumably independent of any behavioral repertoire man possessed at the time.

We shall first discuss the evolution of communicative behaviors in man and then relate this to what is known about the neural mechanisms of communication systems.

Evolution of Communication in Man

Since communication involves behavior, and behavior leaves no direct fossil record, our information concerning the evolution of communication must necessarily be incomplete and inferential. Our inferences are based on three main sources:

1. *Inferences from structure.* The pitfalls here are obvious, in that the mere fact that a structure permits a certain behavior by no means ensures that the behavior was either well developed or even present at the time. However, the absence of an essential structure can be informative.
2. *Inferences from artifacts.* The chief source of pertinent information here is tools, and here again the inferences concerning the related behavior are necessarily indirect.
3. *Inferences from studying the great apes.* Since man and the apes had a common ancestry before our lines diverged about 20 million years

ago, we might expect to see some instructive parallels between our own behaviors and those of our nearest primate relatives. Here one must caution, however, that the great apes *have* diverged from man and have continued to develop; they have not merely stopped at some point in our own past history.

Nevertheless, although each of the above sources of information is imperfect, we can, by a judicious combination of all three, attempt a rough outline of how communication may have evolved.

After the hominid divergence from the great apes, there appears to have been a period of rapid development of several hominid lines (Figure 7.1) related to the formation of savannahs, replacing the forest in Africa. Thus, in all of the hominids—the two *Australopithecus* lines and *Homo erectus*—the upper limbs were freed from locomotion in the course of more upright walking. The teeth were small in all, suggesting that alternate means of defense were available, and there is evidence also of tool use in all early hominids, going back at least 3 million years.

Brain Size

There appears to have been a very rapid increase in the size of the brain in hominids, as estimated from cranial measurements, over the past 4 million years. This is true even when corrections are made for the increasing body size of the various hominid species (Passingham, 1975). Although

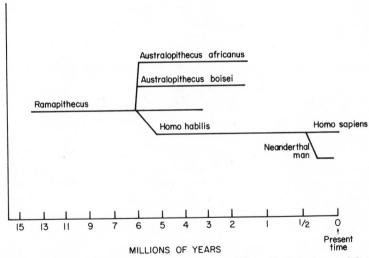

FIGURE 7.1. *Schema of hominid evolution. [Adapted from R. E. Leakey and R. Lewin (1977), Origins, E. P. Dutton, New York.]*

relating cranial capacity to evolutionary development was, and still is, a favorite pasttime of physical anthropologists, Holloway (1972) has pointed out that correlating the evolution of specific functions (e.g., language) with overall brain size is unlikely to be fruitful. Instead, a more sophisticated approach has been to compare across different hominid species those organizational features of the brain that can be estimated from the endocranial casts, for example, the apparent *relative* increase in parieto-temporal regions (Holloway, 1972). That neocortex, and particularly association cortex, increases relative to other brain regions as the primate brain increases in size appears to be well established (Passingham, 1975, and this volume, Chapter 8). It nevertheless remains true that studies on sizes of certain brain areas relative to others, as well as studies on morphological features such as brain asymmetry (Galaburda *et al.*, 1978), have so far failed to provide any new insights into the evolution of communication, although they have helped to round out the picture derived from other sources of information.

Evolution of Speech

An important contribution to our understanding of the evolution of speech has been made by Philip Lieberman (1975). He has addressed himself specifically to the structural requirements of speech and has questioned the extent to which such requirements are met in early hominids and in nonhuman primates. Lieberman contends that, so far as the vocal system is concerned, significant differences between species are to be found not in the larynx, but rather in the structures beyond it, in the supralaryngeal vocal tract.

Lieberman argues that, in order for us to reliably perceive the variety of speech sounds that we do, we must be able to correct for the length of a speaker's vocal tract. The claim is that adult human beings produce three vowels that delimit the frequency range of the sounds each person will produce, and which therefore provide the necessary perceptual framework within which we can perceive variants on these sounds. All other vowels, and therefore at least all stop consonants, are pegged to these vowels. He presents evidence that only vocal tracts of a particular shape (containing a sharp right-angle turn) can produce these sounds. Human infants do not have such a vocal tract and do not produce the three vowels. Whether or not the theory is correct in terms of the importance of the vowel triangle for accurate speech perception, the emphasis on the shape of the supralaryngeal vocal tract appears to have great cogency for differences in vocal ability between species.

Of particular interest for the evolution of speech are the reconstructions of the oral cavities of earlier hominids. *Australopithecus*, by these reconstructions, does not have a right-angled supralaryngeal vocal tract, and Lieberman concludes that *Australopithecus* would therefore not have been able to produce the variety of sounds that *Homo sapiens* could produce, and that his utilization of speech for communicative purposes would consequently be very limited. In fact, a cast of *Australopithecus*'s supralaryngeal vocal tract looks remarkably like that of the chimpanzee (Lieberman, 1975, p. 143).

As recent a hominid as Neanderthal man appears not to have had the requisite vocal tract either, although the contemporary of Neanderthal, *Homo erectus*, does have a right-angled vocal tract. It will be recalled from Figure 7.1 that Neanderthal man (as well as *Australopithecus*) is now assumed to be a separate branch line of hominid that became extinct, whereas *Homo erectus* is on a direct line of descent to man. These studies clearly suggest that the advanced development of speech as a means of communication may have occurred rather late in the course of evolution, appearing less than half a million years ago.

These studies also go far toward explaining why it was so difficult to teach chimpanzees to make human speech sounds (Kellogg, 1968), in contrast to the relative success with which they can be taught a manual sign language (Gardner and Gardner, 1971). The chimpanzee's failure to learn human vocal language has often been cited as evidence for the uniqueness of man's capacity for language, but it now appears that the chimpanzee simply does not have a vocal tract that permits imitation of many human speech sounds. Hindsight suggests that a more rational approach would have been to employ the chimpanzee's spontaneous vocalizations, and to attempt to attach them to objects in the environment, as a means of communication. Yerkes and Learned (1925), however, after attempting to condition certain spontaneous vocalizations in the chimpanzee, concluded that the vocalizations were not conditionable, indicating that even the natural vocal sounds of a chimpanzee may not be available for any but very limited communication. Attempts to arbitrarily condition vocalizations to previously neutral situations in other nonhuman primates have also met with mixed success (Myers, 1976; Sutton, this volume, Chapter 3).

Evolution of Gesture

In comparison with the apparently late and limited development of vocalization as a means of communication in the higher primates, the capacity to use the rest of the body for communication, particularly the upper limbs, appears to have been favored. When we search for the evolutionary development of a gestural form of communication, we again have only indirect evidence. It does appear that all hominids were upright walkers, indicating that the upper limbs were freed from locomotor activity millions of years ago, but this in itself does not give us much information about the complexity of movement possible in the arms.

More compelling evidence for early complex control of brachial movements comes from fossil remains of tools. It appears that *Australopithecus* used bone tools at least, and to some extent modified them for use as early as 3 million years ago. There is even evidence for a right-hand preference in the wielding of such tools as clubs (Dart, 1949). Wooden tools would not of course have survived to provide a record. *Australopithecus'* contemporary, *Homo habilis*, also employed stone tools, many of which were wielded unimanually (Leakey, 1971; Tobias, 1971). We can thus say with some assurance that *Australopithecus* and *Homo habilis* had the precision of upper limb movement control, which would make gesturing possible. This of course does not constitute proof that it was employed as a means of communication.

When we look for parallels to gestural communication in our living primate relatives, we find that, although spontaneous use of the arms for communication does not appear to occur, chimpanzees nevertheless can be taught to employ a manual sign system (Gardner and Gardner, 1971). This is in marked contrast to the failure to teach chimpanzees to employ human speech sounds for communication, and it points up the importance of the behavioral vehicle of communication for utilization in a linguistic system. Apparently, preadaptation for manual signing is better in the chimpanzee than preadaptation for vocalization. This is probably related to the manipulatory and tool-wielding ability manifest in the chimpanzee. When a communication system develops, it presumably builds on an already existing behavioral repertoire, and the chimpanzee repertoire for the upper limbs is apparently more suitable than that for vocalization. It is quite possible that chimpanzees, left to themselves, might eventually have developed a gestural system of communication, although studies by Menzel (1973) suggest that the chimpanzee already has a very effective whole-body system of communication.

These and other considerations have led to the suggestion that a manual system of communication may have preceded the vocal system of communication in man (Hewes, 1973; Stokoe, 1974). The manual system of communication could very easily have been built on manual skills already developed in association with tool use—the independent and precise positioning of the two upper limbs with respect to the tool. It is also clear that employment of most tools requires the asymmetric use of the two arms, and in modern man this asymmetry is systematic. One hand, usually the left, acts as the stable balancing hand; the other, the right, acts as the moving hand in such acts as chopping, for example. When only one hand is needed, it is generally the right that is used. It seems not too farfetched to suppose that cerebral asymmetry of function developed in conjunction with the asymmetric activity of the two limbs during tool use, the left hemisphere, for reasons uncertain, becoming the hemisphere specialized for precise sequential limb positioning. When a gestural system was employed, therefore, it would presumably also be controlled primarily from the left hemisphere. If speech were indeed a later development, it is reasonable to suppose that it would also come under the direction of the hemisphere already well developed for precise motor control.

We may never know for certain whether this is the actual evolutionary sequence or not. That asymmetrical use of the two limbs is not a necessary condition for the development of brain asymmetry is indicated by Nottebohm's (1971) demonstration of asymmetry for vocal control in the chaffinch. To what extent the phenomenon of neural asymmetry in birds is relevant to human brain asymmetry is at present unclear, since man's nearer mammalian relatives apparently show a much lesser degree of asymmetry, both functional and morphological (Webster and Webster, 1975; Yeni-Komshian and Benson, 1976), than the chaffinch.

Neural Mechanisms

This section will be concerned with the probable neural mechanisms involved in man's two major natural systems of communication: speaking and manual signing. (Writing will not be considered here since it is generally based on speech.) It will be suggested that the left cerebral control of both functions in man is based on left-hemisphere specialization for certain motor activities. The nature of this motor control will be discussed in some detail.

We have known for about a hundred years, from the occurrence of speech disorders called "aphasias," that speech is critically dependent on the left hemisphere. A review of eight cases of manual sign language "aphasia" in the literature also indicates that, in all cases, there is evidence of damage to the left hemisphere (Table 7.1). At first glance, therefore, it may appear that it is specialization for symbolic or representational function in the left hemisphere that these two forms of communication have in common. A closer look at the functional characteristics of left-hemisphere control, however, suggests that that is not necessarily the case.

Consider the manual sign language. The movements employed in manual sign language share many structural features with those movements affected in manual "apraxia," a disorder of movement fre-

TABLE 7.1. *Eight Cases of "Aphasia" in the Deaf*

AUTHOR AND REFERENCE	SALIENT FEATURES	CLINICAL LOCALIZING SIGNS	PROBABLE LOCUS OF LESION
Grasset, *Le Progrès Médical*, 1896	Signing affected in right arm only	Incoordination and weakness of right arm	Left hemisphere
Burr, *N. Y. Medical Journal*, 1905	Insufficiently described, general deterioration	Right hemiplegia, hemianopia	Left hemisphere
Critchley, *Brain*, 1938	Not congenitally deaf, manual signs built on speech	Transient right-sided paralysis	Left hemisphere
Leischner, *Arch. f. Psych. u. Nervenkr.*, 1943	Extensive examination, intelligent subject	Transient weakness of right arm and leg	Left posterior (post-mortem data)
Tureen, Smolik, & Tritt, *Neurology*, 1951	Signing difficulty present only in acute phase, recovered although white matter of third frontal destroyed	Left-sided craniotomy of left frontal cyst with increased intracranial pressure and hemorrhage	Left frontal and internal capsule (plus)
Douglass & Richardson, *Brain*, 1959	Basic signs most affected	Right hemiplegia, left-sided headache	Left middle cerebral artery occlusion
Sarno, Swisher, & Sarno, *Cortex*, 1969	Basic signs least affected	Right-sided weakness	Left cerebral infarction
R. Battison & C. Padden, 49th Ling. Soc. Meeting, New York, 1974	All aspects of production affected	Right hemiparesis	Left middle cerebral artery distribution

quently seen after left-hemisphere damage in hearing patients. Apraxia is typically tested by asking the patient to demonstrate how he would use particular objects (in their absence), or how he would perform certain intransitive movements, for example, salute, wave good-bye, etc. (Liepmann, 1908). Typically, the patient who has difficulty performing these movements also has difficulty in imitating them, indicating that it is not merely a language comprehension defect.

Nature of Praxic Control

Despite Liepmann's very clear exposition of apraxia as a movement disorder, it has sometimes been interpreted as a representational disorder. The latter hypothesis was tested very simply by presenting movements for imitation that are explicitly nonrepresentational (Kimura and Archibald, 1974). That is, meaningless movements of the hands or arms were presented to patients with unilateral damage to the left or the right hemisphere. On this task, called Movement Copying, impairment after left-hemisphere damage can be clearly seen (Table 7.2). The impairment is at least as severe as that seen with the more traditional meaningful movements. It appears, then, that the disorders of manual movement observed after left-hemisphere damage are not restricted to representational movement and that the symbolic value of the movements is irrelevant to the occurrence of apraxia. It should also be noted in passing that the defect in Movement Copying is bilateral and roughly equal on the two hands (when both hands can be tested, i.e., in patients without hemiplegia) and that the performance of the two hands is highly correlated.

In deaf patients with sign language defects after central nervous system (CNS) damage, it has frequently been claimed that there is no associated manual apraxia, suggesting that the defect in control of manual movements could be restricted to linguistic movements (see the references in Table 7.1). The data just presented on the copying of meaningless movements in hearing patients with left-hemisphere damage made this an unlikely interpretation. Fortunately, we had the opportunity to study a deaf man who acquired a signing disorder after a left-hemisphere stroke (Table 7.1, last case). The items in the Movement Copying task are also meaningless in the manual sign language; yet he was just as impaired on them as were our hearing patients with left-hemisphere stroke (Kimura, *et al.*, 1976). Thus, in his case, the impairment in execution of manual movements was not limited to linguistic movements. On more traditional apraxia tests requiring demonstration of object use, he was apparently unaffected, as are many hearing patients with left-hemisphere damage, due almost certainly to the fact that these are empirically easier, more practiced movements (see the discussion of task difficulty in the next section, p. 213).

Given that the defects in manual control that arise from left-hemisphere damage are not restricted to meaningful or representational movements, can one generalize this statement to left-hemisphere control over the vocal musculature? To answer this question, Mateer and Kimura (1977) devised a task requiring patients first to imitate a series of single nonverbal oral

TABLE 7.2.   *Mean Percentage Correct in Imitating Meaningless Movements (Movement Copying)*

	LEFT HAND (%)	N	RIGHT HAND (%)	N	PEARSON r	N
Left-hemisphere lesion	59.2	56	60.0	41	.84	41
Right-hemisphere lesion	75.4	23	75.8	35	.88	23

movements (e.g., protrude tongue) and next to imitate a series of three different movements presented one after the other (e.g., lateral movement of tongue, dropping of jaw, and protruding of lips). These oral movements were required of patients with left- or right-hemisphere damage, and, in the case of the left-hemisphere group, patients were further subdivided into aphasic and nonaphasic groups on the basis of their performance on a standard aphasia test employing verbal tasks. For comparison with the nonverbal oral movements, a series of verbal oral movements (spoken syllable sequences) was also presented.

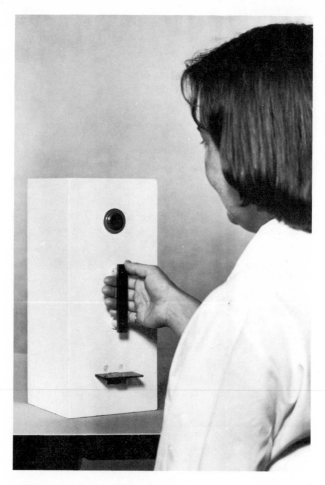

FIGURE 7.2. *The Manual Sequence Box, showing one of the three hand movements required [From Kimura, 1977].*

The results (Table 7.3) clearly show that those patients with speech disorders are also impaired in the reproduction of meaningless nonverbal oral movement sequences. The percentage impairment on such movements is quite comparable to the impairment on verbal movements, even though the two are presented in two different modalities (auditory and visual). We seem, therefore, to have a parallel between the control of oral movements and the control of manual movements by the left hemisphere in that in neither case is control limited to linguistic, verbal, or representational movements but extends to quite meaningless movements as well.

If specialization of function in the left hemisphere (as compared with the right) is not specifically verbal or linguistic, then what can we say about the distinctive motor functions of the left hemisphere that give rise to its importance in speaking and in certain manual activities? Our data suggest that the left hemisphere contains a system important for the selection and/or execution of limb positions and articulatory postures. This system is more than a selection mechanism in the sense that not only is the ordering of movements affected, but efficiency of execution is also impaired. Thus, in the acquisition of another manual skill requiring several changes in hand posture [Figure 7.2, Manual Sequence Box (Kimura, 1977)], it was found that patients with left-hemisphere damage (who had great difficulty with the task) made no more sequencing errors over an equivalent time base than did patients with right-hemisphere damage. The predominant error was a perseverative error, that is, simply repeating the same approximate hand posture as had just been performed (Figure 7.3). This suggested that there was difficulty in moving to a new hand posture, rather than a difficulty in sequencing postures per se. The rather high incidence of unique errors, errors that could not be characterized with reference to the required movements, also indicated that incorrect execution of the movements was an important component of the defect. It should again be noted that, whether the left or the right hand learned the task, the impairment was approximately the same. Mateer (1978), in a follow-up analysis of errors made in reproducing nonverbal *oral* movements, also finds perseverative errors, not sequencing errors, to be most characteristic of patients with left-hemisphere damage.

TABLE 7.3.   *Percentage Correct on Verbal and Nonverbal Multiple Oral Movements*

GROUP	N	NONVERBAL ORAL MOVEMENTS (%)	VERBAL ORAL MOVEMENTS (%)
Right damage, nonaphasic	13	77	99
Left damage, nonaphasic	10	78	99
Left damage, aphasic	13	35	60

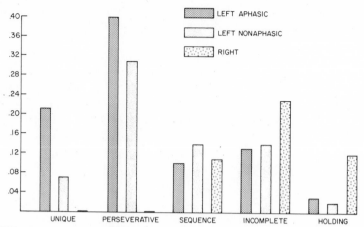

FIGURE 7.3. *Number of different errors per 10 sec of acquisition time on the Box task.*

The left-hemisphere "praxic" system, then, seems to be especially important for the selection and execution of a new posture, but tasks requiring copy of a single hand posture (Kimura and Archibald, 1974), or repetition of the same movement over and over, appear to be relatively less affected by left-hemisphere abnormality. This is true at least of a simple task like repetitive finger tapping (Kimura, 1977). The question is, would it hold for more complex movement, provided the movement is repetitive? The answer appears to be yes, from the findings on a Screw Rotation task (Figure 7.4), where the subject is required to screw a nut up and down a bolt, with coordinated movement of all fingers and the thumb. There is also some slight movement at elbow or shoulder, but it is determined by the hand traveling up and down the bolt. On this task, there is no selective effect of left-hemisphere damage (Figure 7.5). Instead, a defect clearly appears in the *contralateral* hand, and it is approximately equal in cases of either left- or right-hemisphere damage. These data are in good agreement with a study by Pieczuro and Vignolo (1967), employing a slightly different multiple nuts-and-bolts task.

The findings on the Screw Rotation task are in sharp contrast to the Movement Copying and Manual Sequence Box tasks described earlier (Table 7.2 and Figure 7.2), which are selectively under left-hemisphere control, and where a defect, when it is present, is bilateral. That the difference between the latter tasks and the Screw Rotation task is not due simply to the fact that Screw Rotation is a highly speeded task is suggested by the timed performance of the same patients (and same hands) on the post-acquisition trials of the Manual Sequence Box, when patients are also asked to go as quickly as possible (Figure 7.5). Here, although the difference between the groups is not as striking as in the acquisition phase (Kimura,

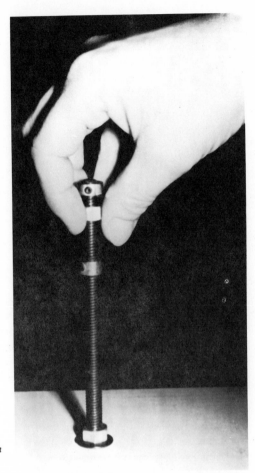

FIGURE 7.4.   *The Screw Rotation task.*

1977), both hands of the left-hemisphere group are slower than both hands of the right-hemisphere group.

Figure 7.5 also suggests that defective sensory information plays a role in the fine finger control required for the Screw Rotation task in that the two-point threshold is raised on the contralateral hand in each group; whereas there appears to be no significant relationship between sensory impairment and performance on the Box task or on Movement Copying. The defect on the latter two limb-positioning tasks (Table 7.2 and Figure 7.5) is in any case bilateral, which again argues against somatosensory control as an important factor. Mateer (1976) also found that defective copying of oral movements after left-hemisphere damage was not explicable on the basis of somatosensory defects.

One can make a further parallel between the praxic defect seen in the

Doreen Kimura

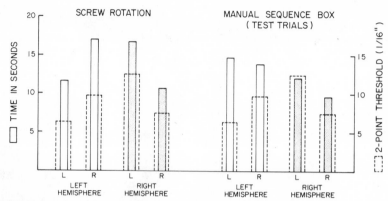

FIGURE 7.5. *Comparison of speed on Screw Rotation and Manual Sequence Box tasks, in patients with left- or right-hemisphere damage. The two-point threshold is indicated by dashed bars.*

hands and arms and that seen in speech disorders in fluent aphasics. In both the vocal and manual modes, individual or single movements can be performed, even repetitively (e.g., finger tapping, Screw Rotation), but errors are made when several different target postures must be organized or when there is a change from one position to another. Thus, patients with fluent aphasia can say ba–ba–ba over and over as quickly as patients with right-hemisphere damage, but when ba–da–ga is the required target they are slower and make more errors (Mateer and Kimura, 1977). This difficulty with multiple articulatory movements is mirrored by a difficulty in nonspeech movements; that is, only multiple oral movements are affected in fluent aphasia while single movements can still be completed (Mateer and Kimura, 1977).

At this point it may be necessary to introduce a caution concerning the use of the terms "single movement" and "multiple movements." Even the reproduction of a single hand posture or a single syllable requires several movements, or a series of movements. Our division of a complex of movements into smaller units is essentially a priori and arbitrary. So far, we have found these arbitrary classifications to be useful, but that may not hold for all analyses.

There are thus a number of statements one can make about the system for praxis. It appears to be critical for changes in posture, but, once a posture is achieved, it can be run off repeatedly without the intervention of this system. It does not appear to depend critically on sensory feedback, and indeed it has been suggested that the system depends on accurate internal representation of moving body parts (Kimura, 1977), somewhat akin to a system that has been suggested for the control of "preprogrammed" movements (Polit and Bizzi, 1978). Finally, it appears to be at least as important for proximal as for fine distal control, in that performing a finger flexion has been found to be unrelated to performance on the Movement Copying task (Kimura and Archibald, 1974), and the Screw Rotation task, requiring rapid movements of the fingers, is not selectively impaired by left-hemisphere damage.

Neuroanatomical Systems in Manual Control

The generally accepted neuroanatomical basis for apraxia is derived from a model originally proposed by Liepmann (1908) and subsequently elaborated by Geschwind (1975) (Figure 7.6). Although there are important differences between the two explanations, they both emphasize cortico-cortical pathways in the control of praxic movements. Consequently, the corpus callosum, the major commissure between the two cortices, figures importantly in both schemas, and particularly in the control of the left hand by the left hemisphere. This chapter will suggest instead that there may be several (at present unknown) complementary systems in praxic control and that such control need not depend exclusively on cortico-cortical or callosal pathways.

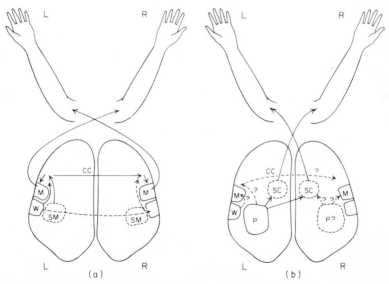

FIGURE 7.6. *Several neuroanatomical schemas for praxic control. CC, corpus callosum; M, motor cortex; SM, supramarginal gyrus; W, Wernicke's area; SC, subcortical (subcallosal) centers; P, praxic system. (a) Liepmann–Geschwind model with emphasis on premotor and callosal pathways. (b) Some suggested alternatives, with the solid lines indicating the most prominent route underpassing the callosum. (See the text for details.)*

Liepmann's proposal can be outlined briefly. The left supramarginal gyrus is the critical region for praxic control; it mediates such control of the right hand through projections to the premotor region in the left frontal lobe, which in turn projects to the left motor hand area, which controls the right hand. Interruptions along this pathway, from damage either to the supramarginal area or to its projection to the frontal area, or to the premotor area itself, will result in apraxia. Damage to motor cortex alone will not result in right-side apraxia, but in loss of fine movement and in sensory deficits (Liepmann, 1908, p. 39).

The control of the left hand is assumed to depend critically on the anterior control system of the *right* hand, through the corpus callosum. Thus, a lesion in the left premotor region will not only result in a right-hand weakness and apraxia but in a so-called "sympathetic" apraxic disturbance of the left hand as well (Liepmann, 1908, p. 39). Apraxia of the left hand alone may result from lesions restricted to the corpus callosum. This rather simplistic, unidirectional model was proposed by Liepmann as a provisional explanation, and he himself suggested modifications to it. It has unfortunately acquired almost the status of a doctrine in the neurological literature.

Geschwind's (1975) variation of the Liepmann model emphasizes, in contrast to Liepmann, the involvement of the pyramidal tract in praxic control and the salience of verbal command, and consequently of Wernicke's area, in eliciting the deficit. The parietal lobe outside of Wernicke's area is not considered in Geschwind's schema, although both Liepmann (1908) and von Monakow (1914) considered it a major area for praxic control.

The restriction of the deficit to the condition of verbal command is a rather special case and should perhaps be dealt with first. There are several cases of disconnection of the two hemispheres by section of the corpus callosum, and it has been reported in these patients that the left hand is defective in carrying out verbal commands, while the right hand has no such difficulty (Gazzaniga, *et al.*, 1967; Geschwind, 1975; Zaidel and Sperry, 1977). The same authors report that no difficulty is experienced by the left hand when the movements are demonstrated and the patient is required to imitate them. Unfortunately, the degree of difference between the left-hand performance on verbal command and its performance on imitation tasks, and how each compares with the right hand's performance, is not clear in any of the above papers, since no quantitative data on the question are presented. Taking the claims at face value, however, the usual explanation for these findings is that verbal command must be analyzed in Wernicke's area, an analysis unavailable to the right hemisphere in the absence of commissural systems (Geschwind, 1975). That the explanation is probably more complicated than this is suggested by Zaidel and Sperry, in that the auditory verbal comprehension capacity of the right hemisphere is quite high as tested on verbal comprehension of the commands, where no manual response other than pointing to the appropriate picture was required.

In any event, the special case of defective left-hand control to verbal command in the presence of a commissural section may not be particularly relevant to the mediating of praxic movements in general, which is impaired by left-hemisphere damage. For one thing, most apraxics have difficulty in performing the required acts, not only to command, but also to imitation (Goodglass and Kaplan, 1963; Kimura and Archibald, 1974; Liepmann, 1908). When there is improvement on imitation, the difference can usually be attributed to the lesser difficulty of imitating. Least difficult of all, and therefore much less likely to be impaired, is the ability to demonstrate object use with the object actually in hand. This scaling of difficulty was suggested many years ago by Hugo Liepmann (1908, p. 15), on the basis of the sensitivity of these measures of apraxia to the severity of the disorder. The most severe apraxic will fail at all three, while the least affected may fail only at the tracing out of practiced movements to command, that is, from memory.

The apparently strong association between right hemiplegia and left-hand dyspraxia has been emphasized by Geschwind (1963, 1975). He suggests that left-hand apraxia is due to loss of control via the corpus callosum from the pyramidal system arising in cortical motor and premotor areas on the left. He therefore contends that the apraxia is particularly evident in the control of individual finger movements (1975, p. 194). There are several problems with this interpretation. While it is true in our own series of patients that those who have a right hemiplegia are more impaired on tests of praxis than patients with left-hemisphere damage without hemiplegia, these two groups also differ greatly in probable extent of CNS damage. Many of the former have hemisphere-wide damage, both cortical and subcortical, often extending into the internal capsule, so that the right hemiplegia is not necessarily an index of focal frontal damage, but rather of widespread damage. A more direct measure of the dependence of apraxia on the involvement of the pyramidal system would be the degree of relation between finger flexion and praxis (Kimura and Archibald, 1974), or hand strength and praxis, in nonhemiplegic limbs. The relationship is weak in each case. Table 7.4 shows the low correlation between right-hand strength and overall praxic ability, as measured by the Movement Copying task. In contrast, the various limb-positioning tasks intercorrelate highly. More critical, perhaps, for the significance of right hemiparesis for left-hand

TABLE 7.4. *Correlation of Movement Copying with Other Motor Tasks*

	MOVEMENT COPYING	df	P
Acquisition time on box	− .74	50	< .01
Praxis to verbal command	.78	23	< .01
Right-hand strength	.28	54	NS

dyspraxia is the correlation between right-hand strength and left-hand Movement Copying. This correlation (not shown in the table) is .16, which is not statistically significant with 54 degrees of freedom.

The claim that manual apraxia is highly associated with Broca's aphasia (Geschwind, 1975, p. 190), and therefore presumptively a common consequence of left anterior damage, is not in agreement with other observations that apraxia is more commonly associated with Wernicke's or "sensory" aphasia than with Broca's or motor aphasia (von Monakow, 1914; de Renzi *et al.*, 1968). We also find that degree of apraxia is, if anything, more highly correlated with "receptive" aphasia scores than with "expressive" scores, suggesting a more posterior locus for praxic control (Kimura, 1977).

The contention that the left frontal lobe anterior to motor cortex is a critical area for control of praxic movements of the hands (Geschwind, 1975; Liepmann, 1908) also receives little support from von Monakow's series of patients (1914, p. 537). Only a small minority of his apraxic patients had left frontal damage, and there are several cases of extensive damage to the frontal lobe on the left with no significant apraxic disturbance. One difficulty in comparing the incidence of apraxia after various lesions, across different authors, is that the term is usually only vaguely defined and may refer to anything from a full-blown bilateral incapacity to produce any standard movements either to command or to imitation to a combination of weakness and clumsiness present only unilaterally. Frontal–callosal systems may indeed be critical for control of fine movements of the distal musculature, but much less important for control of the more molar limb positioning function usually referred to by the term "praxis." Ethelberg (1951) found this distinction to be useful in a series of patients with pathology of the anterior cerebral artery, which supplies wide areas of cortex in the frontal lobe (and the corpus callosum as well). His patients had difficulty, restricted to the contralateral arm, in performing rapid alternating movements or movements requiring fine motor control, such as buttoning a shirt or opening and closing a safety pin, whereas "saluting, handshaking, combing of the hair and aiming an imaginary gun . . . are performed promptly and satisfactorily" (Ethelberg, 1951, p. 105). They thus present a picture of distal clumsiness, rather than apraxia, very much akin to the contralateral hand performance on the Screw Rotation task (Figure

7.5). This clumsiness is probably attributable to Rolandic damage, since it is highly correlated with a raised sensory threshold, and Corkin *et al.* (1970) have shown that such sensory deficits are related to Rolandic lesions.

It is conceivable that, when a hand movement must be initiated through verbal command in a patient with a split callosum, the more posterior praxic system may be bypassed. The pathway for movement control may be directly from a speech comprehension system on the left to the hand area of the left motor cortex. If that were so, it would be understandable that left-hand dyspraxia in the collosal sections would be restricted to verbal command and would be more prominent for distal than for proximal movements (Zaidel and Sperry, 1977). Why verbal command specifically (as compared to imitation) should have more restricted access to the praxis system within the same (left) hemisphere in a patient with callosal section than in one with an intact callosum is rather a puzzle for any schema. Nevertheless, the inference that it is the right hemisphere that mediates the left hand's purportedly accurate imitation of movements in a split-brain patient (Geschwind, 1975) is difficult to reconcile with the fact that the intact right hemisphere in a patient with left-hemisphere damage is unable to do so. Zaidel and Sperry's (1977) report of accurate left-hand copies of static hand postures presented exclusively to the right hemisphere corroborates the finding that copying of single (static) hand postures is no more affected by left-hemisphere damage than by right (Kimura and Archibald, 1974), but there are no data in the split-brain literature at present on the right hemisphere's capacity to copy the *moving* limb.

It is clear from our own series of patients that parietal-lobe damage on the left (without callosal involvement) is sufficient to cause severe bilateral apraxia without hemiparesis. The literature as a whole also suggests that praxic movements depend primarily on a system in the posterior peri-Sylvian region (von Monakow, 1914; de Renzi *et al.*, 1968). The exact limits of this system cannot at present be defined, but there is no good evidence that it is strictly cortical or that the premotor area is a critical component. The praxis system clearly exerts bilateral control over left and right upper limbs (and perhaps lower limbs as well), and this control is at least as salient for the proximal as for the distal musculature, suggesting that it is not critically dependent on the pyramidal system, although it may well exert control over the cortical pyramidal system (see Figure 7.6). The bilateral control need not be mediated by the corpus callosum, since what limited evidence there is suggests that imitation of manual movements by the left hand can be achieved in the presence of callosal section. A discussion of the neural basis for the bilateral control mechanism must be largely speculation at this point, but it may conceivably operate via the basal ganglia, since there are known to be bilateral cortico-striate connections in monkeys (Kemp and Powell, 1970), and cooling of basal ganglia has been shown to be followed by inability to perform learned arm movements (Hore *et al.*, 1977). An alternative, or perhaps an addition, to basal ganglia pathways may be other subcortical structures including the thalamus. Certainly, it would be premature at present to conclude that we know the neuroanatomy of the praxis system.

Neuroanatomical Systems in Oral Control

The evidence presented above suggests that control of postural changes in both oral and brachial musculature appears to depend on the left posterior peri-Sylvian region. Fluent aphasics, who can produce speech effortlessly but who make many speech errors, tend to have damage in this region (Benson, 1967). They also have a high incidence of manual apraxia. However, although production of multiple postures is affected, production of single oral and manual movements is relatively unaffected by posterior damage.

In contrast, production of single oral movements, whether verbal or nonverbal, *is* impaired in nonfluent aphasia, a speech disorder in which there are clearly articulatory problems and hesitancies in speaking, at the level even of a single syllable (Mateer and Kimura, 1977). Nonfluent aphasia has been linked to left anterior damage (Broca, 1861; Benson, 1967). This fact suggests that, for oral movements at least, there may be a somewhat separate system that can mediate the execution of individual speech sounds (and individual nonverbal oral movements), but which requires the integrity of the posterior system for the organization of these movements when several must be produced. (It is not clear that there is an analogous system for control of single hand postures, although we know that individual finger movements can be severely impaired by motor cortex damage.)

Reduction in general facial expressiveness and in vocalization has been reported after bilateral frontal damage in monkeys (Myers, 1972). "Loss of affect," which is usually inferred from reduced facial expressiveness, has also frequently been reported after frontal-lobe damage in man, and Knight (1964) reports similar changes after excision of tissue limited to orbital frontal regions. Kolb (1977) has recently reported some systematic observations that show reduced facial expressiveness after damage to either the left or right frontal lobe. It may therefore be that the frontal lobes in man have become specialized for conveying facial expression (among other things) and that the importance of Broca's area in the left inferior frontal region is a product of increased specialization along two dimensions: (*a*) the left hemisphere for certain kinds of motor control and (*b*) the frontal regions for facial expression. The consequence may be that the left orbital frontal region has become especially involved in the oral-articulatory components of facial movement, with, however, important coordination functions being exercised from the left parietal region.

Summary and Conclusions

It appears probable that communication in man evolved from a stage that was primarily gestural, though with vocal concomitants, to one that is primarily vocal but with the capacity for manual communication retained. The early freeing of the upper limbs from locomotion and the capacity for unimanual tool use in early hominids, as well as the relative success with which the great apes can be taught a manual (as contrasted with a vocal) system of communication, all suggest that manual communication has been favored for a very long time. The structural limitation of the vocal tract in the chimpanzee and in some hominids suggests that vocal communication in man may have reached its current level of versatility only in relatively recent times.

Both manual and vocal communication involve a series of complex self-generated movements, which appear to depend minimally, if at all, on sensory feedback. Both forms of communication depend critically on the left hemisphere of the brain for normal functioning. The left hemisphere is apparently not specialized for symbolic or representational function, however, but for certain kinds of motor function, both verbal and nonverbal. The evidence suggests that the left hemisphere contains a system for accurate internal representation of moving body parts, important for the control of changes in the position of both oral and brachial articulators. Once a posture is achieved, however, it can apparently be run off repetitively without further intervention from this system. The function controlling accurate positioning of limb and oral articulatory musculature has been labeled "praxis," from the widely used term for its disruption, "apraxia." The praxis function appears to be dissociable from another function not selectively mediated by the left hemisphere, which involves fine finger control and which may depend critically on somatosensory feedback. It is conceivable that left-hemisphere specialization for praxic function evolved in association with the asymmetric employment of the limbs during tool use.

Evidence concerning the neuroanatomical basis for praxic control is discussed in some detail. It is generally agreed that the left posterior peri-Sylvian region is a critical area for such control, with some schemas incorporating the premotor region as well. The posterior praxis system exercises control over both hands, as well as over the oral musculature, but the specific pathways for such bilateral control are not known. Past explanations have emphasized cortico-cortical and consequently callosal pathways, but it is suggested instead that there may be more than one neuroanatomical pathway and that a major route may involve subcortical systems not requiring callosal transmission. For oral movements at least, and perhaps for manual as well, achievement of single movements or postures can apparently be mediated by a more anterior system on the left, with the coordination of these individual movements into a series being critically dependent on the posterior system.

Acknowledgments

I would like to thank Jonathan Lomas, Ron Kramis, and Jeannette McGlone for their comments on the chapter.

References

Benson, D. F. (1967). *Cortex 3*, 373–394.
Broca, P. (1861). *Bull. Soc. Anat. Paris 6*, 330–357. (Translated into English in: "Some Papers on the Cerebral Cortex," G. Bonin, ed., pp. 49–52. Charles C. Thomas, Springfield, Ill., 1960.)
Chomsky, N. (1967). *In* "Brain Mechanisms Underlying Speech and Language" (C. H. Millikan and F. L. Darley, eds.), pp. 73–88. Grune & Stratton, New York.
Corkin, S., Milner, B., and Rasmussen, T. (1970). *Arch. Neurol. 23*, 41–58.
Dart, R. A. (1949). *Amer. J. Anthrop. 7*, 1–38.
Ethelberg, S. (1951). *Acta Psychiat. Neurol.* Suppl. #75, 3–211.
Galaburda, A. M., LeMay, M., Kemper, T. L., and Geschwind, N. (1978). *Science 199*, 852–856.
Gardner, B. T. and Gardner, R. A. (1971). *In* "Behaviour of Nonhuman Primates," (A. M. Schrier and F. Stollnitz, eds.) Vol. IV, pp. 117–184. Academic Press, New York.
Gazzaniga, M. S., Bogen, J. E., and Sperry, R. W. (1967). *Arch. Neurol. 16*, 606–612.
Geschwind, N. (1963). *Trans. Amer. Neurol. Assoc.*, 219–220.
Geschwind, N. (1975). *Amer. Scient. 63*, 188–195.
Goodglass, H. and Kaplan, E. (1963). *Brain 86*, 703–720.
Hewes, G. W. (1973). *Curr. Anthrop. 14*, 5–24.
Holloway, R. L. (1972). *In* "The Functional and Evolutionary Biology of Primates" (R. Tuttle, ed.), pp. 185–203. Aldini-Atherton, Chicago.
Hore, J., Meyer-Lohmann, J., and Brooks, V. B. (1977). *Science 195*, 584–586.
Kellogg, W. N. (1968). *Science 162*, 423–427.
Kemp, J. M. and Powell, T. P. S. (1970). *Brain 93*, 525–546.
Kimura, D. (1977). *Brain 100*, 527–542.
Kimura, D. and Archibald, Y. (1974). *Brain 97*, 337–350.
Kimura, D., Battison, R., and Lubert, B. (1976). *Brain Lang. 3*, 566–571.

Kolb, B. (1977). Neural mechanisms in facial expression in man and higher primates. Paper presented at meeting of Can. Psychol. Assoc., Vancouver, British Columbia.

Knight, G. (1964). *Brit. J. Surg. 51*, 114–124.

Leakey, M. D. (1971). "Olduvai Gorge." Cambridge Univ. Press, Cambridge.

Lieberman, P. (1975). "On the Origins of Language." Macmillan, New York.

Liepmann, H. (1908). "Drei Aufsätze aus dem Apraxiegebiet." Karger, Berlin.

Mateer, C. (1976). *University of Western Ontario Research Bulletin* #393.

Mateer, C. (1978). *Brain. Lang. 6*, 334–341.

Mateer, C. and Kimura, D. (1977). *Brain Lang. 4*, 262–276.

Menzel, E. W. (1973). *In* "Precultural Primate Behaviour" (E. W. Menzel, ed.), pp. 192–225. Karger, Basel.

von Monakow, C. (1914). "Die Localisation im Grosshirn." Bergmann, Wiesbaden.

Myers, R. E. (1972). *Acta Neurobiol. Exp. 32*, 567–679.

Myers, R. E. (1976). *In* "Origins and Evolution of Language and Speech," *Ann. N. Y. Acad. Sci. 280*, 745–757.

Nottebohm, F. (1971). *J. Exp. Zool. 177*, 229–261.

Passingham, R. E. (1975). *Brain Behav. Evol. 11*, 73–90.

Pieczuro, A. and Vignolo, L. (1967). *Sistem. Nerv. 19*, 131–143.

Polit, A. and Bizzi, E. (1978). *Science 201*, 1235–1237.

de Renzi, E., Pieczuro, A., and Vignolo, L. A. (1968). *Neuropsychol. 6*, 41–52.

Stokoe, W. C. Jr. (1974). *Semiotica 10*, 117–130.

Tobias, P. V. (1971). "The Brain in Hominid Evolution," Columbia Univ. Press, New York.

Webster, W. G. and Webster, I. H. (1975). *Physiol. Behav. 14*, 867–869.

Yeni-Komshian, G. H. and Benson, D. A. (1976). *Science 192*, 387–389.

Yerkes, R. M. and Learned, B. W. (1925). "Chimpanzee Intelligence and Its Vocal Expressions." Williams and Wilkins, Baltimore.

Zaidel, D. and Sperry, R. W. (1977). *Neuropsychol. 15*, 193–204.

8

Specialization and the Language Areas[1]

RICHARD E. PASSINGHAM

Introduction

"The one great barrier between brute and man is language. . . . Language is our Rubicon, and no brute will dare cross it" (Miller, 1871). One hundred years later chimpanzees are having a brave try. Washoe makes comments with gestures of the hand (Gardner and Gardner, 1971), Sarah by writing with plastic symbols (Premack, 1976), and Lana by typing on a keyboard (Rumbaugh, 1977). Of course there have been those who have denied that these chimpanzees have in fact crossed the Rubicon; but it is difficult to deny that they have mastered some of the stages of language development through which human children pass (Brown, 1973; Gardner and Gardner, 1978).

It is puzzling that chimpanzees should have these capacities, because, although they can be demonstrated in the laboratory, they do not appear to be used in the wild. Yet evolution could not build into an animal a capacity of which it has no need, because there can be no selection pressure if there is no value for survival. It is just this problem that Humphrey (1976) has elegantly discussed in relation to other intellectual capacities that chimpanzees can be persuaded to exercise when trained by man. The solution to the dilemma must be either that chimpanzees do in fact use their language capacities in the wild or that their abilities to understand and use

[1] This work was supported by MRC grant 971/1/397/B.

221

some forms of language in captivity must be general ones, allowing them to do other things of importance to them in the wild. Perhaps they can progress so far with language just because they are so intelligent. But, if that is so, it raises the intriguing question of how far our own linguistic abilities are the result of our general intelligence, and how far they depend on specialized mechanisms.

It has usually been assumed that mankind evolved special brain mechanisms that made speech and language possible, and these are referred to as the speech or language areas. These, it has been supposed, are unique to the human brain. If this is so it is not clear why these areas evolved, because chimpanzees appear to be capable of mastering some aspects of language without the benefit of these specialized mechanisms. The answer may be that we have overestimated the extent to which the capacity for learning language depends on areas specialized for this purpose. It is worth considering the matter in the hope of establishing what capacities are indeed unique to man and require specializations not to be found in chimpanzees.

We will discuss first the language capacities that have been claimed for chimpanzees. In this way we can establish what can be achieved without specialized language areas in the brain. We will then compare the brains of man and ape and look particularly for evidence that the human brain has structural specializations not to be found in the brain of a chimpanzee.

The Abilities of Nonhuman Primates

The view that language capacities might be related to the size of the brain or intelligence is not one that is popular with Chomsky. He asserts (Chomsky, 1972) that "as far as we know, possession of human language is associated with a specific type of mental organization, not simply with a higher degree of intelligence." His evidence is that there are, he thinks, rules that are common to all languages, which a child could not appreciate unless it had special inbuilt mechanisms for language. Lenneberg (1967) reviews the empirical evidence on how children learn language and is led to suppose that human infants possess some language acquisition device. It would be a problem for this view if chimpanzees could be shown to learn a language, and Chomsky believes that they cannot.

It is essential to distinguish between speech and language. A chimpanzee that lacked specializations for speech might still be able to acquire a language. Speech and language are therefore treated separately in the discussion that follows.

Speech

PERCEPTION

There are two aspects of speech perception for which specialized mechanisms might be required. The first is the ability to recognize speech sounds as being the same when made by different speakers. This requires that the differences in intensity, pitch, and quality of the voice be ignored. This ability can be demonstrated in animals. Burdick and Miller (1975) found that chinchillas (Chinchilla laniger) could tell the difference between /a/ and /i/ and that they were able to do so even when they heard the vowels spoken by different speakers or heard vowels that were artificially synthesized. Cats trained to tell the difference between /i/ and /u/ when spoken by a man recognize them when spoken by a woman (Dewson, 1964). Old World monkeys trained to discriminate between /ba/ and /da/ can transfer to a discrimination of new examples from the same speaker and to the same sounds synthesized (Sinnott et al., 1976). There is no reason to assume that animals are naturally less proficient than people in this respect, especially given the vast amount of practice that people have.

The second respect in which animals could differ from people is in the way they categorize the speech sounds they hear. Lieberman (1975) points out that the discrimination that people make between plosives such as /b/ and /p/ is based on slight differences in the time at which voicing starts. With short delays the sound is categorized as a /b/ and with long as a /p/. Examples of /b/ and /p/ can be synthesized such that the voice onset time varies in small steps between different examples. People are found to be better at distinguishing between the two phonemes /ba/ and /pa/ than they are at making distinctions between examples varying in voice onset time within either phoneme boundary. It is important not to exaggerate this claim, since they are certainly able to tell the difference between examples within the phoneme boundary (Sinnott et al., 1976).

The chinchilla was found by Kuhl and Miller (1975) to draw the distinction between /da/ and /ta/ in much the same way as English-speaking subjects. This result is perhaps surprising as the phoneme boundaries are not the same in all languages. Morse and Snowdon (1975) tested rhesus monkeys on /bae/, /dae/, and /gae/ and Waters and Wilson (1976) on /ba/ and /pa/. Both groups found better discrimination between than within phonemes; but their results cannot readily be compared with data on people, because the methods of testing were very different from those used for human subjects. A more satisfactory study is reported by Sinnott et al. (1976) in which Old World monkeys and people were given the same task with the same apparatus. Different versions of /ba/ and /da/ were synthesized, varying in formant transitions with different onsets of the second and third formant. The subjects were given a standard and required to release a key if it changed to another stimulus. The monkeys were less sensitive than people to differences in formant transition. People took longer to make decisions when making discriminations within the phoneme boundary, but monkeys did not show this effect. The authors conclude that these results may be evidence that people have special mechanisms for speech perception.

There are various reasons for being cautious in accepting this view. One is that people might learn to categorize speech sounds in the way they do. It has been claimed that they do not because babies react to differences between and not within phoneme boundaries (Eimas, 1975; Streeter, 1976); but Sinnott et al. (1976) cite evidence suggesting that if the continuum between /ba/ and /pa/ is varied by altering voice onset time there may be an actual discontinuity in acoustical features between but not within the phoneme boundaries. It is also found that whether adults discriminate within the phoneme boundaries is very much dependent on the method of testing. It is quite possible that monkeys tested in the same way as the babies would not differ in performance. In general, it is very dangerous to conclude that monkeys lack capacities shown by people when, as in the Sinnott et al. (1976) study, the people have had many years of experience in making the relevant judgments, and the monkeys none. It is also dangerous to draw conclusions about differences between man and nonhuman primates on the basis of results obtained with monkeys and not chimpanzees. A final reason for caution is the uncertainty that categorical perception is specific to the perception of speech. Cutting et al. (1976) have shown that people also draw similar distinctions between musical sounds, such as notes that appear to be made by an instrument that is plucked or is played with a bow. Here, too, discrimination is much better between than within the categories. It may be that categorical perception is a feature of our hearing in general and that it does not reflect special mechanisms for the perception of speech.

AUDITORY–VISUAL ASSOCIATIONS

The understanding of speech also depends on the ability to form associations between the sound of names and the sight of the things that they refer to. This point has been stressed by Geschwind (1967), who at that time argued that man might be specialized in possessing cross-modal abilities. Since then we have learned that both monkeys and chimpanzees can match the feel of an object to its shape as seen (Cowey and Weiskrantz, 1975;

Elliott, 1977; Jarvis and Ettlinger, 1977; Davenport *et al.*, 1973). But auditory–visual associations are more directly relevant to Geschwind's (1967) claim with respect to language. Here the issue has not yet been resolved by experimental studies. Dewson and Burlingame (1975) have shown that rhesus monkeys can learn to choose between two colors according to which sound they have just heard. But this conditional task does not demand that the animal learn more than that in the presence of A they should do X; they do not have to form the sort of link that there is between a name and its referent. Davenport (1976) reports a lack of success up to that time in teaching chimpanzees to match Morse code pulses of sound to flashes of light. But the ability to appreciate a correspondence of that sort is not one that is required for language, where the relation between a name and its referent is an arbitrary one.

The best way to establish whether an animal can make the required associations is to expose it to human speech and then see if it can learn to understand it. Warden and Warner (1928) tested a dog called Fellow. It obeyed many different commands even when it could hear but not see its master. The dog's comprehension of the names of objects was tested by commanding it to pick out one of three objects. It could do this with an accuracy well above that expected by chance.

A chimpanzee called Gua was reared by Kellogg and Kellogg (1933) in their home and was thus exposed to human speech for 9 months. She learned to understand 95 words and phrases compared with the 107 understood by their son Donald, who was 3 months older. When shown a card with 4 pictures on it Gua was able to point to the dog and shoe when told to do so. The vocabulary of words that she understood included the names of people, objects, and actions. Her comprehension was not dependent on which person spoke to her or on the intonation used.

Another chimpanzee, called Ally, was also reared in a human household and showed some understanding of speech. Fouts *et al.* (1976) describe an experiment in which five words were chosen that Ally understood. Ally was then taught the gestures used in American Sign Language, or Ameslan, for these words. Although he was taught the gestures without the objects present, Ally was then able to give the correct gestural sign for each of the objects when shown them. To do this the chimpanzee must have understood the English words.

Recently Patterson (1978) has systematically studied the ability of a young lowland gorilla to understand spoken English. The gorilla, Koko, has been taught sign language for several years and has at the same time been exposed to human speech. Her comprehension of words and phrases has been assessed on a standard picture vocabulary test, on which she performed as well as educationally handicapped children 4 to 5 years old.

Though it is clear that she can understand many nouns and adjectival phrases, we do not as yet know whether she can appreciate grammatical rules.

It looks from this evidence as if apes can form the auditory–visual associations required for understanding speech. When Ally heard a word it appears to have reminded him of the object as seen. There is no reason to suppose that this happened any less with Gua than with the child. Apes appear to have one important ability required for comprehension of speech.

SPEECH PRODUCTION

There is no question that it is very difficult to teach apes to speak even a few simple words. A prolonged attempt was made to teach the chimpanzee Viki to talk, and at the end she was only able to say "mama," "papa," "cup," and "up" (Hayes and Nissen, 1971). Even these she produced only with effort and with poor articulation. There are several reasons why chimpanzees might be so handicapped. Their vocal tract might be incapable of producing many of the sounds used in human languages, the animals might be unable to produce sounds at will, or they might have difficulty in imitating the sounds they hear.

Lieberman (1975) has made a thorough study of the vocal tracts of nonhuman primates, comparing them with our own. He finds that the shape of the tract is such in other primates that there can be little change in the cross-sectional area of the pharynx as a result of movements of the tongue. Lieberman (1975) suggests that the tract of a monkey or chimpanzee is unable to produce the full range of vowel sounds, although Jordan (1971) has recorded chimpanzee sounds resembling the five vowels. (see also Chapter 7, this volume.)

It is not obvious, however, that this is the sole reason for the poor performance of chimpanzees. It is not clear that they would be able to learn even those words using sounds within their range. There could, for instance, be some further difficulty in voluntary control of the vocal chords. Yamaguchi and Myers (1972) were unable to teach rhesus monkeys to make a call given one visual stimulus but not given another, although they referred to a study in the same laboratory in which some cebus monkeys had been taught to do so. Sutton *et al.* (1973) claimed to have successfully taught rhesus monkeys to control their calls on the grounds that the monkeys made more calls given a white light, and less when the light had gone out. However, this is not a fully convincing demonstration. It could be that the monkeys call because they expect food, and call less when the light is out because they know that food is then no longer available. A recent experiment by Sutton (this volume, Chapter 3) is more convincing; in this, rhesus monkeys learned to produce a coo to a red light and a bark to a green one. Randolph and Brooks (1967) carried out a similar study on a chimpanzee. The animal was trained to produce a bark to persuade a human companion to play, but only if the person stood in one position. If the person stood in another, the chimpanzee had to reach out and touch him to initiate play, and under these conditions the animal produced fewer barks. It seems that the chimpanzee barked not because it expected play but because it learned that play could be started in this way.

Assuming that chimpanzees do have some voluntary control of their calls, the final question is whether they can modify them according to the pattern of sounds they hear. Some birds, but not all, are able to learn parts of their species-specific song by copying the sounds of others (Nottebohm, 1972). A few species, such as some parrots and mynah birds, can even imitate sounds that are not in the normal repertoire of their species. A well-trained mynah bird can do this so well that its imitations of phrases are hardly distinguishable in the spectrograph from the same phrases spoken by a native speaker (Sebeok, 1965). Unfortunately, there has been little work on vocal imitation in primates. Green (1975) reported that the coo sounds made by Japanese monkeys (*Macaca fuscata*) when being fed differed between animals in three areas. It is certainly possible that these are cultural variants, but it was not possible to rule out the possibility that the differences reflected different motivational states in the three troops. What is required is a laboratory study of monkeys or chimpanzees in which the animal is played calls from its own repertoire and required to produce calls to match them. When played a coo it would have to produce a coo to obtain food, when played a bark to bark, and so on. Until this has been tried, it is too early to conclude that other primates lack the ability to imitate sounds shown by birds and people.

Language

Even if nonhuman primates do lack the ability to imitate sounds, this alone would not prevent them from acquiring forms of language other than speech. Recently chimpanzees have been taught to communicate with signs (Gardner and Gardner, 1971), plastic symbols (Premack, 1976), or designs on panels (Rumbaugh, 1977).

LINGUISTIC ACHIEVEMENTS

Rather than enter the debate on whether performance of these apes fulfills all of the possible criteria for a language, it is perhaps more profitable to consider what basic functions a language must perform. A language must be able to refer to things and events in the world. It must further have some means of distinguishing between different states of the world. While a language may not be sufficiently defined by reference to these functions, it is at least necessary that a language be able to perform them.

Chimpanzees are clearly capable of learning to use symbols to refer to things and events. Gardner and Gardner (1971) tested their first chimpanzee, Washoe, under conditions that leave no room for doubt. Washoe was shown pictures and then required to use the gestures of Ameslan, American Sign Language, to tell someone what she had seen. Although that person could not see the pictures, he was often able to tell which of the many possible objects Washoe had seen in the picture. Washoe further learned many signs for actions, such as "come–gimme" and "open." Finally she could use words for the qualities of things, that is, adjectives such as "red." Premack (1976) and Rumbaugh (1977) report similar demonstrations.

These claims are only convincing if it can be shown that the chimpanzees are not simply learning what to do by rote. When the animals have been taught a gesture or symbol in the presence of one example of the class of things to which it refers, they must then be tested with new examples in order to see if they appreciate its meaning. Such tests are difficult to carry out adequately, because chimpanzees have a very impressive rote memory (Farrer, 1967) and the capacity for learning new items very quickly (Gardner and Gardner, 1978). It is clear that Washoe can spontaneously use gestures to refer to new situations. When taught the gesture "open" for opening doors, she then used it for opening containers such as jars (Gardner and Gardner, 1971, 1975). She can also spontaneously link words to form new meanings, as in her description of a swan as a "water bird" (Fouts, 1975). Unfortunately, neither Premack (1976) nor Rumbaugh (1977) always provides adequate data on the results of the first few trials of the tests they designed to demonstrate that their chimpanzees have an appreciation of the full meaning of the symbols used (Gardner and Gardner, 1978). Nonetheless, given the demonstrations with Washoe, it seems implausible that chimpanzees taught other symbolic systems do so without any understanding of the meaning of the symbols used.

The next question is whether chimpanzees can be taught to distinguish between different states of the world in their use of symbols. Given symbols for objects and actions, they can clearly point to the existence of something or the occurrence of an event. But to distinguish between different events it is often necessary to link symbols together and to do so according to certain rules. Brown (1970) points out that the two situations of a cat biting a dog and a dog biting a cat can only be distinguished by following some form of grammatical rule. In English the subject precedes the object, except in passive constructions, but there are languages in which the order is reversed (Greenberg, 1966) and others in which the subject is determined not by the order but by case grammar.

In the systems taught to chimpanzees, word order has been the means of drawing the relevant distinctions. There are two interesting cases in which there is information on the word orders used. The first is the use of order to refer to the agent. Gardner and Gardner (1971, 1975) analyzed the orders in which Washoe used gestures and found that she placed "you" or the name of her companion first on most occasions. The young chimpanzees with which they are now working use the order agent–action and action–object in the same way as their teachers (Gardner and Gardner, 1978). However, none of these animals has been specifically tested in situations such as that suggested by Brown (1970), where the subject and object are both specified by the word order.

A second important case in which word order can be used in referring to different states of the world is its use when describing the relative positions of objects (Lenneberg, 1971). Chown *et al.* (in press) tested a chimpanzee, Ally, to see if he could correctly describe the positions of objects in, on, or under various objects. Ally could do this for objects other than those he was trained with. He always used the order subject–preposition–location.

It would seem that chimpanzees can master the simplest rules of word order and are not clearly to be distinguished from young children in this respect (Brown, 1973; Gardner and Gardner, 1975, 1978). Gardner and Gardner (1975, 1978) also compared Washoe and young children in their ability to give the appropriate replies to "Wh" questions. These are questions of the sort, "Who are you?"; "Who is smoking?"; and "Where is the shoe?" Appropriate answers would be, respectively, "Washoe," "you," and "there." Like young children, Washoe rarely gave words in the inappropriate category, thus showing an appreciation of what would count as a grammatically correct reply.

That chimpanzees cannot match the later achievements of children is of relevance only if these are thought to be essential for the possession of language. At the least chimpanzees can make use of symbols to perform some of the functions of language. If they can go no further, this might reflect low intelligence and not necessarily the lack of some special language mechanism.

While the philosophical issue of the definition of language may be evaded, one important scientific issue may not. Chimpanzees can be successfully taught to communicate with gestures or simple visual symbols. The question is whether our own ability to communicate in the same way depends on mechanisms of the dominant hemisphere in adults, or whether it is independent of the language mechanism of this hemisphere. The evidence is not clear-cut. Sarno *et al.* (1969) review the literature for gestural signs. They report the case of a congenitally deaf man who became aphasic after a left hemisphere stroke. He was also impaired in understanding gestural signs and in making signs with the left hand. This suggests that the same hemisphere controlled gestural and natural language.

On the other hand, Gardner *et al.* (1976) found that they could teach global aphasics a simple system of visual symbols, much like that used by Premack (1976). It could be that the system did not constitute a language, but at least one patient spontaneously used it for requests, and the authors acquired the impression that the patients regarded the system as other than a simple game. It seems plausible to suppose that it was the right hemisphere that was able to acquire the system, and, if so, it could be argued that language mechanisms were not required in learning it. But it should be remembered that the system was very simple and that the right hemisphere has some language abilities in adult man. These same abilities may permit the learning of a simple visual language.

SYMBOLS

We can now reconsider the problem posed at the beginning. Chimpanzees can learn some important aspects of language, but are not required to do so in the wild. Yet abilities of no use to the animal should not have been given them by evolution. The most plausible solution is that the abilities that enable them to learn language are not ones that are specific for language acquisition but general ones enabling them to learn any tasks intelligently. If so, it is worth examining these abilities further.

The first issue is why chimpanzees can learn to use symbols. Languages use a set of symbols, and symbols are often distinguished from signs of the sort that animals use in the wild. A symbol is strictly a substitute for its referent, whereas a sign is not. The word "lion" refers to the animal and can be used in talking about it in the absence of the real lion. The lion's roar, on the other hand, signals the approach of a lion; it is a sign that something is to happen rather than a token of that event that could be used in a language. This distinction is a logical one and appears sound.

It is not so obvious that the distinction is important in understanding how signs and symbols can be learned. Both the sign and the symbol evoke some mental representation of the thing that is signified. In both cases there must be some neural mechanism for retrieving the stored representation of the object or event that is predicted by the signal or referred to by the symbol. In principle this retrieval could be done in the same way for both. If so, a mechanism that evolved to allow animals to make predictions on the basis of spatial or temporal conjunctions of objects and events could also allow them to learn symbolic relations. Whether any particular species could do so might depend on their level of general intelligence.

Chimpanzees can further be shown to be able to appreciate the relation in which one thing is a token or substitute for another. Wolfe (1936) trained chimpanzees to pull a bar against a heavy weight in order to bring grapes within reach. The animals would also do this to obtain tokens that could be substituted for grapes only after a certain number of trials. Cowles (1937) found that the same animals would work at visual discrimination problems and other tasks when rewarded with tokens to be exchanged at the end of the daily session. The notion that a token is convertible into something else is one that makes sense to chimpanzees.

Symbols may be iconic or not. The simplest symbols are parts of the thing represented or pictures of the original. Premack (1976) found that chimpanzees can be taught to pick out one of two fruits on the basis of a sample consisting of just part of one of them, for example, a seed or the stalk. They can also interpret pictures. Davenport *et al.* (1975) taught chimpanzees to feel an object and then pick out from two pictures the one that represented that object. The animals could do this even when shown small photographs or line drawings. Line drawings are degraded pictures, not so very different from pictograms.

If chimpanzees have the basic equipment that would enable them to learn to appreciate symbols, it must still be shown directly that they can use it. The evidence comes from studies of chimpanzees using sign language and from studies of their use of plastic symbols. The demonstration with Ally, referred to earlier, provides one example. Taking five spoken words that Ally understood, Fouts *et al.* (1976) taught the animal the gestures for these words in sign language. When Ally was presented with the objects to which these words referred, he could give the appropriate sign. Since the objects were not present when the gestures were taught, Ally must have had access at that time to mental representations of them on hearing the spoken words of English.

The other demonstrations come from Premack's (1976) work with Sarah. In the first demonstration, Sarah was asked to describe an apple by choosing between those features on pairs of cards that best represented an apple to her. When shown the plastic symbol for "apple," a blue triangle, she gave the same description, which would be appropriate for an apple, though not for the triangle itself. Similarly she gave the same description when shown the symbol for "caramel" as she gave for an actual caramel. The simplest explanation is that when she saw the symbol it evoked a representation of the object symbolized. In the second demonstration, Sarah was taught two symbols for colors. She was presented with two

sentences in which the only symbols she did not know were those representing colors. The sentences were "brown color of chocolate" and "green
color of grape." Although at this stage real chocolate bars or grapes were
not present, she was then able to correctly name the colors of chocolate and
grapes when presented with the real things. Again the simplest explanation
is that the symbols for "chocolate" and "grape" served to remind her of actual chocolate and grapes. However, more demonstrations of this kind are
needed, taking particular care to ensure that the results on transfer tests
cannot be due to rapid rote learning.

RULES

The second issue is why chimpanzees appear to be able to learn certain
elementary rules of language. This achievement must reflect their general
intelligence. This intelligence is easily demonstrated on tests carried out in
the laboratory.

Many tests of human intelligence, such as Letter Series and Matrices, require the person to detect the rule followed by some sequence and then to
apply it to generate the next item in the series. They are specifically designed to test the ability to comprehend new rules. The best tests of animal
intelligence also require them to learn rules. On visual discrimination learning sets the improvement found across a series of problems must be accounted for by supposing that the animals come to appreciate the basic rule
that applies to all problems. This rule is the simple one that on each trial a
piece of food is found under one object and that it will remain under that
one for the remaining trials on that problem. The ability to learn this rule is
correlated with intelligence in children (Harter, 1965).

Chimpanzees improve very quickly on learning sets, though less quickly
than children (Kintz *et al.*, 1969). They are also quicker to improve on
visual discrimination learning sets if they have previously been given a
series of visual discrimination reversal problems (Schusterman, 1964). This
indicates that they perceive the general rule that applies to both types of
problems (Warren, 1974).

There is no problem in understanding how this ability could be of value
to chimpanzees in the wild. The chimpanzee would be well fitted for working out the rules that apply in the world about it. It also comes to the
laboratory quick to learn about such things as the actions that can be performed with an object or by some agent (Premack, 1976). It should not be
surprising if it can learn a simple linguistic rule, such as the rule that subjects must precede objects, as it knows about agents and actions and is intelligent enough to be able to learn the relevant rules. There is no need to

suppose that a language acquisition device is necessary when learning a rule as simple as this.

We must conclude that chimpanzees are indeed able to master some of the elements of language and that they are able to do so through the use of general abilities of value to them in the wild. This conclusion forces us to reconsider the extent to which people require special abilities to learn language. The main reason for supposing that they do has been the belief that there are specialized areas for language in the human brain but not in the ape brain. We must now therefore turn to the brain and see what we know of the differences between the brain of a man and the brain of an ape or monkey.

The Brain

Size and Proportions

Our brain could differ from those of other primates in several respects. Apart from the obvious difference in gross size, it could differ in the relative proportions of the different areas. The anatomical asymmetries between the two hemispheres of the human brain might not exist in the brains of other primates. Finally, there might be specialized areas unique to the brain of man. Each possibility will be discussed in turn.

METHODS

It is not as easy as it might seem to make valid comparisons between the size and proportions of the brains of man and those of the brains of other primates. Certain technical details must first be reviewed before discussing the results of such comparisons.

The reason for considering the size of the brain and of its component parts is that it is hoped that measures of this sort are related to intelligence. Intelligence might be thought to depend, in part, on the amount of brain tissue in excess of that required for analyzing incoming sensory information and for controlling the muscles of the body. The most direct measure of the inputs and outputs of the brain is the size of the spinal cord and cranial nerves. A less direct measure is the size of the medulla, since many of the inputs and outputs of the brain pass through this region, although it also contains many nuclei. The size of the human brain may therefore be compared with the size of the spinal cord (Passingham 1978) or of the medulla (Passingham, 1975a). But the most commonly used index of the information the brain must handle is the weight of the body; the rationale is that, the larger the body, the greater the number of sensory receptors and the larger the muscles that the brain must control. Comparisons of brain to body size give orderly data with which to rank different species, unless they are very closely related (Jerison, 1973; Passingham, 1975a).

If the relationship between the size of the brain and the weight of the body is known for a series of nonhuman primates, it is easy to estimate the brain size that would be expected for a primate of the same body weight as man. This is done by plotting log brain weight against log body weight and fitting a least-squares regression line through the points for the primates other than man (Figure 8.1). This line gives the value of brain weight to be expected for a primate of any particular body weight. The difference between the weight of the human brain and the value predicted for a primate of the same body size is a measure of how many times greater our brain is than would be expected given our build. The same calculations can be performed for parts of the brain to establish which parts are disproportionately large or small for a primate of our size.

It could be that as the brain increases in size it maintains the same inter-

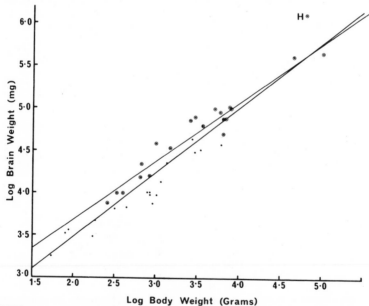

FIGURE 8.1. *Least-squares regression lines for brain on body weight. Lower line is for nonhuman primates; upper line is for simians. Dots are for prosimians, and circled dots are for simians. For the nonhuman primates, log brain weight (milligrams) = 1.96 + 0.76 × log body weight (grams). For simians, log brain weight (milligrams) = 2.31 + 0.69 × log body weight (grams). H = Homo. Data are from Stephan et al. (1970). [Figure from Passingham and Ettlinger, 1974.]*

nal proportions, that is, that each area remains a constant proportion of the whole. In fact, this is not the case in primates. Comparisons of a series of brains of increasing size tend to show that some areas expand more than others. This can be seen, for example, by plotting log neocortex volume against log brain volume (Figure 8.2). The slope of the regression line is greater than 1, indicating that primates with larger brains have a greater proportion of neocortex to total brain volume; this is simply demonstrated by comparing the ratios of neocortex to brain volume in primates differing in brain size (Passingham, 1975a).

It is essential to appreciate that different questions are asked when relating an area of the brain to body size and the same area to brain size. Given that the human brain is larger than would be expected for a primate of our weight, it is obvious that many, though not necessarily all, of the parts of the human brain will also be larger than expected given our build. It is nonetheless an open question, given that information, whether the size of any part of our brain is or is not greater than would be expected for a hypothetical primate with a brain as large as ours. While the brain, and

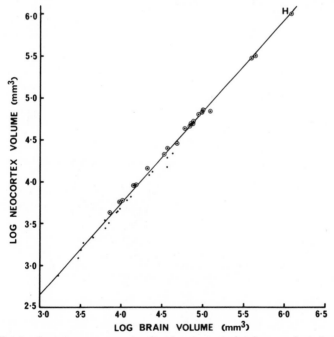

FIGURE 8.2. *Least-squares regression line for neocortical volume on brain volume. Log neocortex (cubic millimeters)* = 0.63 + 1.1 × *log brain (cubic millimeters). Dots are for prosimians, and circled dots are for simians. H = Homo. Neocortical volumes include white matter. Data are from Stephen et al. (1970). [Figure from Passingham, 1975b.]*

thus the brain parts, may be unusually well developed for a primate, the changes in proportions within the brain could be wholly predictable from information on the proportions of the brain for primates differing in brain size. If the latter were true, it would show that although the brain had expanded in size it had done so in a way that fits the primate pattern. This point will be illustrated in the discussion that follows of the size and proportions of the human brain.

COMPARISONS BY BODY SIZE

The human brain is 3.1 times as big as would be expected for a primate of our size, and 3.0 times as big as expected for a monkey or ape (excluding prosimians) (Figure 8.1). When various areas of the brain are considered, and comparisons made with the values predicted for nonhuman primates matched for weight, it is apparent that some parts of our brain have expanded more than others (Figue 8.3). The medulla is no bigger than would be expected, whereas a considerable expansion is shown for all cortical areas, hippocampus, paleocortex, cerebellum, and neocortex. The greatest increase is in the neocortex and in the cerebellum, presumably mainly in the neocerebellum. Within the neocortex the best information that we have on the volume of different cytoarchitectonic areas comes from Shariff

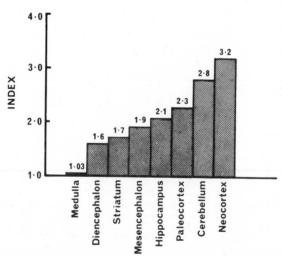

FIGURE 8.3.   Indexes showing how many times greater is each area of man's brain compared with the values predicted for a nonhuman primate of the same body weight. Least-squares regression lines were fitted to the data given by Stephan et al. *(1970) for body weight and volumes of each brain area in the nonhuman primates. [Figure from Passingham, 1975b.]*

(1953), and even these data are not wholly reliable (Passingham, 1975b). Nonetheless, using these data, it appears that, compared with body size, the areas that show the greatest increase are motor cortex (agranular) and association areas (eulaminate) (Figure 8.4).

There are no numerical data at all on the relative sizes of specific association areas, and this is particularly unfortunate because no statement can be made about parietal and temporal association cortex, in which the posterior speech area is represented in man.

An indirect, and not fully satisfactory, way to consider the issue is to examine the volumes of different nuclei of the thalamus, since each projects to a known and delimited area of neocortex. Hopf (1965) gives data for man and three nonhuman primates, an ape, monkey, and prosimian. By fitting least-squares regression lines to the data on nonhuman primates for each nucleus, plotted against body weight, the value for any nucleus in the human brain can be compared with that predicted. The indexes in the first column of Table 8.1 show how much larger or smaller any nucleus is than would be expected for a nonhuman primate matched for weight. The increase in the size of the pulvinar, which projects in man to temporal and parietal association areas (Walker, 1966), is no greater than the increase in the size of the dorsomedial nucleus projecting to prefrontal cortex. There is as yet no evidence that any particular association area has expanded more than any other.

COMPARISON OF PROPORTIONS

As not all of the areas of the human brain have expanded equally, it is of interest to know whether the changes in proportions are ones that could be

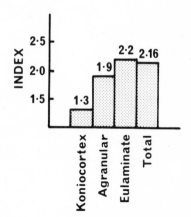

FIGURE 8.4. *Indexes showing how many times greater is each subarea of the human neocortex compared with the values predicted for a nonhuman primate of the same body weight. Least-squares regression lines were fitted to the data given by Shariff (1953) for body weight and cytoarchitectonic areas of the neocortex. The indexes are statistically significant for agranular cortex (equivalent to motor and premotor cortex) and for eulaminate cortex (roughly equivalent to association cortex) but not for konicocortex (including much though not all of the primary sensory areas). [Figure from Passingham, 1975b.]*

predicted from information we have on the brains of other primates. To do this, comparisons may be made between the size of any area and the size of the brain as a whole. Since the greatest expansion has been in the neocortex (Figure 8.3), it is particularly important to consider plots of neocortex volume against brain size. It can be seen from Figure 8.2 that man's neocortex is no greater than would be expected for a nonhuman primate matched for brain size. In Figure 8.5, the association areas of the neocortex, defined on cytoarchitectonic criteria, are plotted against neocortical volume. Again the value obtained for man does not differ from that predicted on the basis of data from other primates.

A similar result is obtained when the volumes of different thalamic nuclei are related to the total volume of the gray matter of the thalamus. The data of Hopf (1965) have been analyzed, by fitting least-squares regression lines to the values for nonhuman primates for each nucleus plotted against volume of the thalamus as a whole. In this way, indexes are derived for each nucleus in the human thalamus. The indexes are given in the second column of Table 8.1. The only values that are markedly high or low are those for the anteroprincipal nucleus, which is larger than expected, and for the lateral geniculate, which is smaller. The data on the anteroprincipal nucleus are suspect since the values given for this are the same for a prosimian and an Old World monkey. The low index for the lateral geniculate is striking. It means that the lateral geniculate is 2.1 times as small as it should be, and this matches well with the finding reported by Passingham and Ettlinger (1974) that striate cortex in man is 2.3 times as small as it should be for a primate matched for size of neocortex. The result for striate cortex is reliable as it is based on a very much larger sample of primates than is

TABLE 8.1
Indexes by Body and by Thalamus[a]

NUCLEUS	INDEX BY BODY	INDEX BY THALAMUS
Lateral geniculate	.85	.48
Medial geniculate	2.02	1.24
Lateral	1.60	.87
Center-median and parafascicular	2.43	1.23
Anteroprincipal	3.45	2.56
Anterodorsal	2.18	1.59
Pulvinar	2.04	.91
Dorsomedial	2.79	1.39

[a] The index by body gives how many times greater or smaller the nucleus is in man compared with the value predicted for a nonhuman primate of the same body weight. The index by thalamus gives how many times greater or smaller the nucleus is in man compared with the value predicted for a nonhuman primate with the same volume of thalamus. Data for gray substance of the thalamus and for particular nuclei are from Hopf (1965).

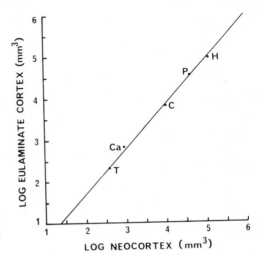

FIGURE 8.5. *Least-squares regression line for eulaminate cortex on
neocortex. Log eulaminate cortex
(cubic millimeters) = 0.41 + 1.07 ×
log neocortex (cubic millimeters).
Eulaminate cortex is roughly equivalent to "association" cortex. Data are
from Shariff (1953). [Figure from
Passingham, 1973.]*

available for comparisons of thalamic nuclei (Stephan, 1969). But the low
indexes for the lateral geniculate and for striate cortex must not be taken to
imply that in the human brain all sensory receiving areas are disproportionately small. The medial geniculate, which projects to primary auditory
cortex, is of the size expected. Shariff (1953) gives data on all koniocortex
in man and four other primates; koniocortex includes all of the primary
visual and auditory and much of the primary somatosensory receiving
areas. The value for man does not differ significantly from that predicted
for a nonhuman primate matched for neocortical volume (index = .8). It
seems, then, that the decrease in the proportion of primary visual cortex
may be offset by a slight increase in the auditory cortex as judged by the index for the medial geniculate. More extensive data are needed to confirm
this.

The aspects of the human brain already discussed are remarkable, in
that, apart from its very large size, in these other respects it is fully predictable given information about other primates. A further striking example of
this predictability is the number and spacing of the cells in the neocortex.
The number of cells in the human neocortex fits the relationship found between number of cells and neocortical volume for nonhuman primates
(Passingham, 1973) and for mammals as a whole (Jerison, 1973). Rockel *et
al.* (1974) have demonstrated that the number of neurons in a cylinder,
taken vertically through the thickness of the neocortex from the surface, is
the same from one mammal to another and from one area to another, except in striate cortex. This is due to an increase in the depth of the
neocortex as cell size increases and cell density decreases. The human brain
has the same number of cells in such a cylinder as other primate and non-

primate mammals, although man is similar to nonprimate mammals, but not to other primates, in the number of cells through the depth of striate cortex. It is apparent that our brains are constructed with the same building blocks as in other primates.

It is now clear that, given the size of the human brain, its proportions are in general those that would be expected for a hypothetical nonhuman primate that had expanded its brain to the same size as ours. That is to say that our brains, though very large indeed, have changed according to a pattern that can be demonstrated in other primates. To say that a change is predictable is not to say that it is unimportant. The human brain has more neocortex to total brain volume than does a chimpanzee, the latter has more than a monkey, and a monkey more than a prosimian (Passingham, 1975a). The fact that the relation between these proportions is lawful does not imply that the changes have no significance for determining intellectual capacities.

BRAIN SIZE AND LANGUAGE

When a brain reaches a certain size, and also changes in proportions, the result is not simply that improvements become possible in capacities available to a smaller brain. It can also be that totally new capacities become available. A chimpanzee has a brain roughly three times as large as a small prosimian, such as a mouse lemur (*Microcebus murinus*), even after controlling for the effects of body size (Stephan, 1972), and its proportion of neocortex (including white matter) to brain volume is 76% compared with only 46% in the mouse lemur (Passingham, 1975a). Yet the chimpanzee not only learns more quickly, but can learn things of which the mouse lemur is not capable, such as the use and manufacture of tools. The difference between the size of our brain and that of the chimpanzee is as large as that between chimpanzee and mouse lemur, so that this difference could well account for capacities possessed by man but not by chimpanzee. The obvious analogy is that of the computer, in which the number of logic elements, as well as amount of storage space, is one of the determinants of what the machine can and cannot do.

To what extent, then, can the size of the human brain account for the fact that only people naturally have language? Jerison (1976) has argued that there may indeed by a Rubicon, a size the brain must reach if language is to be possible. Lenneberg (1967) had previously denied this, pointing to the small brains of nanocephalic dwarfs who could nonetheless speak. In answer, Passingham and Ettlinger (1974) showed that, when account is taken of the small body size of these dwarfs, their brains may be better developed than those of the great apes. Although this is true for nanocephalic dwarfs, it is not the case that all microcephalics are dwarfs, and, as Holloway (1968) points out, even those that are not may nonetheless have some capacity for speech. But the fact that a person with a brain little larger than that of a chimpanzee can *speak* does not prove that brain size is irrelevant for *language*. Certainly we may have specialized mechanisms for speaking for which a particular brain size may not be critical.

Although chimpanzees cannot speak they have some capacity for language. In other words, the size of the brain essential to master the elements of language appears to be much less than that possessed by most people. But that is not to say that the impressive size of the human brain is unimportant for language. It must be responsible, in part, for the complexity of the language acquired by normal children. The human brain may also have had to achieve a critical mass to permit the intellectual achievement of inventing language as a tool for thought and communication. Deaf children have been reported to spontaneously invent an elementary system of gestures for communication (Goldwin-Meadow and Feldman, 1977), an achievement not yet matched by an ape.

Cerebral Dominance

Given the considerable difference in brain size, the second respect in which it is thought that the human brain might differ from that of a great ape is in the specialization of function between the hemispheres. Three aspects of cerebral dominance must be considered: the distribution of handedness, the relation of language dominance to handedness, and anatomical asymmetries in size between the two hemispheres.

DOMINANCE IN MAN

If people are asked which hand they prefer to use on a variety of common tasks it is found that they tend to use their right hand on more tasks than their left. Hand preference as assessed over many tasks is normally distributed in a population, but the mean is shifted toward greater preference for the right hand (Annett, 1972). The same is true for efficiency on a skilled task, the right hand tending to be more agile than the left (Annett, 1972).

There is a clear relation between hand preference and the hemisphere that is dominant for language in adults. The best evidence comes from a study reported by Milner (1975) on 262 epileptic patients whose epilepsy had been caused by lesions acquired after infancy. Sodium Amytal was injected into the blood supply of either hemisphere and the effects on speech were noted. Speech was found to be controlled by the left hemisphere in 96% of right-handers and 70% of non-right-handers (ambidextrous or left-handed), by the right hemisphere in 4% of right-handers and 15% of non-right-handers, and by both in 15% of non-right-handers, but in none of those who were right handed. Language comprehension was not fully tested, but patients can still obey commands with the ipsilateral hand after injection of sodium Amytal into the blood supply of the dominant hemisphere (Milner *et al.*, 1964). The distribution of cerebral dominance for the understanding of spoken language has still to be properly established.

Recently it has been shown that the two cerebral hemispheres of the human brain are not entirely symmetrical. Small differences have been found between the left and right hemispheres (Galaburda *et al.*, 1978; Witelson, 1977a). Of these some favor the left. Several studies have confirmed the earlier report by Geschwind and Levitsky (1968) that the planum temporale, behind Heschl's gyrus, tends to be larger on the left than the right and have found the same to be true in infant brains (Wada *et al.*, 1975; Witelson, 1977a). There are related asymmetries for the length of the Sylvian fissure, the left tending to be longer than the right (Connolly, 1950; Rubens, 1977), and for differences in the posterior course of the Sylvian fissure, which tends to remain lower on the left than on the right (LeMay, 1976; Rubens, 1977). But asymmetries have also been reported that appear to favor the right hemisphere. Using X-ray tomography, LeMay (1976) has shown that, for instance, the frontal lobe tends to be wider at the level of the frontal horns of the lateral ventricles in the right hemisphere than in the left. Taking measurements from photographs of brains, Wada *et al.* (1975) found that the operculum of the right third frontal gyrus tends to be larger on the right than the left in infants and adults.

These asymmetries are only of much interest if they can be shown to relate to handedness or to cerebral dominance for language. Only a few studies provide data on handedness. LeMay (1976) has shown that the back of the Sylvian fissure tends to be lower in the left hemisphere in 67% of right-handed patients compared with only 21.4% of left-handers; in contrast, this point was higher in the left hemisphere of 71.4% of left-handers but in only 25.5% of those who were right handed. Similarly, she found a clear relationship between handedness and other asymmetries, such as the width of the right frontal region.

All of these measures are very crude. It has been assumed that the asymmetries reported represent differences in cortical volumes. Yet widths and lengths of fissures or hemispheres may be very misleading unless the cortex in fissures is estimated. As Wada *et al.* (1975) point out, a greater frontal opercular surface on the right might tell us little of value if the cortex was more fissured on the left.

In only one study have cortical volumes been estimated from coronal sections. Galaburda *et al.* (1978) studied three human brains and found an area of temporoparietal cortex to be roughly seven times as large on the left as the right. Though based on a small sample this result is clearly intriguing.

DOMINANCE IN NONHUMAN PRIMATES

We can now compare the evidence for handedness, dominance, and asymmetries with what we know of these in other primates. Hand preferences have been studied most in monkeys. Data are available on the hand used for one task or several closely related tasks (Milner, 1969; Gautrin and Ettlinger, 1970; Lehman, 1978) and for preference as shown across many tasks (Warren *et al.*, 1967; Beck and Barton, 1972). For comparison with Annett's (1972) data on people, evidence must be examined for consistency over different unrelated tasks, and the data of Warren *et al.* (1967) have therefore been analyzed. The distribution of hand preference across five tasks has been plotted in Figure 8.6 for 33 rhesus monkeys. The distribution

HANDEDNESS

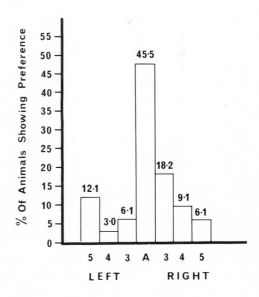

FIGURE 8.6. *Handedness in rhesus monkeys, based on an analysis of the data from the experiment described in Warren* et al. *(1967). An animal was judged as having a hand preference on a task if it used one hand 75% or more times on that task. A, ambidextrous. It includes those animals that did not show a significant hand preference on more than 2 out of the 5 tasks, and those that showed a significant hand preference for the right on 2 and for the left on 2. On the abscissa the numbers 3, 4, and 5 refer to animals showing the same hand preference on 3, 4, and 5 tasks.*

is clearly a normal one, and there is no evidence of a shift of the mean toward the right as reported by Annett (1972) for people. The best data for apes are those of Finch (1941) on chimpanzees. Forty chimpanzees were tested on four unrelated tasks. Thirty percent of the animals used their right hand on 90% or more of the trials across all tasks, and 30% used their left. Again there is no tendency for a shift toward a preference for the right hand.

On the issue of cerebral dominance, it is important to appreciate the distinction between demonstrating that a lesion of one hemisphere alone has a disruptive effect on some capacity and that the effect differs according to which hemisphere is damaged. Hemispheric specialization is demonstrated only if the latter is found to be the case. Unfortunately, there is as yet no information on specialization for the functions most relevant to language, because, understandably, those chimpanzees that have acquired some aspects of language have not been used for experiments involving neurosurgery. As a result it is only possible to discuss evidence for capacities that might have indirect relevance to language.

Unilateral lesions of association areas can disrupt the capacity of rhesus monkeys to learn certain tasks: This is true for inferotemporal cortex (Ettlinger and Gautrin, 1971; Warren and Nonneman, 1976), superior temporal cortex (Dewson et al., 1970; Dewson, 1977), parietal cortex (Ettlinger and Dawson, 1969), and prefrontal cortex (Warren et al., 1969). In none of these studies was the animal's hand preference found to be relevant. Of these findings the most interesting with respect to language capacities is the disruption by lesion of superior temporal cortex of the ability to interpret sounds (Dewson, 1977) and sequences of sounds (Dewson et al., 1970; Cowey and Weiskrantz, 1976).

But, although unilateral lesions have an effect, there is only one study in which the effect was found to be greater after a lesion on one side than after a lesion on the other. This is the study that was still in progress at the time of the report given by Dewson (1977). Rhesus monkeys were tested for their ability to discriminate a tone from a short burst of white noise. A deficit on the task was not found after removal of superior temporal cortex on the right in two animals, but was found after removal of the same area on the left in four animals, one of which had previously had a right-side lesion without effect. Unfortunately, the results are not easy to interpret, as all but one of the animals had previously had its cochlea destroyed on one side. Confirmation of the findings is required with further animals without cochlea destruction.

An alternative strategy is to test the capacities of each hemisphere separately by splitting the cerebral commissures, and the optic chiasm as well, when testing visual performance. This has been done by Hamilton

(1977). When the monkeys were taught visual discriminations prior to surgery, separate testing of each hemisphere did not reveal better performance by the one than the other. Nor did the animals learn tasks after surgery more efficiently with one hemisphere than the other.

There is a third way of tackling the issue. The auditory paths are arranged such that the crossed projection is stronger than the uncrossed, and, therefore, sounds presented to one ear will have a more potent representation in the contralateral than the ipsilateral hemisphere. There are many studies of normal human subjects in which the technique of dichotic listening has been used to demonstrate differences between the hemispheres in their way of analyzing sounds. By presenting sounds either to the left or to the right ear of Japanese monkeys, Petersen *et al.* (1978) have claimed to find that in these animals the left hemisphere is better at discriminating calls than the right.

It is perhaps too early to make strong claims about lateralization of function in monkeys. But it would be as premature to deny such lateralization as it would be to take it as proven. Furthermore, the brains of apes have yet to be examined in this way.

It is all the more important that apes be studied, given the evidence for anatomical asymmetries between the hemispheres in apes. The length of the Sylvian fissure was measured by Yeni-Komshian and Benson (1976) in 25 people, 25 chimpanzees, and 25 rhesus monkeys. There was no difference between the mean lengths for the left and right hemispheres in rhesus monkeys. But in chimpanzees, as in people, there was a difference. The left fissure was longer than the right in 80% of the chimpanzees and 84% of the people, and the right longer in 8% of the chimpanzees compared with 16% of people. A similar result has been reported by LeMay (1976) for the height of the back of the Sylvian fissure. In 30 monkeys of different species, left and right fissures were found to be equal in height, whereas the right fissure was higher in 57.1% of great apes and the left in 3.5% of the apes. These figures compare with 57.5% of people in whom the right fissure is higher, and 7.5% in whom the left is higher (LeMay, 1976).

The similarities between the figures for man and apes are striking. They provoke two alternative speculations. The first assumes that the asymmetries in man are indeed related to cerebral dominance. If so, perhaps in chimpanzees, also, there is some form of dominance; but this could only be for general intellectual capacities of use in the wild, although in captivity they permit the learning of language. It is not clear why dominance for such abilities should evolve. The second speculation is that the asymmetries found in man and apes are not related to language at all, although in man they do relate to handedness. There are many bodily asymmetries (Morgan, 1977) just as there are many different brain asymmetries (LeMay,

1976), and it may be wrong to assume that they all indicate functional specializations.

LANGUAGE CAPACITIES OF THE
NONDOMINANT HEMISPHERE

As yet we have not found anatomical asymmetries between the two hemispheres that are unique to the human brain. It must be admitted that we have so far considered only gross measures. There may well be differences in fine structure between the two hemispheres in the human but not in the chimpanzee brain. But for the moment we have no direct evidence that this is so. No one denies that the left hemisphere controls speech in all but a few right-handers and that there must be some mechanism that selects which hemisphere is to be the dominant one for speech. But there is no reason to suppose that the dominant hemisphere is constructed differently from the nondominant hemisphere, unless we can show that the nondominant hemisphere is totally incapable of language.

Two separate issues must be considered. The first is whether the nondominant hemisphere normally carries out any linguistic functions. The second is whether it can take over such functions from the dominant hemisphere if the latter is damaged.

Dominance is established early in life. Reviewing studies of dichotic listening, Witelson (1977b) concludes that there is evidence of left hemisphere specialization for language in children as young as 3 years old. Wada (1977) has measured evoked potentials for clicks and flashes from the two hemispheres and claims asymmetries in auditory processing even in babies as young as 5 weeks or so. But we are not entitled to conclude that in young children the right hemisphere has no linguistic function.

This question is best answered by looking at the effects on language of right hemisphere lesions in children. Dennis and Whitaker (1977) have summarized the studies of children with hemiplegia. Only in the study by Basser (1962) were children with left hemiplegia as likely to suffer language disturbance as children with right hemiplegia. There are four other recent studies, in all of which the incidence of speech problems was found to be much lower in left than in right hemiplegics. Annett (1973), for example, found such problems in 40.7% of children with right hemiplegia but in only 14.7% of children with left hemiplegia. Even in children the nondominant hemisphere usually plays a lesser role in controlling speech than the dominant hemisphere.

The language abilities of the nondominant hemisphere can be directly measured in adults in two ways. The first is to examine patients who have had the dominant hemisphere removed because of a tumor acquired after early childhood. The second is to study patients who have the cerebral commissures sectioned to relieve epilepsy; in these patients material can be separately presented to the nondominant hemisphere. In both cases, of course, we may be led to overestimate the role normally performed by the nondominant hemisphere, because functional reorganization could have occurred as a result of the long-standing pathology of the brain.

Smith (1966) reported on a man who had the whole of the left hemisphere removed. He was nonetheless able to understand 82 of 112 spoken words as assessed on the Peabody Picture Vocabulary test. Comprehension of spoken words was also tested by Zaidel (1977) in one patient with a left dominant hemispherectomy and two patients with section of the commissures. Using a new and ingenious system for showing pictures to the right hemisphere in patients with commissure section, he required the patients to pick out pictures corresponding to the words they heard. He found that the comprehension vocabulary of the right hemisphere was surprisingly good on a series of picture vocabulary tests such as the Peabody; this was true for action words and verbs as well as nouns, and for low as well as for high frequency words.

Levy and Trevarthen (1977) also used patients with commissure section to study the ability of the right hemisphere to interpret written rather than spoken words. If written words were presented to both hemispheres at the same time, the patients usually chose pictures corresponding to the word shown to the left hemisphere. But if the words were presented to the right hemisphere alone, the patients were able to match them to pictures. In a further test, the patients were shown pictures and asked to pick others out on the basis of correspondence in sound between the names of the objects shown in the two sets of pictures. If, for instance, they were shown a picture of a bee, they had to pick out a picture of a key. The patients were only able to do this if the pictures were shown to the left hemisphere. From this it appears that only the left hemisphere in adults can evoke the sound of words, as in the names of pictures.

There is a little evidence that the right hemisphere may not be totally incapable of speech in adults. The patient with a hemispherectomy studied by Smith (1966) was occasionally able to produce whole sentences. Kinsbourne (1971) reported that three aphasic patients were able to continue speaking even after injections of sodium Amytal into the blood supply of the dominant hemisphere; in two cases they spoke as clearly as before. The right hemisphere appears even to have some capacity to write. Levy *et al.* (1971) asked two patients with commissure sections to write with the left hand short words shown to the right hemisphere, and one of them could write 12 of 39 nouns correctly.

It appears that, even though aphasia is not often produced by lesions of the nondominant hemisphere in adults or even in children, this hemisphere may have a greater capacity for understanding speech than was once supposed. But it must be admitted, nonetheless, that the role normally played by this hemisphere in subserving speech and language is probably still a minor one.

But what it normally does and what it can do are different matters. The capacity of this hemisphere to take over speech and language is well documented in young children. Early left-side lesions tend to shift the pattern of dominance toward the right hemisphere as assessed by the sodium Amytal test (Rasmussen and Milner, 1977). In such cases the right hemisphere is more likely to play a major role in controlling speech than it is in patients without early damage to the left hemisphere. The right hemisphere is more likely to become the dominant one if the damage to the left hemisphere includes one or both speech areas (Rasmussen and Milner, 1977).

The capacity of the right hemisphere to take over language functions in children can be demonstrated in two other ways. First, children with right hemiplegia tend to recover quickly from any dysphasia caused by the lesion (Basser, 1962). Second, removal of the left hemisphere in childhood or adulthood may have little or no effect on speech and language if the lesion causing the initial hemiplegia is incurred early in life (Basser, 1962). In such cases the right hemisphere is able to assume full direction of the comprehension and production of speech.

We find then that the nondominant hemisphere is anatomically equipped in young children for the control of speech and language. We have no reason to assume that if the left hemisphere takes control of speech the right now loses the relevant fine circuitry. If in adults it is less efficient at taking over language functions, that is presumably the result of a prior commitment to other functions, which now leaves less scope for handling language. We certainly do not have to believe that in adults the two hemispheres are constructed differently from each other.

THE FUNCTION OF DOMINANCE

Even if the left hemisphere is not as specialized in function as once supposed, no one denies that speech and language comprehension are usually more dependent on one hemisphere than the other. If so, what is the advantage of dominance?

There are three issues. The first is the tendency of people, but not of other primates, to make more use of their right hand than their left. The second is the greater role of the dominant than the nondominant hemisphere for language. The third is the strong association between handedness and cerebral dominance. We have to decide which came first, handedness or language dominance, and why, given one, the other should have followed.

There is an obvious reason why it might pay individuals to use one hand in preference to the other when tackling skilled tasks. If the same hand is used on different tasks, the skills acquired by the hemisphere controlling that hand may be of value on new tasks. It seems not unlikely that the development of our considerable abilities at using and making tools could have led to pressure for a strong hand preference. It is less obvious why there should have been pressure for different individuals to use the same hand, the right.

Given handedness for the skilled use of tools, ·one of the ways in which dominance could have evolved becomes apparent. If our ancestors first invented communication with gestures of the hand, as Hewes (1973) and some others suppose, they might be expected to have used the same hand they used for the other manual tasks. Kimura (1976) points out that left hemisphere lesions impair the sign language of the deaf (Sarno *et al.*, 1969). On this view, when spoken language was later invented, the hemisphere came to have dominant control over the muscles of the speech apparatus.

The advantage of this theory is that it explains the association between hand preference and speech dominance; but it may still not be correct. An alternative view is that there are reasons for evolving cerebral dominance for speech that have nothing to do with handedness. The best evidence that there may be such reasons is the existence of dominance for song production in some song birds. Nottebohm (1977) summarizes the evidence he has collected on asymmetries in the neural control of song. Section of the nerve innervating the left syrinx in canaries markedly impairs singing, but section of the nerve innervating the right syrinx has little or no effect. This has been found to be true for 49 canaries, 16 chaffinches, 2 white-crowned sparrows, and 14 white-throated sparrows. Only in one bird, a white-throated sparrow, was the dominance not clear-cut. If a central controlling area, the hyperstriatum ventrale pars caudale, is removed in canaries, the number of syllables of song that the bird loses is much greater for left than right lesions, although recovery takes place after some months. Here then is a case of dominance for sound production, and it does not result from any preference between use of different limbs. It is very tempting to believe that the reason must have something to do with the abilities of these birds, but not all others, to learn songs by imitation, although that would not explain why parrots, which are also able to imitate sounds, appear to lack dominance (Nottebohm, 1977). This phenomenon suggests that we should at least investigate the advantages of hemispheric specialization between the hemispheres for the production of learned sounds.

Specialized Areas

We are now ready to ask whether the areas controlling speech and language in the left hemisphere of man differ only in size from similar areas to be found in other primates or whether these areas are unique to the human brain. Comparisons can be made on the basis of cytoarchitecture and anatomical connections, and of electrophysiology, and on the effects of lesions.

In considering anatomical criteria, one embarrassing problem is that there is little agreement on the extent of Wernicke's area in the human brain (Bogen and Bogen, 1976). The reason is that impairments in the comprehension of language can result from lesions over a large area of parietal and temporal cortex (Conrad, 1954; Russell and Espir, 1961; Luria, 1970). The chances of a patient having a problem in understanding speech sounds are greatest with lesions of the superior temporal gyrus, are still considerable with lesions of inferior parietal cortex, are less with damage to inferior temporal cortex, and are even less when the lesion is in other areas (Luria, 1970).

Superior temporal cortex may be identified with the area TA and inferior parietal cortex with PG and PF of Von Economo (1929). The same areas have been identified by Von Bonin and Bailey (1947) in the rhesus monkey and by Bailey *et al.* (1950) in the chimpanzee. However, by using pigment preparations, Braak (1978) identified an area of the temporal plane with very large pyramidal cells in man, and he was unable to find a similar area in a hamadryas baboon (*Papio hamadryas*). Von Economo (1929) labeled Broca's area FCBm, and a similar area of premotor cortex has been described in the rhesus monkey (Von Bonin and Bailey, 1947) and chimpanzee (Bailey *et al.*, 1950). Von Bonin and Bailey (1961) later said that the area was less distinctive than they had previously suggested.

Almost nothing is known of the cortico-cortical connections in the human neocortex. Geschwind (1965) supposes that the superior temporal and inferior parietal region receives information directly from several senses, thus making possible the association of the sounds of speech with the sight and touch of the things to which the words refer. In the rhesus monkey it has been shown that the superior temporal sulcus and the inferior parietal region receive projections from areas concerned with sound, vision, and touch (Pandya and Kuypers, 1969; Jones and Powell, 1970). The arcuate fasciculus has been thought to be important in man in connecting Wernicke's to Broca's area. Projections from superior temporal cortex have also been described in the rhesus monkey both into the arcuate sulcus and on to the inferior prefrontal convexity (Chavis and Pandya, 1976). These connections may correspond to arcuate fasciculus. A similar projection was indicated in the chimpanzee using the technique of strychnine neuronography (Bailey *et al.*, 1950).

Electrical stimulation of Broca's area in man disrupts but does not evoke speech (Penfield and Roberts, 1959). Unfortunately, the analogous experiment has never been carried out in other primates, that is, stimulation of anterior neocortex while the animal is calling. Instead, there have been several attempts to evoke calls by electrical stimulation, in the squirrel monkey by Jürgens and Ploog (1970; Jürgens, this volume, Chapter 2) and in the rhesus monkey by Robinson (1967, 1972). No success has been reported from sites on the lateral frontal surface of the brain. The area called FCBm in the rhesus monkey and chimpanzee by Von Bonin and Bailey (1961) nonetheless clearly plays some role in the control of vocalization. Electrical stimulation here produces movements of the muscles of the larynx in the rhesus monkey (Hast *et al.*, 1974) and of the larynx and pharynx in the chimpanzee (Bailey *et al.*, 1950). Stimulation here might interfere with calls the animal was already making.

By electrophysiological recording in the squirrel monkey, Newman and Wolberg (1973a, 1973b; Newman, this volume, Chapter 4) have shown that there are cells in the supratemporal plane that respond differently to different calls played to the animal. Newman and Lindsley (1976; Newman, this volume, Chapter 4) also studied cells in frontal cortex, in the arcuate sulcus and around the principal sulcus. They found cells which responded to calls, but which distinguished less between different calls. It seems likely that in monkeys there are neocortical mechanisms for the analysis of calls as of other sounds.

The most critical experiment is the effect of removing cortex from the superior temporal area or from different areas of frontal cortex. People suffer dysphasia after removal of Wernicke's or Broca's area (Penfield and Roberts, 1959). They may be impaired not just in the perception of speech sounds but also in the identification of meaningful nonspeech sounds (Faglioni *et al.*, 1969). Lesions of superior temporal cortex, which exclude the primary auditory area, produce an impairment in the ability of rhesus monkeys to discriminate sounds, including speech sounds (Dewson *et al.*, 1969; Newman, this volume, Chapter 4). Such animals are impaired in a further respect: Even when they have been retrained adequately to distinguish between two sounds, they are poor at telling apart different sequences of these sounds (Cowey and Weiskrantz, 1976). The ability to discriminate patterns of sounds is of obvious relevance for language.

Removal of the area FCBm on both sides in rhesus monkeys does not appear to disrupt the production of calls they have been trained to give for food (Sutton *et al.*, 1974; Sutton, this volume, Chapter 3). The addition of a lesion of superior temporal and inferior parietal cortex was not found to have any further effect. Myers (1972, 1976) also reports that lesions of FCBm, superior temporal, or inferior parietal cortex did not alter the pattern of spontaneous calls. On the other hand, prefrontal lesions led to a considerable decrease in spontaneous calling, but it was not shown whether this was a direct effect of the lesion or an indirect effect of changes in emotionality. It seems that FCBm may not, then, exert direct control over vocalization in monkeys. Either there is no neocortical area that does so or the area may be found elsewhere, for example, in front of premotor cortex. These findings in monkeys do not, of course, show that there is no area analogous to Broca's area in chimpanzees.

To conclude: The evidence for specialized mechanisms for language unique to the human brain is not as conclusive as some have supposed. The most plausible such mechanism is the specialization of part of the frontal cortex for the production of speech sounds. We have not been able to show that the dominant hemisphere differs in its fine structure from the nondominant hemisphere, or that either differ qualitatively from the hemispheres of the ape brain. This is not to say that there are no specializations of the human brain, only to admit that we are as yet uncertain as to their extent and nature.

Conclusion

Chimpanzees learn some forms of language although they have no specialized abilities and presumably no specialized brain areas for this purpose. If this is so, we must reconsider the extent to which man has specialized linguistic abilities and specialized language areas of the brain. It may be that Chomsky (1972) has underestimated the contribution of general intelligence to the learning of language by children.

It would be absurd to suppose that man has no specializations for language. The fact that chimpanzees can learn some forms of language should only force us to consider what these specializations are. But we should not underestimate the chimpanzee. Bertrand Russell once wrote that "no matter how eloquent a dog may be he cannot tell me that his father is poor, but honest." Perhaps not a dog . . . but a chimpanzee?

Acknowledgments

I am grateful to J. M. Warren for providing the data on which Figure 8.6 is based and to L. Weiskrantz for his comments on the manuscript.

References

Annett, M. (1972). *Brit. J. Psychol. 63*, 343–358.

Annett, M. (1973). *Cortex 9*, 4–33.

Bailey, P., Von Bonin, G., and McCullogh, W. S. (1950). "The Isocortex of the Chimpanzee." Univ. of Illinois Press, Ill.

Basser, L. S. (1962). *Brain 85*, 427–460.

Beck, C. H. M. and Barton, R. L. (1972). *Cortex, 8*, 339–363.

Bogen, J. and Bogen, G. M. (1976). *Ann. N.Y. Acad. Sci. 280*, 834–843.

Braak, H. (1978). *Anat. Embryol. 152*, 141–169.

Brown, R. (1970). "Psycholinguistics." Macmillan, New York.

Brown, R. (1973). "A First Language: The Early Stages." Allen and Unwin, London.

Burdick, C. K. and Miller, J. D. (1975). *J. Acoust. Soc. Amer. 58*, 415–427.

Chavis, D. A. and Pandya, D. W. (1976). *Brain Res. 117*, 369–386.

Chomsky, N. (1972). "Language and Mind." Harcourt Brace, New York.

Chown, W., Couch, J. B., Fouts, R. S., and Kimball, G. H. (in press), *J. Exp. Psychol.*

Connolly, C. J. (1950). "The External Morphology of the Primate Brain." Charles C. Thomas, Springfield, Ill.

Conrad, K. (1954). *Brain 77*, 491–509.

Cowey, A. and Weiskrantz, L. (1975). *Neuropsychologia 13*, 117–120.

Cowey, A. and Weiskrantz, L. (1976). *Neuropsychologia 14*, 1–10.

Cowles, J. T. (1937). *Comp. Psychol. Mon. 14*, 1–96.

Cutting, J. E., Rosner, B. S., and Fuard, C. F. (1976). *Quart. J. Exper. Psychol. 28*, 361–378.

Davenport, R. K. (1976). *Ann. N.Y. Acad. Sci. 280*, 143–149.

Davenport, R. K., Rogers, C. M., and Russell, I. S. (1973). *Neuropsychologia 11*, 21–28.

Davenport, R. K., Rogers, C. M., and Russell, I. S. (1975). *Neuropsychologia 13*, 229–235.

Dennis, M. and Whitaker, H. A. (1977). *In* "Language and Development and Neurological Theory" (S. J. Segalowitz and F. A. Gruber, eds.), pp. 93–106. Academic Press, New York.

Dewson, J. H. (1964). *Science 144*, 555–556.

Dewson, J. H. (1977). *In* "Lateralization in the Nervous System" (S. Harnad, R. W. Doty, L. Goldstein, J. Jaynes, and G. Krauthamer, eds.), pp. 63–71. Academic Press, New York.

Dewson, J. H. and Burlingame, A. C. (1975). *Science 187*, 267–268.

Dewson, J. H., Cowey, A., and Weiskrantz, L. (1970). *Exp. Neurol. 28*, 529–548.

Dewson, J. H., Pribram, K. H., and Lynch, J. C. (1969). *Exp. Neurol. 24*, 579–591.

Eimas, P. D. (1975). *In* "Infant Perception: from Sensation to Cognition" (L. B. Cohen and P. Salpatek, eds.), pp. 193–231. Academic Press, New York.

Elliott, R. C. (1977). *Neuropsychologia 15*, 183–186.

Ettlinger, G. and Dawson, F. R. (1969). *Neuropsychologia 7*, 161–166.

Ettlinger, G. and Gautrin, D. (1971). *Cortex 7*, 317–331.

Faglioni, P., Spinler, H., and Vignolo, A. (1969). *Cortex 5*, 366–389.

Farrer, D. N. (1967). *Perc. Motor Skills 25*, 305–315.

Finch, G. (1941). *Science 94*, 117–18.

Fouts, R. S. (1975). *In* "Socioecology and Psychology of Primates" (R. H. Tuttle, ed.), pp. 371–390. Mouton, The Hague, Paris.

Fouts, R. S., Chown, W., and Goodwin, L. (1976). *Learning and Motiv. 7*, 458–475.

Galaburda, A. M., LeMay, M., Kemper, T. L., and Geschwind, N. (1978). *Science 199*, 852–856.

Gardner, B. T. and Gardner, R. A. (1971). *In* "Behavior of Nonhuman Primates" (A. M. Schrier and F. Stollnitz, eds.), Vol. IV, pp. 117–184. Academic Press, New York.

Gardner, B. T. and Gardner, R. A. (1975). *J. Exp. Psychol. 104*, 224–267.

Gardner, H., Surif, F. E. B., Berry, T., and Baker, E. (1976). *Neuropsychologia 14*, 275–292.

Gardner, R. A. and Gardner, B. T. (1978). *Ann. N.Y. Acad. Sci. 309*, 37–76.

Gautrin, D. and Ettlinger, G. (1970). *Cortex 6*, 287–292.

Geschwind, N. (1965). *Brain 88*, 237–294; 585–644.

Geschwind, N. (1967). *In* "Progress in Learning Disabilities" (D. R. John and H. R. Myklebust, eds.), pp. 182–198. Grune & Stratton, New York.

Geschwind, N. and Levitsky, W. (1968). *Science 161*, 186–187.

Goldwin-Meadow, S. and Feldman, H. (1977). *Science 197*, 401–403.

Green, S. (1975). *Zeit. Tierpsychol. 38*, 304–314.

Greenberg, J. H. (1966). "Language Universals." Mouton, The Hague, Paris.

Hamilton, C. R. (1977). *In* "Lateralization in the Nervous System" (S. Harnad, R. Doty, L. Goldstein, J. Jaynes, and G. Krauthamer, eds.), pp. 45–62. Academic Press, New York.

Harter, S. (1965). *J. Child Psychol. 2*, 31–43.

Hast, M. H., Fischer, J. M., Wetzel, A. B., and Thompson, V. E. (1974). *Brain Res. 73*, 229–240.

Hayes, K. J. and Nissen, C. H. (1971). *In* "Behavior of Nonhuman Primates" (A. M. Schrier and F. Stollnitz, eds.), Vol. IV, pp. 59–115. Academic Press, New York.

Hewes, G. (1973). *Curr. Anthropol. 14*, 5–24.

Holloway, R. L. (1968). *Brain Res. 7*, 121–172.

Hopf, A. (1965). *J. Hirnforsch 8*, 25–38.

Humphrey, N. K. (1976). *In* "Growing Points in Ethology" (P. P. G. Bateson and R. A. Hinde, eds.), pp. 303–317. Cambridge Univ. Press, Cambridge, London.

Jarvis, M. J. and Ettlinger, G. (1977). *Neuropsychologia 15*, 499–506.

Jerison, H. J. (1973). "Evolution of the Brain and Intelligence." Academic Press, New York.

Jerison, H. J. (1976). *Ann. N.Y. Acad. Sci. 280*, 370–382.

Jordan, J. (1971). *Folia Morphol. 30*, 322–340.

Jones, E. G. and Powell, T. P. S. (1970). *Brain 93*, 793–820.

Jurgens, U. and Ploog, D. (1970). *Exp. Brain Res. 10*, 532–554.

Kellogg, W. N. and Kellogg, A. (1933). "The Ape and the Child." McGraw-Hill, New York.

Kimura, D. (1976). *In* "Studies in Neurolinguistics" (H. Avakian-Whitaker and H. A. Whitaker, eds.), Vol. 2, pp. 145–156. Academic Press, New York.

Kinsbourne, M. (1971). *Archiv. Neurol. 25*, 302–306.

Kintz, B. L., Foster, M. S., Hart, J. O., O'Malley, J. J., Palmer, E. L., and Sullivan, S. L. (1969). *J. Gener. Psychol. 80*, 189–204.

Kuhl, R. N. and Miller, J. D. (1975). *Science 190*, 69–72.

Lehman, R. A. (1978). *Neuropsychol. 16*, 33–42.

LeMay, M. (1976). *Ann. N.Y. Acad. Sci. 280*, 349–366.

Lenneberg, E. H. (1967). "Biological Foundations of Language." Wiley, New York.

Lenneberg, E. H. (1971). *J. Psycholing. Res. 1*, 1–28.

Levy, J., Nebes, R. D., and Sperry, R. W. (1971). *Cortex 7*, 49–58.

Levy, J. and Trevarthen, C. (1977). *Brain 100*, 105–118.

Lieberman, P. (1975). "On the Origins of Language." Macmillan, New York.

Luria, A. R. (1970). "Traumatic Aphasia." Mouton, The Hague, Paris.

Miller, F. M. (1871). "Lectures in the Science of Language," Vol. I. Longmans and Green, London.

Milner, A. D. (1969). *Neuropsychologia* 7, 375–378.

Milner, B., Branch, C., and Rasmussen, T. (1964). *In* "Disorders of Language" (A. V. S. De Rueck and M. O'Connor, eds.), pp. 200–214. Churchill, London.

Milner, B. (1975). *In* "Advances in Neurology" (D. P. Purpura, J. K. Penry and R. D. Walter, eds.), Vol. VIII, pp. 299–321. Raven, New York.

Morgan, M. (1977). *In* "Lateralization in the Nervous System" (S. Harnad, R. W. Doty, L. Goldstein, J. Jaynes, and G. Krauthamer, eds.), pp. 173–194. Academic Press, New York.

Morse, P. A. and Snowdon, C. T. (1975). *Perc. Psychophysics* 17, 9–16.

Myers, R. E. (1972). *Acta Neurobiol. Exp.* 32, 568–579.

Myers, R. E. (1976). *Ann. N.Y. Acad. Sci.* 280, 745–757.

Newman, J. D. and Lindsley, D. F. (1976). *Exp. Brain Res.* 25, 169–181.

Newman, J. D. and Wollberg, Z. (1973a). *Exp. Neurol.* 40, 821–824.

Newman, J. D. and Wollberg, Z. (1973b). *Brain Res.* 54, 287–304.

Nottebohm, F. (1972). *Amer. Naturalist* 106, 116–140.

Nottebohm, F. (1977). *In* "Lateralization in the Nervous System" (S. Harnad, R. W. Doty, L. Goldstein, J. Jaynes, and G. Krauthamer, eds.), pp. 23–44. Academic Press, New York.

Pandya, D. N. and Kuypers, H. G. J. M. (1969). *Brain Res.* 13, 13–26.

Passingham, R. E. (1973). *Brain Beh. Evol.* 7, 337–359.

Passingham, R. E. (1975a). *Brain Beh. Evol.* 11, 1–15.

Passingham, R. E. (1975b). *Brain Beh. Evol.* 11, 73–90.

Passingham, R. E. (1978). *In* "Recent Advances in Primatology: Evolution" (D. J. Chivers and K. A. Joysey, eds.). Academic Press, London.

Passingham, R. E. and Ettlinger, G. (1974). *Int. Rev. Neurobiol.* 16, 233–299.

Patterson, F. G. (1977). Paper presented at symposium of Amer. Assoc. for Adv. Sci., Denver, Col.

Penfield, W. and Roberts, L. (1959). "Speech and Brain Mechanisms." Princeton Univ. Press, Princeton, N.J.

Petersen, M. R., Beecher, M. D., Zoloth, S. R., Moody, D. B., and Stebbins, W. C. (1978). *Science* 202, 324–327.

Premack, D. (1976). "Intelligence in Ape and Man." Erlbaum, N.J.

Randolph, M. C. and Brooks, B. A. (1967). *Folia Primatol.* 5, 70–79.

Rasmussen, T. and Milner, B. (1977). *Ann. N.Y. Acad. Sci.* 299, 355–369.

Robinson, B. W. (1967). *Physiol. Beh.* 2, 345–354.

Robinson, B. W. (1972). *In* "Perspectives on Human Evolution" (S. L. Washburn and P. Dolhinow, eds.), pp. 438–443. Holt, Rinehart and Winston, New York.

Rockel, A. J., Hiorns, R. W. and Powell, T. P. S. (1974). *J. Anat.* 118, 371.

Rubens, A. B. (1977). *In* "Lateralization in the Nervous System" (S. Harnad, R. W. Doty, L. Goldstein, J. Jaynes, and G. Krauthamer, eds.), pp. 503–516. Academic Press, New York.

Rumbaugh, D. M. (ed.) (1977). "Language Learning by a Chimpanzee" Academic Press, New York.

Russell, W. R. and Espir, M. L. E. (1961). "Traumatic Aphasia." Oxford Univ. Press, Oxford.

Sarno, J. E., Swisher, P. L., and Sarno, M. T. (1969). *Cortex* 5, 398–414.

Schusterman, R. J. (1964). *J. Comp. Physiol. Psychol.* 58, 153–156.

Sebeok, T. A. (1965). *Science* 147, 1006–1014.

Shariff, G. A. (1953). *J. Comp. Neurol. 98,* 381–400.

Sinnott, J. M., Beecher, M. D., Moody, D. B., and Stebbins, W. C. (1976). *J. Acoust. Soc. Amer. 60,* 687–695.

Smith, A. J. (1966). *J. Neurol. Neurosurg. Psychiat. 29,* 467–471.

Stephan, H. (1969). *Proc. 1st Int. Congress Primatol.,* Vol. 3, 34–42.

Stephan, H., Bauchot, R., and Andy, O. J. (1970). *In* "The Primate Brain" (C. R. Noback and W. Montagna, eds.), pp. 289–297. Appleton-Century-Crofts, New York.

Streeter, L. A. (1976). *Nature 259,* 39–41.

Sutton, D., Larson, C., and Lindeman, R. C. (1974). *Brain Res. 71,* 61–75.

Sutton, D., Larson, C., Taylor, E. N., and Lindeman, R. C. (1973). *Brain Res. 52,* 225–231.

Von Bonin, G. and Bailey, P. (1947). "The Neocortex of Macaca mulatta." Univ. of Illinois Press, Urbana, Ill.

Von Bonin, G. and Bailey, P. (1961) *Primatologia II 2* Lief 10, 1–42.

Von Economo, C. (1929). "The Cytoarchitecture of the Human Cerebral Cortex." Oxford Univ. Press, London.

Wada, J. A. (1977). *Ann. N.Y. Acad. Sci. 299,* 370–379.

Wada, J. A., Clarke, R., and Hamm, A. (1975). *Archiv. Neurol. 32,* 239–246.

Walker, A. E. (1966). *In* "The Thalamus" (D. P. Purpura and M. D. Yahr, eds.), pp. 1–11. Academic Press, New York.

Warden, C. J. and Warner, L. H. (1928). *Quart. Rev. Biol. 3,* 1–28.

Warren, J. M. (1974). *J. Hum. Evol. 3,* 445–454.

Warren, J. M. (1977). *In* "Lateralization in the Nervous System" (S. Harnad, R. W. Doty, L. Goldstein, J. Jaynes, and G. Krauthamer, eds.), pp. 151–172. Academic Press, New York.

Warren, J. M., Ablanalp, J. M. and Warren, H. B. (1967). *In* "Early Behavior: Comparative and Developmental Approaches" (H. W. Stevenson and E. H. Hess, eds.), pp. 73–101. Wiley, New York.

Warren, J. M., Cornwell, P. R., and Warren, H. B. (1969). *J. Comp. Physiol. Psychol. 69,* 498–505.

Warren, J. M. and Nonneman, A. J. (1976). *Ann. N.Y. Acad. Sci. 280,* 733–744.

Waters, R. S. and Wilson, W. A. (1976). *Perc. Psychophysics 17,* 285–289.

Witelson, S. (1977a). *Ann. N.Y. Acad. Sci. 299,* 328–354.

Witelson, S. (1977b). *In* "Language Development and Neurological Theory" (S. J. Segalowitz and F. A. Gruber, eds.), pp. 213–287. Academic Press, New York.

Wolfe, J. B. (1936). *Comp. Psychol. Mon. 12,* 1–72.

Yamaguchi, S. and Myers, R. E. (1972). *Brain Res. 37,* 109–114.

Yeni-Komshian, G. H. and Benson, D. A. (1976). *Science 192,* 387–389.

Zaidel, E. (1977). *Cortex 12,* 191–211.

9

Behavioral and Neurobiological Aspects of Primate Vocalization and Facial Expression

HORST D. STEKLIS   MICHAEL J. RALEIGH

Introduction

Comparative and evolutionary investigations of human and nonhuman primate communication frequently utilize three types of information: data derived from (a) studies of cognitive capacities; (b) accounts of overt communicative behavior; (c) investigations of neural correlates of these behaviors. Recently several investigators have described the cognitive capabilities of chimpanzees and discussed the evolutionary implications of these observations (e.g., Harnad et al., 1976; Rumbaugh, 1977; Premack, 1976). In this chapter we discuss selected aspects of primate communicative behavior and its underlying neural basis. This discussion is designed to supplement recent accounts of primate cognitive capacity by critically examining two related issues. One involves the interpretation of certain similarities in and contrasts between the communicative abilities of human and nonhuman primates. The other concerns a purported dichotomy between the neural systems mediating nonhuman primate social communication and those implicated in the control of human communication.

A common view is that a fundamental contrast between human and nonhuman primates is that the latter lack significant voluntary control over their vocalizations and facial expressions (e.g., Campbell, 1974; Erwin, 1975; Myers, 1976). Ethological studies have emphasized certain invariant, species-typical aspects of primate communicative displays (e.g., Andrew,

257

NEUROBIOLOGY OF
SOCIAL COMMUNICATION IN PRIMATES

1963; van Hooff, 1976). This invariance is often regarded as supporting the view that nonhuman primate communication conveys information primarily about the sender's motivational state [e.g., anger or fear (Lancaster, 1968)]. This functional interpretation is thought to be strengthened by evidence suggesting that nonhuman primate vocalization and facial expression are mediated largely by parts of the central nervous system associated with the integration and expression of affective states (Apfelbach, 1972; Robinson, 1976; Myers, 1976). These include the cingulate gyrus, orbital frontal cortex, septum, amygdala, and other phylogenetically ancient components of what is commonly termed the limbic system (MacLean, 1958; Livingston and Escobar, 1971; Yakovlev, 1972). Thus, behavioral and neurological data have led to the view that nonhuman primate vocalizations and facial expressions are largely involuntary behavioral manifestations of underlying emotional states. Furthermore, this perspective emphasizes that an important behavioral difference between human and nonhuman primates is that the former are able voluntarily to communicate information that is largely dissociated from the sender's emotional state. This peculiarly human type of communicative ability is supposedly mediated by more recently evolved neocortical and other nonlimbic structures associated with "higher" activities, such as planning, foresight, and fine perceptual and motor skills. Thus, in its extreme form, this point of view maintains that there is a behavioral and neurological dichotomy between humans and other primates in terms of the communicative function and neural basis of vocalization and facial expression.

As a number of authors (Robinson, 1976; Hewes, 1977), including ourselves (Steklis and Harnad, 1976; Raleigh and Ervin, 1976), have noted, if such qualitative human–nonhuman contrasts exist, a major question concerning the evolution of human communication would be how our ancestors moved from an involuntary, limbically mediated communicative system to a voluntary, neocortically organized system.

In our own view, however, neither the behavioral nor the neurological data actually support this type of dichotomy. There are clearly tremendous contrasts between human and nonhuman primate communication (e.g., see Lieberman, 1975; Harnad *et al.*, 1976). Our intention is not to minimize important differences. Rather, we suggest that these contrasts can more reasonably be regarded as end products of quantitative rather than qualitative differentiation.

To support this, we shall discuss evidence from behavioral studies for the voluntary control of vocalization and facial expression in nonhuman primates. Conditioning studies, observations of free-ranging and captive primates, and developmental investigations will be examined. We will then review data on the neurobiology of primate communication. Perspectives on the interactions between the limbic system and neocortical structures in the mediation of complex behavior, various conceptions of encephalization, and data from stimulation and ablation studies are discussed.

The scope of both the behavioral and neurological discussions is limited. In the behavioral section, attention will be confined to vocalization and facial expression. These behaviors are most readily recognizable and are least ambiguous in terms of their role in social communication. They fulfill common ethological criteria for communicative behavior, in that they are employed in social interactions dependent upon a shared code that is of mutual benefit to members of the species (Smith, 1977). Furthermore, they are the only communicative behaviors whose neural correlates have been investigated. Our neurological section is limited largely to a discussion of primate species of the two genera that have been extensively investigated: *Saimiri* and *Macaca*.

The absence of data on the great apes, especially the chimpanzee, is striking. Clearly this paucity of information limits the generality of any conclusions concerning differences between human and other primates. Data from human studies, treated more extensively in this volume by Ojemann and Mateer (Chapter 5), Brown (Chapter 6), and Kimura (Chapter 7), are included only in the context of specific comparisons with nonhuman data.

Behavioral Data

In this section we review evidence from conditioning studies, ethological and other behaviorally oriented investigations, as well as developmental data related to the issue of voluntary control of nonhuman primate vocalization and facial expression.

Conditioning Studies

A problem inherent in determining whether nonhuman primates can voluntarily control communicative behavior has to do with how volition can be operationally defined. This problem has long plagued psychologists, and for many types of investigations volition is not a useful concept (Kimble and Perlmuter, 1970). Nevertheless, the notion permeates discussions of primate communication, and it is important to ascertain whether any objective, reproducible criteria for "volition" can be established with reference to an organism's behavior. Traditionally, behaviors that could be conditioned to arbitrary stimulus cues have been regarded as being under voluntary control. As Bindra (1976, p. 85) has stated, "The development of 'voluntary control' is essentially the process by which stimulus control of an act is transformed from the initial species-specific, unconditioned eliciting stimuli of movements to extraneous individual-specific conditioned stimuli."

In several experimental procedures, nonhuman primate vocalizations have been conditioned to arbitrary visual stimulus cues. Both prosimians and anthropoids appear to be capable of exerting considerable learned control over their vocalizations. For instance, lemurs (*Lemur catta*) can modify their rate of vocalization in response to arbitrary (on/off) light stimuli (Wilson, 1975). Reversal of reward cues resulted in the expected shifts in vocalization. In contrast to the findings of earlier investigations (Yamaguchi and Myers, 1972), Sutton (Chapter 3) has demonstrated that rhesus monkeys can be conditioned to produce a specific vocal response (coo or bark) to an arbitrary visual cue (red or green light). Changes in reward contingencies alter the rate and type of vocalization. Several physical parameters of the coo calls (duration and amplitude) were also successfully altered by this procedure.

Similar experimental procedures have not yet been utilized to assess whether apes can voluntarily control vocalization. Nonetheless, as noted by Passingham (Chapter 8) and Sutton (Chapter 3), Randolph and Brooks (1967) did succeed in modifying the vocal behavior of a chimpanzee by social reinforcement (play). The frequency and latency of vocalization varied directly with the acquisition and extinction training.

Thus, the available evidence suggests that nonhuman primate vocalization is conditionable. This suggests that under appropriate experimental conditions they are indeed capable of exerting voluntary control over their vocalizations.

Whether facial expression can be similarly conditioned has yet to be investigated experimentally. The absence of relevant studies persists in part because the experimental paradigms employed in vocal conditioning studies are inappropriate for investigating facial expression. These paradigms require that a behavior (e.g., coo vocalization) appear at some initial (minimal) spontaneous rate in order for it to be differentially reinforced. In our experience, without appropriate (social) eliciting stimuli, facial expression, unlike vocalization, is unlikely to appear with sufficient frequency to be amenable to such shaping procedures. Obviously, this difficulty needs to be surmounted before facial expression can be successfully conditioned.

Ethological and Other Behavioral Observations

Since vocalization can be brought under voluntary control, it is reasonable to anticipate that there may be analogous behavioral patterns in free-ranging settings. As Humphrey (1976) has noted, cognitive capacities manifested in laboratory situations are likely to be of some functional (or adaptive) importance in the naturalistic conditions under which the species evolved these abilities. However, this is not always apparent. For example, Rumbaugh (1977), Premack (1976), Gardner and Gardner (1969), and others have shown that the great apes exhibit exceedingly creative (even language-like) abilities in the lab; yet these capacities are not so readily apparent in naturalistic settings. It is therefore of interest to review the evidence for volitional control of communicative behavior provided by naturalistic studies.

In this discussion, the problem of specifying operational criteria for "volition" again emerges. As a minimal criterion in order to be considered a voluntary act, a behavioral pattern must vary in form between members of the same species. That is to say, for a given species, the form of a behavior under voluntary control is less predictable on the basis of general knowledge about that species than is the form of an involuntary act (Bindra, 1976, p. 85). Conversely, the form of an involuntary action is likely to be less variable, presumably because it is determined primarily by momentary stimulus factors and the organism's general physiological condition (Hebb, 1949, p. 145). Thus, our discussion of the naturalistic data focuses on the degree to which primate facial expressions and vocalizations vary between individuals or subgroups of a species.

As was mentioned, most descriptions of primate communicative behavior have emphasized the stereotyped, species-typical nature of the signals (Kummer, 1970). Functional analyses have suggested that these signals are intimately linked to the sender's motivational or emotional state (Rowell and Hinde, 1962; Green, 1975). However, linkage to motivational or emotional states does not exclude their being subject to voluntary control.

Several investigations have documented individual and intergroup differences in communicative behavior. For example, Bertrand (1969) has reported that both confined and free-ranging stump-tailed macaques (*Macaca arctoides*) exhibit individual variability in several calls (e.g. coo, chirp, and whistle) emitted in identical contexts. Earlier studies of rhesus monkey (*Macaca mulatta*) calls also drew attention to such individual variation (e.g., Rowell and Hinde, 1962). Free-ranging chimpanzees, too, show individual variation in their (pant-hoot) calls (Marler and Hobbett, 1975).

It is unlikely that this type of variability is solely an artifact imposed by the investigator's analytical devices (see Snowdon and Pola, 1978); rather, it appears to be meaningful to the animals. This is shown by several recent studies employing experimental (including playback) techniques. Japanese macaque (*Macaca fuscata*) mothers show selective responses to playbacks of recorded vocalizations of juveniles. They respond much more vigorously to the coos of their own offspring than to those of unrelated juveniles (Pereira and Bauer, 1978). Similarly, squirrel monkey (*Saimiri sciureus*) mothers are more responsive to their own infant's vocalizations than to those of other infants (Kaplan *et al.*, 1978). Differences in the calls of rhesus monkey mothers are also responded to selectively by their juvenile offspring (Hansen, 1976). A playback experiment using chimpanzee pant-hoot calls and control sounds showed that chimpanzees discriminate between calls of familiar and strange animals as well as between male and female pant-hoots (Bauer, 1978). Studies of free-ranging forest mangabeys (*Cercocebus albigena*) have shown, through use of playback techniques, that groups respond selectively to differences in the long-distance calls ("whoopgobble") of adult males from different groups (Waser, 1977).

It must be noted at this point that, while such functionally meaningful variation exists in the calls of monkeys and apes, it is not clear how this variation arises (see Lillehei and Snowdon, 1978): Differences in call structure may reflect genetic variation, age and/or social maturational differences, experiential idiosyncrasy, vocal apparatus morphology, or some combination of these factors. Thus, while suggestive, these data alone do not permit conclusions about the degree of voluntary control of vocalization among primates.

More convincing evidence exists in cases where imitational processes are involved. For example, Van Lawick-Goodall (1973) described the acquisition of a facial expression in a novel context by a chimpanzee. While playing, a 2-year-old female produced a "sucking in cheeks" expression instead of the usual play face. The animal's peers readily imitated this facial expression, and a short-lived tradition developed. The gesture was not unique, however, as other infants utilized it in other contexts. What was novel was its utilization in a play context. Individual idiosyncracies of facial expression also occur, as Van Lawick-Goodall's (1968) account of a male chimpanzee who produced a novel "mock smile" face in a variety of contexts showed.

These examples illustrate that the form and function of primate communicative signals are not necessarily invariant. Rather, arbitrary associations between a signal's form and its function (context) can be acquired and propagated among group members, presumably through observational learning. The existence of such group-specific communicative patterns suggests that nonhuman primates can modify their social signals to some degree (Stephenson, 1973a). Both vocalizations and facial expressions have undergone such modification. Green (1975) has reported that Japanese macaques at three different sites imposed a locale-specific quality on the shared tonal theme of their coo calls. The variants were emitted in identical contexts (at each group's provisioning site); the remainder of the repertoire was shared between the groups; old animals lacked such locale-specific modifications. Green concluded that these variants represented learned dialects. He further speculated that a coo variant was originally emitted in the new context (provisioning) by a single group member and was subsequently imitated by others.

In a related study, Stephenson (1973b) identified similar intergroup variation in several nonvocal gestures among the same Japanese macaques. He reported that the males of only two of the three groups directed a facial expression involving "lip quivering" to a consort female. In all three groups this facial expression occurred during encounters between a male and the progeny of a female with whom he was consorting. Because this study examined group-specific communicative behavior patterns, excellent control for context was achieved. Stephenson concluded that these macaques have the capacity to acquire and to propagate arbitrary communicative signs.

This evidence is compatible with the view that primates possess the capacity for voluntarily modifying their vocal and facial expressions, and that imitation may be one means by which novel communicative behavior is propagated. As Passingham (Chapter 8) and Sutton (Chapter 3) note, little can be said regarding the extent or limits of this imitative capacity due to the paucity of direct experimental data. A promising context for the investigation of such abilities is the ape language learning paradigm. Suggestive evidence for the ability to imitate facial gestures comes from Patterson's work (1978a,b) on language acquisition in the gorilla. She reported that, while viewing photographs of ape facial expressions, the gorilla called Koko imitated the portrayed gestures. Chevalier-Skolnikoff (1976) observed Koko and other gorillas imitate a human model making a new facial expression not in the normal repertoire of the gorilla. Such expressions were also performed in the absence of the human. Some imitation of vocalization was also reported. To our knowledge, no one has explored systematically the extent of this apparent capacity in the language learning context, where, for example, the ape could be supplied with signs denoting particular facial gestures (or vocalizations) and then asked to perform these on command. Similarly, systematic observations could be made on the ape's variety and accuracy of facial expression, as well as associated signs of affect when shown pictures of conspecific expression.

Some indirect evidence of imitational capacities comes from observations on heterospecific monkey groups. Available data suggest that such capacities are more readily manifested in copying conspecifics, for in these mixed species groups vocalizations and facial expressions of other species are not imitated (Maple and Lawson, 1975). It has also been known for some time that even with considerable training human-reared chimpanzees are not successful in imitating human speech sounds (Kellogg and Kellogg, 1967).

Additional, although weaker, evidence regarding the degree of voluntary control may be provided by noting the extent to which behaviors are goal-directed or purposeful (Hebb, 1949) or, in naturalistic terms, the amount of modulation possible under social circumstances. As Gautier and Gautier (1977) have pointed out, primatologists are frequently struck by the rich subtlety of communicative events, which may be indicative of learned control. For instance, Bertrand (1969) has maintained that many of the gestures and vocalizations of stump-tailed macaques resemble "deliberate goal-directed actions." Particularly interesting is the animal's ability to inhibit communicative behavior when it is socially appropriate or beneficial to do so (e.g., "feigned indifference"). Of course, it has long been known (Yerkes and Yerkes, 1929) that the great apes are supremely endowed with such capacities, to the point of being able to practice deception (e.g., Menzel, 1974; Chevalier-Skolnikoff, 1976).

A final example illustrates the degree to which behaviors generally regarded as involuntary can be modified through social reinforcement. Bertrand (1976) described a fully toilet-trained hand-reared macaque (*Macaca nemestrina*), capable of urinating on (verbal) command, and able to utilize facial expressions instrumentally (as illustrated by signaling her intention of urinating in order to achieve some other end, such as leaving the room). While Bertrand concedes that she could have been observing a genius or mutant monkey, we concur with her inclination to discount this as unlikely.

Developmental Investigations

Developmental studies indicate that unlearned (or innate) factors contribute significantly to the production and recognition of communicative displays in humans and other primates. Present at birth and (presumably largely) genetically specified, unlearned factors appear to determine the form of vocalization and facial expression in many primate species. While underscoring the importance of unlearned factors in primate communication, this section also emphasizes that the presence of these components does not preclude primates from voluntarily controlling their communicative displays.

Innate factors in the production and recognition of speech in man have been extensively reviewed (e.g., Lenneberg, 1967; Kavanagh and Cutting, 1975) and will not be discussed in this chapter. We will instead focus on the development of vocalization and facial expression in nonhuman primates.

Human facial expressions are also discussed, since they share a number of features with the expressions of other primates (van Hooff, 1976).

There appear to be species differences in the relative contribution of innate factors to the development of vocal production. In squirrel monkeys (*Saimiri sciureus*), neither deafening at birth nor rearing by muted mothers alters vocal production in later life (Newman, Chapter 4). Talapoin monkeys (*Miopithecus talapoin*), removed from their mothers at birth, display virtually all of the basic calls of the species (Gautier and Gautier, 1977). In contrast, pigtailed macaques (*M. nemestrina*) deafened at birth exhibit vocal abnormalities (Sutton, Chapter 3). Similarly, rhesus monkeys reared in partial social isolation (deprived of physical contact) show idiosyncratic structural abnormalities in their calls (Newman and Symmes, 1974); abnormalities in form and frequency have also been reported for isolation-reared Japanese macaques (*Macaca fuscata*) (Kawabe, 1973).

Unlearned components also contribute substantially to the development of vocal perception. Isolation studies indicate that in rhesus monkeys a "tuned vocal perceptual mechanism" exists at birth, which is maximally responsive to the frequencies most characteristic of adult rhesus vocalizations (Sackett, 1973).

While both the production and perception of vocalization depend on innate mechanisms, perception may be more readily modified by experience than is production. For instance, in heterospecific groups, monkeys respond appropriately to the vocal signals of other species; however, the forms of the signals they send are not altered. When modification in signaling occurs, it is in the timing and reciprocity (e.g., acquisition of novel sequences) rather than in the form of the signal itself (Maple and Lawson, 1975).

Unlearned factors are also involved in the production and recognition of facial expression in primates. Crab-eating macaques (*Macaca fascicularis*) blinded postnatally before Day 19 show all of the normal facial expressions (with the possible exception of threat), although these expressions are produced less often (Berkson and Becker, 1975). Congenitally blind people are not impaired in the ability to produce facial expressions when associated with emotional states such as rage, fear, surprise, and delight (Eibl-Eibesfeldt, 1973). However, the blind may be less skilled in voluntarily producing such expressions (Charlesworth and Kreutzer, 1973).

Sackett (1966) has demonstrated that, between 2.5 and 4 months, isolation-reared rhesus monkeys respond to pictures of threatening conspecifics by a tremendous increase in disturbance behaviors (e.g., rocking, self-clasping, and vocalizing) and a decrease in the rate at which they chose to view these slides.

Among humans, the rules governing the circumstances under which it is appropriate to express emotion vary cross-culturally; however, whenever emotional states are expressed spontaneously, the same facial expressions are employed in all studied cultures. Furthermore, many facial expressions are recognized cross-culturally (e.g., those communicating fear) (Ekman, 1973). These observations suggest that imitative learning may be less of a factor than innate mechanisms in the spontaneous production and recognition of facial expression in both humans and other primates.

These data suggest that neither humans nor other primates are free from the innate components, although humans are obviously more capable than are other primates of voluntarily altering the form of communicative signals and modifying the contexts in which they are utilized (Ekman, 1977). Nonetheless, in all primates, the proper timing and assignment of appropriate cue functions to these signals depend on adequate social experience (Mason *et al.*, 1968). Apparently the sequence in which displays are manifested is more labile and context dependent than the form of the display itself. By altering the sequence of displays, primates can voluntarily control the type of information communicated. In some primate species, changes in the ordering of displays may give rise to learned innovations (see Ethological and Other Behavioral Observations, p. 263). In other species the actual form of the vocalization or facial expression may be modified (Redican, 1975; Chevalier-Skolnikoff, 1974). In all primates, innate and learned components contribute to the development of communicative behavior and in no primate does the presence of unlearned factors reduce the likelihood of voluntary control over communicative behavior.

Summary and Conclusions

Evidence commensurate with the view that nonhuman primates can voluntarily control vocalization and facial expression comes from conditioning studies, naturalistic observations, and developmental investigations. These observations suggest that the differences between human and nonhuman primates in this capacity appear to be of a quantitative nature. However, alternative interpretations of these data are possible. The conditioning studies may be representative of particular species rather than of the entire primate order. Investigations utilizing comparable procedures in a variety of species are required before the distribution and extent of volitional control can be specified.

The hypothesis that communicative behaviors are directly reflective of emotional states (or largely involuntary) is exceedingly difficult to test with naturalistic data. For example, a facial expression like "lip smacking" may occur in several contexts, including grooming, approach, or submission. In these situations the animal's overt arousal level may vary from calm to fearful. Nonetheless, it is still possible to maintain that the animal experiences the same emotional state in all three contexts. As a result of this type of dilemma, some students of animal communication have abandoned the notion that a communicative act provides information about the motivational or physiological state of the signaler. Rather, it is maintained that messages convey information about what the sender is likely to do next (Beer, 1977; Smith, 1977).

The documentation of individual and group differences in the form of communicative displays may also be variably interpreted. It is our view that these contrasts are likely to represent learned acquisitions. However, the differences may also be due to genetic dissimilarities, particularly in cases where groups are geographically isolated (e.g., Vogel, 1971). It is also possible to postulate that group differences in motivational states account for the locale-specific variations, as, for example, in the Japanese macaque coo vocalizations reported by Green (1975).

Finally, developmental data indicate that the appropriate production and recognition of communicative signals depend on the interactions between learned voluntary control and innate factors. Nonetheless, it is possible to maintain that this type of modification of innate mechanisms is qualitatively different in humans from that occurring in other primates.

At present, then, perhaps the strongest behavioral support for the view that primates exercise some significant volitional control over their vocalizations is that this hypothesis is consistent with a wide diversity of data. Clearly this type of volitional control is likely to vary with the species: For example, apes may possess a larger capacity for such control than monkeys. Similarly, different ecological or social pressures may select for these abilities in some but not all species (see Nottebohm, 1972). Many more descriptive, analytical, and experimental studies are required in order to establish the distribution and nature of such capacities.

Neurobiological Data

As described earlier, several investigators have maintained that there are qualitative differences between the neural mechanisms mediating nonhuman primate vocalization and facial displays and those underlying human speech and facial expression (see Introduction, p. 258). In this section, we evaluate this perspective by reviewing traditional and more recent perspectives on the functional interplay between limbic system and neocortical structures in the mediation of behavior. We then examine the concept of encephalization and its bearing on evolutionary changes in neocortical and limbic system function. Finally, we discuss the effects of central nervous system manipulations (primarily ablations) on vocalization and facial expression in primates. Comparisons are made with corresponding human clinical data.

Limbic System, Neocortical Structures, and Behavior

Methodological advances have compelled investigators to revise many of the previous conceptions of the interplay between the limbic system and neocortical structures in the mediation of communicative behavior. Descriptive and experimental neuroanatomy have been transformed by the recent perfection of staining, histofluorescent, and other techniques (Nauta and Ebbesson, 1970; Ungerstedt, 1971; Kater and Nicholson, 1973; Cowan and Cuénod, 1975). Recently developed behavioral tests have enabled investigators to assess more accurately the behavioral consequences of manipulating the central nervous system (Weiskrantz, 1968, 1977). Utilization of more refined techniques has enabled researchers to specify more precisely the relationship between nervous system structure and activity and species differences in behavior [see Masterton *et al.* (1976a,b) and Passingham and Ettlinger (1974) for reviews].

Prior to the development of modern neuroanatomical techniques, pioneering investigators had described many of the changes undergone by the primate brain during its evolution. Changes in the size, fissural patterns, course of myelination, and cytoarchitectonic features of the primate brain were described by Brodmann (1909), Elliot-Smith (1902), Tilney (1928), LeGros Clark (1930), Edinger (1929), and others. As these investigators noted, the progressive expansion and elaboration of the six-layered neocortex constitutes a striking evolutionary trend. This progressive enlargement of the forebrain is readily apparent when the brains of living primates are compared. With increasing phylogenetic relatedness to man, the neocortex becomes larger, more richly fissured, and histologically more clearly defined. Accompanying this enlargement is the displacement of phylogenetically older "limbic" cortex from a ventrolateral to a medial position in the cerebrum. In anthropoid primates, this cortex becomes a "limbus" (hence, limbic cortex) encircling the brainstem.

The early anatomists noted that this limbic cortex received direct input from the olfactory bulb and possessed extensive reciprocal connections with the brainstem areas involved in the regulation of autonomic activities. Such anatomical observations, experimental studies, and clinical data were compatible with the view that the limbic system regulates emotional and motivational states critical for preservation of the individual and the species (Herrick, 1933; Papez, 1937; Klüver and Bucy, 1939). Subsequently, the term "limbic system" or "visceral brain" was applied to include a variety of structures, including several interconnected cortical structures (e.g., orbitofrontal, anterior temporal, parahippocampal, and cingulate regions) and subcortical formations (e.g., amygdaloid, anterior thalamic, hippocampal, and septal nuclei) (MacLean, 1958). Some investigators continue to assert that the primary function of the limbic system is to mediate emotional behavior (MacLean, 1973). However, many investigators have pointed out that the diversity of the structures included in this system makes it highly probable that they mediate a wide variety of behavioral functions (Pribram and Kruger, 1954). Experimental evidence has implicated the limbic system in functions ranging from motivation to memory to cognition (see, e.g., Isaacson, 1974).

Just as the limbic system has traditionally been connected with affect, so the neocortex has been associated with the mediation of skilled sensorimotor abilities, refined perceptual capacities, and elaborate intellectual achievements. Since the nineteenth century, the voluntary control of communicative abilities has been regarded as one such function of the neocortex [see, e.g., Geschwind (1970) for a review]. Pioneering cytoarchitectonic and electrophysiological studies have demonstrated that, in contrast to other mammals, in man and other anthropoids considerably less of the neocortex is directly devoted to the mediation of sensorimotor functions (Brodmann, 1909; Woolsey, 1947). Much of the frontal, temporal, and parietal cortex was found to be either unresponsive to sensory stimuli, as measured electrophysiologically, or devoid of direct primary sensory or motor projection systems. As a consequence, these areas were designated as "silent" or "association" cortex, that is, cortex that did not directly process sensory information of a particular modality but was involved in the association, regulation, or modulation of output from primary cortical areas and subcortical structures. Hughlings Jackson, for example, maintained that the human prefrontal cortex was primarily concerned with voluntary control of movement. This control was achieved principally by modulating the motor commands originating in ("lower") subcortical areas.

As in the case of the limbic system, the currently available clinical and experimental data indicate that association cortex is involved in a variety of functions. These range from visual perception and memory (in the temporal cortex) (Gross, 1973; Dean, 1976) to planning and establishment of behavioral strategies (in parts of the prefrontal cortex) (Pribram and Luria, 1973) [for a general review, see Masterton and Berkley (1974)]. However, the notion that voluntary, goal-directed behaviors are primarily regulated by neocortical association areas continues. It is often maintained, for example, that behaviors mediated by limbic system structures are less subject to voluntary control than those mediated by the neocortex (Geschwind, 1965; Campbell, 1974).

Encephalization and Central Nervous System Evolution

The dissociation between neocortical and limbic system structures with regard to volition and "higher intellectual" functions and involuntary emotional activities persists, in part, because of the influence of the doctrine of encephalization on theories of brain organization. As originally described by Hughlings Jackson (1887–1888), this doctrine maintains that structural and functional evolution of the vertebrate central nervous system proceeds from a caudal to rostral direction. According to this perspective, phylogenetically more recent species are more reliant on neocortical mechanisms to mediate abilities previously (in more primitive species) controlled by subcortical structures. In this view, human speech, for example, may be regarded as a product of increasing encephalization in the control of phonatory mechanisms that were at one time controlled by subcortical means. The concept of encephalization was heuristically useful in that, in contrast to strict localization perspectives, it emphasized cortical–subcortical relationships and suggested that functions were unlikely to be represented in discrete "centers." In the past 20 years, however, more sophisticated approaches to the functional and structural reorganization of the primate brain have appeared.

Recent comparative neuroanatomical investigations have led to a substantially different perspective regarding the nature of structural differences among vertebrate groups. For example, it was traditionally argued that the nonmammalian forebrain was "dominated" by olfactory inputs (Herrick, 1933). Apparently, however, only a fraction of the area traditionally associated with olfaction actually receives olfactory input. Visual, somatic, and auditory projections also reach these parts of the forebrain through the thalamus in all vertebrate classes (Ebner, 1976). Furthermore, some thalamocortical projection systems, present in reptiles and mammals, are homologous (Ebner, 1976). Thus, there appears to be considerable organizational similarity between "lower" and "higher" vertebrate groups. This suggests that the evolution of the brain did not only involve accretion to existing structures; rather, the system has become reorganized as a whole.

Modern comparative allometric investigations have shown that during primate evolution both subcortical and neocortical structures manifest changes in absolute and relative size, histological complexity, and pattern of afferent and efferent connections. Passingham (Chapter 8) has quantitatively analyzed changes in the relative proportion of cortical area and thalamic nuclei in prosimians and anthropoids. He has demonstrated that primates with large brains (e.g., apes and humans) possess larger neocortices relative to total brain size than do primates with smaller brains (e.g., monkeys and prosimians). For instance, humans possess the largest proportion of neocortex relative to total brain volume. However, this proportion is no greater than what is expected for any primate with a similarly sized brain. Sensorimotor and association areas are likewise no larger in humans or apes than would be anticipated on the basis of total neocortical size. These data suggest that changes in relative size and proportion among cortical areas occurring during primate evolution follow an organizational scheme that does not differ appreciably between man and other primates. Alterations in size and proportion appear to be primarily a consequence of the progressive increase in overall brain volume.

Allometric investigations also demonstrate that there have been substantial changes in paleocortical and subcortical formations. Stephan and his colleagues (e.g., Stephan and Andy, 1970) have made detailed volumetric measurements of these structures in extant primates and insectivores. They have demonstrated that, as is the case with neocortex, these parts of the central nervous system evolve at different rates. Compared to a hypothetical insectivore of similar body weight, the greatest increases in size have occurred in the striatum, hippocampus, septum, and amygdala. The paleocortex has remained relatively unchanged, while the olfactory bulbs have decreased in size (Stephan and Andy, 1969). A recent detailed study (Stephan and Andy, 1977) indicates that more subtle reorganization within these structures also occurs: While the entire amygdaloid complex enlarges during primate evolution, the small-celled part of the cortico-basolateral group is the most enlarged and the nucleus of the lateral olfactory tract has actually decreased in size.

In summary, then, structural modifications occur at all levels of the neuraxis. Both neocortical and limbic system structures undergo enlargement and shifts in relative size. It is clearly the neocortex of the anthropoids that manifests the most dramatic expansion. This undoubtedly accounts, at least in part, for human special abilities. It is equally apparent, however, that non-neocortical structures have also undergone substantial evolutionary modification. These changes have altered corresponding functional capabilities. The continuing involvement of the limbic system in emotional actions does not preclude it from also contributing significantly to "higher cognitive functions" (see Riklan and Levita, 1969), including language (Penfield and Roberts, 1959; Brown, this volume, chapter 6; Ojemann and Mateer, this volume, chapter 5). From both a behavioral and an anatomical perspective, the evolution of the primate brain is more reasonably viewed in terms of dynamic interaction among different components (Noback and Shriver, 1969) than as "imperialism" within brains (see Weiskrantz, 1977).

Lesion and Stimulation Studies

In this section we review data on the effects of brain lesions and electrical stimulation on facial expression and vocalization in humans and other primates. As mentioned under Conditioning Studies (p. 259) and under Ethological and Other Behavioral Observations (p. 260), nonhuman primates can learn to modify their vocalizations and perhaps also their facial expressions. One of the objectives of this section is to ascertain whether the neural correlates of these behaviors are the same as those of voluntary facial expression and vocalization in humans. In addition, we evaluate critically the hypothesis that nonhuman primate facial expression and vocalization rely more heavily on limbic system mechanisms than do those of man. This hypothesis predicts that limbic system lesions have more devastating consequences on nonhuman primate communicative behavior than on human facial expressions or speech.

FACIAL EXPRESSIONS

Years ago Bard (1934) and Bard and Mountcastle (1948) demonstrated that in cats extensive bilateral forebrain ablations (including all tissue rostral to the hypothalamus) did not alter facial expressions or body postures accompanying rage and other induced affective states. More recent work has shown that the marsupial opposum's defensive postures, which include facial threat, remain intact following ablation of the entire neocortex (Hara and Myers, 1973). These observations suggest that the motor patterns of facial expression accompanying aggression and other affective states depend on diencephalic and lower brainstem structures acquired during the early phases of mammalian evolution. This does not imply, however, that facial movements in primates are similarly organized. During primate evolution, facial musculature became considerably more complex (Huber, 1931), permitting a degree of subtlety in facial expression matched by that in few other mammals (Chevalier-Skolnikoff, 1973). As a consequence, the neural control of primate facial expression is likely to be substantially more complex than that in nonprimate mammals.

Only a few studies have investigated the effects of lesions on primate facial expression, and comparative data are lacking since only the rhesus monkey has been studied in any depth. Some of the early investigations of cortical function in mammals, although not addressing specifically the problem of the neurology of species-typical facial expression, contain relevant observations. Of particular interest are the studies by Green and Walker (1938) of the inferior portion (or face area) of the precentral gyrus in the rhesus macaque and baboon (species unidentified). These investigators found that bilateral ablation of motor and premotor facial area resulted in long-term (up to 2 months) paresis of the entire lower facial musculature. Unilateral ablations of the entire facial motor area produced a similar but contralateral paresis.

This facial weakness was described primarily in reference to feeding movements or other responses (e.g., withdrawal of the corners of the mouth to manual stimulation of the face). These movements were referred to as "volitional." Occasional reference was made to "grimacing" behavior, which Green and Walker observed to occur during emotional arousal (e.g., as a consequence of painful stimulation). It is noteworthy that the bilateral precentral lesions abolished this display. However, following unilateral ablations of the face area, emotional grimacing was less affected than the "volitional" movements, a dissociation in the use of the face that, as noted by Green and Walker, is well known clinically.

A more recent ablation study has focused directly on the question of the neurology of facial expression in primates. Unilateral ablations confined to the precentral motor gyrus of rhesus monkeys in some cases produced mild deficits in contralateral facial expressiveness (Myers, 1969). In contrast, bilateral anterior temporal or prefrontal-orbitofrontal lesions led to marked long-term reductions in the variety and frequency of most facial expressions in socially living rhesus monkeys (Franzen and Myers, 1973). In man, lesions involving the precentral gyrus commonly produce an inability to initiate (voluntary) facial movements upon command, while the capacity for spontaneous facial expression (e.g., smiling in response to a joke) is left unimpaired (Myers, 1976; Kolb and Milner, in press). These data have been interpreted as demonstrating that rhesus monkeys, unlike man, do not control facial expression through neocortical mechanisms involved in volitional movements (Myers, 1976).

This conclusion may be premature for several reasons. In assessing these lesion data, it is useful to examine the degree to which the behavioral deficits are qualitative (i.e., eliminating motor abilities) or quantitative (e.g., in terms of frequency). In making human–nonhuman comparisons it is useful to distinguish, where possible, between impairments of a voluntary or involuntary nature. Available evidence suggests that extensive bilateral ablations of prefrontal or temporal limbic regions do not interfere with motor mechanisms of facial expression but instead alter affective or motivational processes, resulting in less frequent and socially inappropriate facial expression. In the Franzen and Myers (1973) study, operated subjects were described as "typically poker-faced"; however, the same animals actually produced submissive "grimaces" at significantly elevated frequencies. Although sparse, the data on facial expression following such lesions in man (e.g., Goldstein, 1948; Blumer and Benson, 1975) also show that, while quantitative decrements occur, patients are nevertheless capable of producing the full range of expression spontaneously (Kolb and Milner, in press). Thus, the neural basis for spontaneous facial expression appears to be similar in man and other primates.

As reviewed above, precentral motor and premotor cortical areas do appear to make a contribution to the actual motor organization of facial expression in primates. The nature of this contribution, however, remains unclear, as do species differences in the importance of precentral gyrus to facial expression. Systematic investigations of these questions have yet to be carried out, and the few existing relevant ablation studies have produced variable results (i.e., from mild to severe effects on facial expression). Similarly, the question whether in nonhuman primates the motor areas contribute, as they do in man, more to volitional than to involuntary (or spontaneous) expression cannot be decided on present evidence. Experimental data regarding the existence of voluntary control over facial expression in monkeys or apes are required. If, for example, expressions prove to be conditionable, it would then be of interest to document the effects of precentral lesions on conditioned expression. In the absence of such data, no conclusions can be reached regarding species differences in the neocortical motor mechanisms of voluntary facial expression.

Stimulation experiments may also contribute information about the neural representation of facial expression. As is the case for ablation effects, it is often very difficult to interpret stimulation data. For example, it is often unclear whether an evoked behavior is secondary to an induced emotionally motivated state or represents a directly elicited motor response (see Jürgens, chapter 2). Stimulation (particularly of subcortical sites) has elicited facial gestures in primates (e.g., Delgado and Mir, 1969; Perachio and Alexander, 1975). However, most studies do not distinguish between emotion-induced and primary motor activity, thereby leaving unclear the role of stimulated brain structures in facial expression per se [for a review of this problem in relation to agonistic patterns, see Plotnick (1974)].

While brain regions responsible for producing integrated facial expression in primates have not been delineated, it has been apparent for nearly a century that the facial musculature has abundant representation in sensorimotor cortex. Many of the early cortical stimulation studies revealed that in humans (e.g., Cushing, 1909) and other primates (e.g., Beevor and Horsley, 1890; Walker and Green, 1938) movements of individual or groups of facial muscles could be elicited from circumscribed sensorimotor fields. Later work has shown that the amount of cortex devoted to the representation of the face relative to other body parts is quite large in all of the higher primates (Woolsey, 1958). Yet, the significance of this complex representation in relation to patterned facial expression is unknown. Furthermore, it is unclear whether significant species differences exist in the size or organizational detail of the motor face area. As a starting point, an interesting nonexperimental approach would be to make quantitative comparisons of this somatotopic map [similar to what Passingham (1973) has done for hand and leg areas] to determine whether the relative size of the face area differs among monkeys, apes, and humans.

VOCALIZATIONS

A frequent observation in clinical neurology is that severely aphasic patients are nevertheless capable of spontaneous outbursts of words (often profanities) and of nonverbal vocalizations (e.g., crying, laughing). This observation suggests that there may be a dissociation between the neural control of speech and nonverbal vocal expression. In normal speech, nonverbal elements (e.g., intonation, pitch) are intimately associated with verbal elements. However, it may be heuristically useful for cross-species studies to analyze the verbal and nonverbal aspects of vocal patterning separately. Recent ablation studies on rhesus monkeys indicate that it may also be useful for comparative purposes to distinguish between volitional and spontaneous control of phonation (see Sutton, chapter 3, and Conditioning Studies, p. 259). Thus, in discussing the neural control of vocal abilities, it is appropriate to compare human speech and learned (e.g., conditioned) vocalizations in monkeys.

Sutton (chapter 3) demonstrated that large bilateral ablations of inferior frontal or parietotemporal neocortex in rhesus monkeys do not affect the quality or quantity of spontaneous vocalization. Furthermore, discriminative (or voluntary) vocal responding in a conditioning task was not altered. On the other hand, bilateral anterior cingulate cortex ablations initially abolished both types of vocalizations. It appears, however, that learned vocalizations were more dramatically affected by cingulate lesions, for the spontaneous vocalizations returned more rapidly. While vocal abilities in both contexts reappeared, the normal dynamic range of vocalization was not apparent in the postoperative test situation. It is noteworthy that coo calls were the type most severely affected, for these may be subject to learned modification among Japanese macaques (Green, 1975).

Myers and his colleagues [see Myers (1976) for a review] have performed a series of ablation studies. They report that in rhesus monkeys bilateral removal of any one of the four neocortical areas thought to correspond to the human "speech areas" (e.g., inferior precentral gyrus, supramarginal gyrus) failed to produce any well-defined changes in spontaneous vocal behavior. However, sequential ablation of all four areas reduced spontaneous vocal behavior to 50% of the preoperative level. Myers' further studies showed that bilateral orbitofrontal lesions led to the largest quantitative reduction in the frequency of vocalization. Cingulate and anterior temporal lesions resulted in less severe and more temporary reductions. However, it is noteworthy that an earlier study by Green and Walker (1938) had shown that in rhesus monkeys bilateral ablation of the entire facial motor cortex impaired spontaneous phonation. Postoperatively, vocalization was described as "a feeble chirp or low pitched and husky," and "the characteristic noisy chatter and variation of intonation were absent" (Green and Walker, 1938, p. 266). In one subject such abnormalities were apparent as late as the 49th postoperative day.

These varying results may be due to a number of methodological differences, including variations in lesion size or observational methods. Sutton *et al.* (1974) quantified vocal responding, but their precentral lesions appear to be smaller than those produced by Green and Walker (1938). Since Myers (1969) presents neither quantitative behavioral data nor histological information, it is difficult to compare his findings with those of other investigators.

In the Franzen and Myers (1973) study, not all vocalizations were affected equally by frontal lesions. In some subjects the frequency of "scream" vocalization increased postoperatively. As with facial expression, rather than affecting the capacity to vocalize, the prefrontal lesions may exert their behavioral effects by altering the motivational states (e.g., increased fear) associated with particular calls (e.g., "scream").

Collectively, the ablation studies suggest that cingulate and orbital (limbic) cortices are involved in the regulation of spontaneous vocalization. However, the anterior cingulate region appears to be of particular importance in the mediation of learned vocal responding. The evidence for neocortical (i.e., motor face area) involvement is less conclusive since large ablations have produced conflicting results. Nevertheless, the importance of these structures in vocalization is also supported by stimulation work (reviewed by Jürgens, Chapter 2). This work indicates that anterior cingulate cortex plays a critical role in call production, while the cortical larynx area functions in the motor coordination of voluntary vocalization.

This review has indicated that there are no data supporting the suggestion that a dichotomy exists in regard to the neural correlates of either voluntary or spontaneous vocalizations between humans and other primates. Human speech and other vocalizations are organized in neocortical, limbic, and subcortical (e.g., thalamic) brain regions (see Chapters 5 and 6). Some of these areas seem to play similar roles in the mediation of nonhuman primate vocalization (see Chapter 2). For example, lesions or stimulation of inferior frontal cortex (larynx and face area) interfere with speech. Cingulate and supplementary speech area lesions reduce spontaneous verbal and nonverbal expression, resulting in mutism in severe cases. In humans, additional regions (e.g., those in the temporoparietal cortex) subserve speech, while in other primates these areas apparently do not contribute to call production. This additional neural representation may be indicative of the greater complexity of speech when compared to nonhuman primate vocalization. Nevertheless, voluntary control of vocalization is neither dependent on nor a consequence of the evolution of these neocortical systems. Rhesus monkeys (and probably other anthropoids) possess mechanisms for volitional control at the level of limbic (cingulate) and (perhaps) premotor cortex.

Summary and Conclusions

In this chapter we examined the view that a fundamental contrast between man and other primates is that the latter lack significant voluntary control over their vocalizations and facial expressions. Both behavioral and neurobiological data were reviewed.

Conditioning studies, developmental observations, and field and laboratory behavioral investigations indicate that nonhuman primates can voluntarily control their communicative displays. These investigations have utilized objective criteria (such as instrumental conditioning, behavioral variability, and imitative ability) as evidence for volitional control. Vocalizations can be conditioned in both prosimians and anthropoids (see Conditioning Studies, p. 259). In combination with other observations (e.g., on the imitative abilities of apes, p. 263), naturalistic data suggest that nonhuman primates voluntarily modify their facial expressions and vocalizations. In some cases this results in the development of group traditions. Although in many primate species the essential form of calls and facial gestures is largely unlearned, these are still subject to substantial voluntary control (see Developmental Investigations, p. 264). This is most apparent in humans, where much of spontaneous expression is innate, yet clearly subject to learned modification. In some other primates the forms of their communicative gestures are not ordinarily modified. Still, these animals manifest voluntary control by combining these displays into socially appropriate behavioral sequences.

Neurobiological evidence shows that in all anthropoid primates vocalizations and facial expressions are mediated by neocortical and by limbic system structures (see pp. 272–277). In different species these mechanisms may contribute differentially to the production and recognition of communicative behaviors. However, this interspecific variability does not appear to be indicative of the relative degree of involuntary or voluntary mediation. It is therefore no longer tenable to maintain that the evolution of volitional control of communicative behavior was contingent upon the expansion and reorganization of the neocortex in human evolution (see p. 269). Comparative anatomical investigations reveal that during evolution all levels of the primate CNS have been modified. Functional capacities, such as significant learned control of communicative displays, are likely to be mediated by neocortical limbic, and subcortical structures (pp. 272–277). Among monkeys, the anterior limbic cortex plays such a role in the regulation of vocal behavior.

In primates, the nature and extent of volitional control of communicative gestures have not been systematically investigated. This is particularly apparent in behavioral studies, where considerable attention has been directed to the "species-typical" aspects of primate communication. Some neurobiological studies have directly examined the potential flexibility and modifiability of primate communicative behavior (e.g., Sutton, Chapter 3). Still, conclusions are at best tentative since only a few species have been investigated. Particularly noticeable is the lack of relevant data from the African apes. This makes conclusions regarding human–nonhuman differences tenuous, since these animals are biochemically, morphologically, and behaviorally considerably closer to humans than are monkeys. While rudimentary in monkeys, abilities (such as voluntary control of facial expression) may be well developed in the great apes. In our view, many of the differences between humans and other primates in behavioral capacities can best be specified by utilizing the language learning paradigms of Rumbaugh (1977), Premack (1976), and others (see p. 263).

Finally, we have not intended to minimize important contrasts between humans and other primates. There are striking differences in both communicative abilities and their underlying substrates. In both its behavioral manifestations and neurobiological basis, human speech is dramatically more complex than nonhuman primate vocal communication. However, the data reviewed in this chapter indicate that the differences are quantitative, not qualitative. Nonhuman primates possess (in at least rudimentary form) the neurobiological mechanisms (e.g., neocortical sensorimotor regions) and behavioral capacities (e.g., volitional control of vocalization and facial expression) that were elaborated during the evolution of human speech. As discussed in Chapter 10, this suggests that speech may have evolved during the early phases of human evolution through selection for the increased development of the neural substrates underlying learned control of vocal behavior.

References

Andrew, R. J. (1963). *Science 142*, 1034–1041.

Apfelbach, R. (1972). *Z. Tierpsychol. 30*, 420–430.

Bard, P. (1934). *Psychol. Rev. 41*, 309–329, 424–449.

Bard, P. and Mountcastle, J. B. (1948). *Research Publications of the Association in Nervous and Mental Disease 27*, 362–404.

Bauer, H. R. (1978). Paper presented at Second Annual Meeting of Primatologists, Atlanta, Ga.

Beer, C. G. (1977). *Amer. Zool. 17*, 155–165.

Beevor, C. E. and Horsley, V. (1890). *Philos. Trans. Roy. Soc. B 181*, 120–158.

Berkson, G. and Becker, J. D. (1975). *J. of Abnormal Psychology 84*, 519–523.

Bertrand, M. (1969). "The Behavioral Repertoire of. the Stumptail Macaque." Bibliotheca Primatologica No. 11. Karger, Basel.

Bertrand, M. (1976). *Z. Tierpsychol. 42*, 139–169.

Bindra, D. (1976). "A Theory of Intelligent Behavior." Wiley, New York.

Blumer, D. and Benson, D. F. (1975). *In* "Psychiatric Aspects of Neurologic Disease" (D. F. Benson and D. Blumer, eds.), pp. 151–170. Grune & Stratton, New York.

Brodmann, K. (1909). "Vergleichende Lokalisationslehre der Gros-shirnrinde," Leipzig.

Campbell, B. G. (1974). *In* "Frontiers of Anthropology" (M. J. Leaf, ed.), pp. 290–307. Van Nostrand, New York.

Charlesworth, W. R., and Kreutzer, M. A. (1973). *In* "Darwin and Facial Expressions" (P. Ekman, ed.), pp. 91–168. Academic Press, New York.

Chevalier-Skolnikoff, S. (1973). *In* "Darwin and Facial Expression" (P. Ekman, ed.), pp. 11–89. Academic Press, New York.

Chevalier-Skolnikoff, S. (1974). "The Ontogeny of Communication in the Stumptail Macaque," Contributions to Primatology, Vol. 2. Karger, Basel.

Chevalier-Skolnikoff, S. (1976). *In* "Origins and Evolution of Language and Speech" (S. Harnad, H. D. Steklis, and J. Lancaster, eds.), pp. 173–211. Ann. N.Y. Acad. Sci., Vol. 280.

Clark, W. E. Le Gros. (1930). *J. Anat. (Lond.) 64*, 371–414.

Cowan, W. M., and Cuénod, W. (1975). "The Use of Axonal Transport for Studies of Neuronal Connectivity." Elsevier, Amsterdam.

Cushing, H. (1909). *Brain 32*, 44–53.

Dean, P. (1976). *Psychological Bulletin 83*, 41–71.

Delgado, J. M. R. and Mir, D. (1969). *Ann. N.Y. Acad. Sci. 159, 3*, 731–751.

Ebner, R. R. (1976). *In* "Evolution of Brain and Behavior in Vertebrates" (R. B. Masterton, M. E. Bitterman, C. B. G. Cambell, N. Hotton, eds.), pp. 147–168. Wiley, New York.

Edinger, T. (1929). *Ergeb. Anat. Entwicklungsgesch. 28*, 1–249.

Eibl-Eibesfeldt, I. (1973). *In* "Social Communication and Movement" (M. Cranach, and I. Vine, eds.) pp. 163–194. Academic Press, New York.

Ekman, P. (1973). *In* "Darwin and Facial Expression" (P. Ekman, ed.), pp. 169–222. Academic Press, New York.

Ekman, P. (1977). *In* 'The Anthroplogy of the Body" (J. Blacking, ed.) pp. 30–84. Academic Press, New York.

Elliot-Smith, G. (1902). *J. Anat. Physiol. (Lond.) 23*, 309–319.

Erwin, J. (1975). *In* "The Rhesus Monkey" (G. H. Bourne, ed.), Vol. 1, pp. 365–379. Academic Press, New York.

Franzen, E. G. and Myers, R. E. (1973). *Neuropsychologia 11*, 141–157.

Gardner, R. A. and Gardner, B. T. (1969). *Science 165*, 664–672.

Gautier, J. P. and Gautier, A. (1977). *In* "How Animals Communicate" (T. A. Sebok, ed.), pp. 890–964. Indiana Univ. Press, Bloomington.

Geschwind, N. (1965). *Brain 88*, 237–294; 585–644.

Geschwind, N. (1970). *Science 170*, 940–944.

Goldstein, K. (1948). "After Effects of Brain Injuries in War." Grune & Stratton, New York.

Green, H. D. and Walker, A. E. (1938). *J. Neurophysiology 1*, 262–280.

Green, S. (1975). *In* "Primate Behavior" (L. Rosenblum, ed.), Vol. 4, pp. 1–102. Academic Press, New York.

Gross, C. G. (1973). *In* "Progress in Physiological Psychology," Vol. 5, pp. 77–123. Academic Press, New York.

Hansen, E. W. (1975). *Developmental Psychobiology 1*, 83–88.

Hara, K. and Myers, R. E. (1973). *Brain Research 52*, 131–144.

Harnad, S. R., Steklis, H. D., and Lancaster, J. (1976). "Origins and Evolution of Language and Speech." Ann. N.Y. Acad. Sci. Vol. 280.

Hebb, D. O. (1949). "Organization of Behavior." Wiley, New York.

Herrick, C. J. (1933). *Proc. Nat. Acad. Sci. Vol. 19*, 7–14.

Hewes, G. W. (1977). *In* "Language Learning by a Chimpanzee. The Lana Project" (D. M. Rumbaugh, ed.), pp. 3–54. Academic Press, New York.

Hockett, F. C. and Altmann, S. A. (1968). *In* "Animal Communication" (T. A. Sebeok, ed.), pp. 61–72. Indiana Univ. Press, Bloomington.

Huber, E. (1931). "Evolution of Facial Musculature and Facial Expression." Johns Hopkins Univ. Press, Baltimore.

Humphrey, N. K. (1976). *In* "Growing Points in Ethology" (P. P. G. Bateson and R. A. Hinde, eds.), pp. 303–318. Cambridge Univ. Press, New York.

Isaacson, R. L. (1974). "The Limbic System." Plenum, New York.

Jackson, J. H. (1887–1888). *In* "Selected Writings of John Hughlings Jackson" (J. Taylor, ed.), 1931–1932, Vol. II, pp. 72–91. Hodder and Stoughton, London.

Kaplan, J. N., Winship-Ball, A., and Sim, L. (1978). *Primates 19*, 187–193.

Kater, S. B. and Nicholson, C. (1973). "Intracellular Staining in Neurobiology." Springer-Verlag, New York.

Kavanaugh, J. F. and Cutting, J. E. (1975). "The Role of Speech in Language." M.I.T. Press, Cambridge, Mass.

Kawabe, S. (1973). *In* "Behavioral Regulators of Primate Behavior" (C. R. Carpenter, ed.), pp. 164–184. Associated University Presses, N.J.

Kellogg, W. N. and Kellogg, L. A. (1967). "The Ape and the Child: A Study of Environmental Influence on Early Behavior." Hafner, New York.

Kimble, G. A. and Perlmuter, L. C. (1970) *Psych. Rev. 77*, 361–385.

Klüver, H. and Bucy, P. C. (1939). *Arch. Neurol. Psychiat., Chicago, 42*, 979–1000.

Kolb, K. and Milner, B. (in press). *Neuropsychologia*.

Kummer, H. (1970). *In* "Old World Monkeys" (J. R. Napier and P. H. Napier, eds.), pp. 25–38. Academic Press, New York.

Lancaster, J. B. (1968). *In* "Primates" (P. C. Jay, ed.), pp. 439–457. Holt, Rinehart and Winston, New York.

Lenneberg, E. H. (1967). "Biological Foundations of Language." Wiley, New York.

Lieberman, P. (1975). "On the Origins of Language." MacMillan, New York.

Lillehei, R. A. and Snowdon, C. T., (1978). *Behaviour 65*, 270–281.

Livingston, K. E. and Escobar, A. (1971). *Arch. Neurol. 24*, 18–21.

MacLean, P. D. (1958). *J. Nervous Mental Disease 127*, 1–11.
MacLean, P. D. (1973). *In* "A Triune Concept of the Brain and Behavior," The Clarence M. Hincks Memorial Lectures, 1969 (T. J. Boag and D. Campbell, eds.), pp. 4–66. Univ. of Toronto Press, Toronto.
Maple, T. and Lawson, R. (1975). *Primates 16*, 99–101.
Marler, P. and Hobbett, L. (1975). *Z. Tierpsychol. 38*, 97–109.
Mason, W. A., Davenport, R. K., and Menzel, E. W. (1968). *In* "Early Experience and Behavior" (G. Newton and S. Levine, eds.), pp. 440–480. Charles C. Thomas, Springfield.
Masterton, R. B. and Berkley, M. A. (1974). *Ann. Rev. Psychol. 25*, 277–312.
Masterton, R. B., Bitterman, M. C., Campbell, C. B. G., and Hotton, N. (1976a). "Evolution of Brain and Behavior in Vertebrates." Wiley, New York.
Masterton, R. B., Hodos, W., and Jerison, H. (1976b). "Evolution, Brain, and Behavior." Wiley, New York.
Menzel, E. W. (1974). *In* "Behavior of Nonhuman Primates" (A. M. Schrier and F. Stollnitz, eds.), Vol. 5, pp. 83–153. Academic Press, New York.
Myers, R. E. (1969). *Proc. 2nd. Intern. Congr. Primate. 3*, 1–9.
Myers, R. E. (1976). *In* "Origins and Evolution of Language and Speech" (S. R. Harnad, H. D. Steklis, and J. Lancaster, eds.), pp. 745–757. Ann. N.Y. Acad. Sci. Vol. 280.
Nauta, W. J. H. and Ebbesson, S. O. E. (1970). "Contemporary Research Methods In Neuroanatomy." Springer-Verlag, New York.
Newman, J. D. and Symmes, D. (1974). *Developmental Psychobiology 7*, 351–358.
Noback, C. R. and Shriver, J. E. (1969). *Ann. N.Y. Acad. Sci. 167, 1*, 118–128.
Nottebohm, F. (1972). *The American Naturalist 106*, 116–140.
Papez, H. W. (1937). *Arch. Neurol. Psychiat. 38*, 725–743.
Passingham, R. E. (1973). *Brain, Behav. Evol. 7*, 337–359.
Passingham, R. E. and Ettlinger, G. (1974). *Int. Rev. Neurobiol. 16*, 223–299.
Patterson, F. (1978a). *National Geographic 154*, 440–465.
Patterson, F. (1978b). *Brain and Language 5*, 72–97.
Penfield, W. and Roberts, L. (1959). "Speech and Brain-Mechanisms." Princeton Univ. Press, Princeton.
Perachio, A. A. and Alexander, M. (1975). *In* "The Rhesus Monkey" (G. H. Bourne, ed.), Vol. 1, pp. 381–409. Academic Press, New York.
Pereira, M. and Bauer, H. (1978). Paper presented at Second Annual Meeting of the American Society of Primatologists, Atlanta, Ga.
Plotnick, R. (1974). *In* "Primate Aggression, Territoriality, and Xenophobia" (R. L. Holloway, ed.), pp. 389–416. Academic Press, New York.
Premack, D. (1976). "Intelligence in Ape and Man." Wiley, New York.
Pribram, K. H. and Kruger, L. (1954). *Ann. N.Y. Acad. Sci., Vol. 58*, 109–138.
Pribram, K. H. and Luria, A. R. (1973). "Psychophysiology of the Frontal Lobes." Academic Press, New York.
Raleigh, M. J. and Ervin, F. R. (1976). *In* "Origins and Evolution of Language and Speech" (S. R. Harnad, H. D. Steklis, and J. Lancaster, eds.), pp. 539–541. Ann. N.Y. Acad. Sci. Vol. 280.
Randolph, M. C. and Brooks, B. A. (1967). *Folia Primatol. 5*, 70–79.
Redican, W. K. (1975). *In* "Primate Behavior" (A. Rosenblum, ed.), Vol. 4, pp. 103–194. Academic Press, New York.
Riklan, M. and Levita, E. (1969). "Subcortical Correlates of Human Behavior." Williams and Wilkins, Baltimore.
Robinson, B. W. (1976). *In* "Origins and Evolution of Language and Speech" (S. R. Harnad, H. D. Steklis, and J. Lancaster, eds.), pp. 761–771. Ann. N.Y. Acad. Sci. Vol. 280.
Rowell, T. E. and Hinde, R. A. (1962). *Proc. Zool. Soc. Lond. 138*, 279–294.

Rumbaugh, D. M. (1977). "Language Learning by a Chimpanzee. The Lana Project." Academic Press, New York.

Sackett, G. P. (1966). *Science 154*, 1471–1473.

Sackett, G. P. (1973). *In* "Behavioral Regulators of Primate Behavior" (C. R. Carpenter, ed.), pp. 56–67. Associated University Presses, N.J.

Smith, W. J. (1977). "The Behavior of Communicating." Harvard Univ. Press, Cambridge, Mass.

Snowdon, C. T. and Pola, Y. V. (1978). *Anim. Behav. 26*, 192–206.

Steklis, H. D. and Harnad, S. R. (1976). *In* "Origins and Evolution of Language and Speech" (S. R. Harnad, H. D. Steklis, and J. Lancaster, eds.), pp. 445–455. Ann. N.Y. Acad. Sci. Vol. 280.

Stephan, H. and Andy, O. J. (1969). *Ann. N.Y. Acad. Sci. 167, 1*, 370–386.

Stephan, H. and Andy, O. J. (1970). *In* "The Primate Brain" (C. R. Noback and W. Montagna, eds.), Advances in Primatology, Vol. 1, pp. 109–135. Appleton-Century-Crofts, New York.

Stephan, H. and Andy, O. J. (1977). *Acta anat. 98*, 130–153.

Stephenson, G. R. (1973a). *In* "Behavioral Regulators of Primate Behavior" (C. R. Carpenter, ed.), pp. 34–55. Associated University Presses, N.J.

Stephenson, G. R. (1973b). *In* "Symp. 4th Inter. Congr. Primat.," Vol. 1, pp. 51–75. Karger, Basel.

Sutton, D., Larson, C., and Lindeman, R. C. (1974). *Brain Res. 71*, 61–75.

Tilney, R. (1928). "Brain from Ape to Man," 2 Vols. London.

Ungerstedt, U. (1971). *Acta. Physiol. Scand., Suppl. 367*, 1–48.

van Hooff, J. A. R. A. M. (1976). *In* "Methods of Inference from Animal to Human Behavior" (M. Von Cranach, ed.), pp. 165–196. Aldine, Chicago.

Van Lawick-Goodall, J. (1968). "The Behavior of Free-Living Chimpanzees in the Gombe Stream Reserve." Anim. Behav. Monogr., Vol. 1/3. Balliére, Tindall and Cassell, London.

Van Lawick-Goodall, J. (1973). *In* "Symp. 4th Inter. Congr. Primat.," Vol. 1, pp. 144–184. Karger, Basel.

Vogel, C. (1971). *Proc. 3rd Inter. Congr. Primatol. 3*, 41–47.

Walker, A. E. and Green, H. D. (1938). *J. Neurophysiol. 1*, 152–165.

Waser, P. M. (1977). *Behaviour 60*, 28–74.

Weiskrantz, L. (1968). "Analysis of Behavioral Change." Harper and Row, New York.

Weiskrantz, L. (1977). *Brit. J. Psychol. 68*, 431–445.

Wilson, W. A., Jr. (1975). *Animal Behavior 23*, 432–436.

Woolsey, C. N. (1958). *In* "Biological and Biochemical Bases of Behavior" (H. F. Harlow and C. N. Woolsey, eds.), pp. 63–81. Univ. of Wisconsin Press, Madison.

Woolsey, C. N. (1947). *Fed. Proc. 6*, 437–441.

Yakovlev, P. I. (1972). *In* "Limbic System Mechanisms and Autonomic Function" (C. H. Hockman, ed.), Charles C. Thomas, Springfield, Ill.

Yamaguchi, S. and Myers, R. E. (1971). *Brain Res., 37*, 109–114.

Yerkes, R. M. and Yerkes, A. W. (1929). "The Great Apes." Yale Univ. Press, New Haven.

10

Requisites for Language: Interspecific and Evolutionary Aspects

HORST D. STEKLIS   MICHAEL J. RALEIGH

Introduction

Several contributors to this volume (especially Passingham, Jürgens, and Sutton) have discussed language- and speech-related abilities in humans and other primates. Although attempting to minimize overlap with their chapters, our primary objective is to relate much of the information presented in this volume to the problem of language evolution. In our discussion we distinguish between speech and language, in that speech represents one means (utilizing the vocal–auditory channel) of expressing language, for the latter is a supermodal phenomenon [as in the "central language system" of Whitaker (1969)]. However, no attempt is made to characterize language definitively. Rather, we examine some of the cognitive, productive (motor), and perceptual abilities that are either associated with language or regarded as critical prerequisites to linguistic performance (e.g., see Lieberman, 1975). After considering some of the previous accounts of the differences between humans and nonhuman primates in both language-related abilities and their neural correlates, we will set forth several new speculations regarding the origin and evolution of language.

While the distinction between cognitive, productive, and perceptual aspects of language is at least partially arbitrary, for in the normal adult they are intimately intertwined, clinical and other observations suggest that

283

NEUROBIOLOGY OF
SOCIAL COMMUNICATION IN PRIMATES

they may be dissociable. For example, inferior prefrontal cortical lesions are likely to impair the expressive aspects of language more severely than the receptive ones, while the converse is the case for more posteriorly located lesions [see Luria (1970) for a review]. When the temporoparietal language areas and their subcortical projection systems are isolated from other brain areas, linguistic ability appears to be restricted to its automatic perceptual, productive, and grammatical aspects. Such patients repeat what they hear and engage in some phonological, grammatical, and semantic encoding, but there is an absence of propositional utterances and other manifestations of higher cognitive functioning (Geschwind *et al.*, 1968; Whitaker, 1976). Comparable dissociations are present in children acquiring language, where perceptual aspects precede productive and cognitive components (McNeill, 1970; Clark, 1977). This potential dissociability suggests that these three aspects of language may have had separate evolutionary histories.

Cognition

Can cognitive abilities (such as memory, symbolic, and other representational capacities not directly dependent upon mode of sensory input or motor output) be compared between humans and other closely related primate species? If, as Chomsky (1957) and Lenneberg (1967, 1971) have proposed, human linguistic competence is firmly enmeshed in a peculiarly human cognitive matrix, then the cognitive abilities of other species have little relevance to those underlying human language. However, the cognitive attributes of language may have evolved from capacities shared with other primates, and precursors to human linguistic abilities may exist in these species (Bever, 1970; Morton, 1970). This proposal is supported by the plethora of data on the humanlike representational and symbolic communicative abilities of African apes (e.g., Mason, 1976; Fouts, 1976; Premack, 1976; Rumbaugh, 1977a; Gallup *et al.*, 1977; for review).

While humans share many cognitive processes with apes, it is possible that some faculties, more closely linked to language, are uniquely human. Since this issue has been evaluated (e.g., McNeill, 1973; Mounin, 1976; Terrace and Bever, 1976; Hill, 1978), we will restrict our discussion to two topics: (*a*) the ability to form cross-modal associations and (*b*) symbolic and propositional abilities. Until recently the existence of these capacities in apes was controversial. Where comparable data are available, we will also discuss the neural correlates of these abilities.

Cross-modal Abilities

Cross-modal perception (i.e., the ability to coordinate information received in one sense modality with information received in other modalities) has frequently been regarded as either dependent on or prerequisite to the evolution of human language. The association between cross-modal performance and language was apparent in Ettlinger's (1967) proposal that language may function as a bridge or mediator in cross-modal tasks. More than a decade ago, Geschwind (1964) maintained that the capacity to make nonlimbic cross-modal associations was a uniquely human ability, contingent upon the evolution of a rich cortico-cortical association fiber network in the angular gyrus. In his view, this development permitted the formation of visual–auditory associations, which were fundamental to the evolution of human speech: For example, a particular sound could become associated with (or evoke) a particular visual image (or referent). Subsequently, this purportedly unique human intermodal system was regarded as critical to the evolution of abilities such as "object naming" or "environmental reference" (Lancaster, 1968; Hewes, 1973a), "concept formation," and even more elaborate cognitive processes, including the recognition of "causality" and the formation of "binary oppositions" (Laughlin and D'Aquili, 1974).

The cross-modal abilities of primates have recently been reviewed by Ettlinger (1977), and their relation to brain anatomy has been discussed by Passingham and Ettlinger (1974). As noted by Ettlinger (1973, 1977), many of the controversies in discussions of cross-modal performance result from the utilization of different training schedules and noncomparable tasks. Consequently, species differences in these abilities are not easily delineated. Nonetheless, available data suggest that humans and other primates do not differ qualitatively in either ability or their purported neural substrates.

The two most frequently utilized tasks are cross-modal matching-to-sample and cross-modal transfer of training. In the latter procedure cross-modal abilities are assessed by measuring the enhancement of learning in a second modality relative to the acquisition rate in the first. This procedure is less direct than the cross-modal matching-to-sample task in assessing cross-modal abilities (Davenport, 1976). Cross-modal matching tasks are readily mastered by adult humans. Matching between touch and vision is also accomplished by prelinguistic infants (Bryant *et al.*, 1972), apes (Davenport, 1976), Old World monkeys (Cowey and Weiskrantz, 1975; Norris and Ettlinger, 1978), and New World monkeys (Elliott, 1977). However, Davenport (1976) reported that apes and Old World monkeys failed to transfer from auditory to visual modalities. Interestingly, prosimians, rabbits, and rats can achieve significant auditory–visual cross-modal transfer of an intensity or pulse pattern discrimination [see Davenport (1977) for a review]; however, these species may achieve this through use of neurological systems other than those employed by monkeys and apes (Ettlinger, 1973).

Premack (1976) suggested that the tests used by Davenport (i.e., matching long and short pulses of sound to long and short flashes of light, respectively) may have been excessively demanding. In other paradigms chimpanzees do show evidence of this type of intermodal transfer. They are, for example, capable of associating spoken English words with the corresponding gestures in sign (Ameslan) language (Fouts, 1974) or with plastic tokens representing the spoken words (Premack, 1976).

These observations suggest that humans do not differ qualitatively from other closely related primates in the ability to form either visual–auditory or other cross-modal associations. Furthermore, while these abilities may well have been prerequisite to the development of human language, they are not likely dependent on the latter. It is apparent, nevertheless, that the development of language in the child facilitates cross-modal performance (e.g., Blank and Bridger, 1964) as does the acquisition of naming in the chimpanzee (Rumbaugh and Gill, 1977).

Since apes and humans are capable of matching information between sense modalities, neural mechanisms underlying these abilities are of interest. As was mentioned, the angular gyrus of the human brain purportedly subserves these integrative abilities. On the basis of anatomical organization, the human inferior parietal lobe is not qualitatively distinguishable from that of either monkeys or apes (Passingham and Ettlinger, 1974). In monkeys, bilateral extirpation of this area results in deficits on a cross-modal recognition of shape task (Sahgal *et al.*, 1975). It is unknown whether similar damage in humans would interfere with cross-modal tasks.

In any event, it is doubtful that a single brain area mediates all cross-modal abilities. Bilateral ablations of the banks and depths of the arcuate sulcus in the monkey impair tactual–visual cross-modal matching performance (Petrides and Iversen, 1976). As pointed out by the investigators, both parietal and frontal areas (in addition to the depths of the superior temporal sulcus) are foci where information from various sensory modalities converges (Pandya and Kuypers, 1969; Jones and Powell, 1970). Both regions, therefore, may integrate information from different sense modalities. It would be of interest to know if particular cross-modal abilities, such as visual–auditory, depend more on the integrity of some brain areas (e.g., superior temporal sulcus or parietal lobe) than others (e.g., frontal lobe). More critical, however, is the lack of data from humans, which precludes ascertaining whether the neural substrates of cross-modal abilities vary between species.

Symbolic and Propositional Abilities

A fundamental feature of human language is its arbitrariness (see Hockett, 1959; Hockett and Altmann, 1968), in that its meaningful elements (or words) do not necessarily possess any physical resemblance to their referents (e.g., objects, actions). At the cognitive level, arbitrariness requires complex representational (symbolic) and mnemonic capacities, which at one time were regarded as uniquely human (e.g., White, 1959; Langer, 1972). However, currently it is recognized that the African apes exhibit many sophisticated symbolic abilities. A variety of these have been investigated in chimpanzees [e.g., "map reading" (Menzel *et al.*, 1978) and recognition of filmed representations of problems and appropriate solutions (Premack & Woodruff, 1978)]. These observations show that, as well as possessing the capacity for acquiring linguistic symbols, apes appear to share an extensive portion of the human cognitive domain (Rumbaugh, 1977b).

There are at least two criticisms of this interpretation. It has been maintained that the apes do not understand and/or use their acquired symbols in the same fashion as do humans [see Hill (1978) for a review]. In an extreme form, it is asserted that the ape's abilities are merely products of conditioning regimes. Rather than comprehending the nature and function of their symbols, apes learn by rote to associate these with appropriate referents. To some extent this criticism is justifiable because some investigations have failed to eliminate experimenter cuing effects or general associative learning and mnemonic capacities [see Savage-Rumbaugh *et al.* (1978b) for a review].

Less severe reservations address the apes' limited use of their symbolic system for purposive social communication [e.g., the ability to communicate about internal states, negation, prevarication (see McNeill, 1973; Mounin, 1976)]. In this view, the utilization of cognitive abilities enables humans to convey propositional information (see Steklis and Harnad, 1976). In most investigations of ape linguistic abilities, the opportunities for propositional interchanges have been severely limited: Frequent reversal of receiver and transmitter roles, or "interchangeability," an important design feature of human linguistic communication (Hockett and Altmann, 1968), has been precluded in most previous studies. Generally, apes have been placed in passive receiver and socially subordinate roles (Hill, 1978). Thus, it has been suggested that it is premature to characterize human–ape interchanges as "communicative" in a purposive, propositional sense (Mounin, 1976).

A few recent studies address both of these reservations. Chimpanzees will use an acquired symbolic code for purposive intraspecific communication (Savage-Rumbaugh et al., 1978b), and both chimpanzees and gorillas show evidence of propositional utterances (i.e., interrogation, negation, prevarication) and may well have a symbolic concept of self (Terrace and Bever, 1976). We will briefly describe the evidence in support of these claims.

Savage-Rumbaugh et al. (1978a, 1978b) demonstrated that two male juvenile chimpanzees acquired a functional symbolic communication system. Through the use of learned symbols ("Yerkish") projected on a computerized keyboard, the two chimpanzees (a) were able to specify accurately the names of foods to one another when food identity was known to only one and (b) spontaneously requested foods of each other by name (Savage-Rumbaugh et al., 1978a). Subsequently it was found that the chimpanzees reliably named and associated distinct functions with six different tools for obtaining foods and learned to both request from and provide for one another specific tools to obtain foods to be shared by both (Savage-Rumbaugh et al., 1978b). When forced to rely on nonsymbolic means of communication (e.g., bodily postures, facial expression), their joint accuracy declined from 92% to 10%. These experiments provide strong evidence that chimpanzees comprehend both the symbolic and the communicative functions of the symbols they use.

There is additional, though largely anecdotal, evidence that apes use symbols for intentional communicative interchanges. For example, the gorilla Koko (Patterson, 1978a) and the chimpanzee Lana (Rumbaugh and Gill, 1977; Gill, 1977) spontaneously ask questions of their trainers. Thus, Patterson reported that Koko utilizes several forms of gestural "intonation" (e.g., leaving the hands in the sign position, seeking eye contact, partial raising of eyebrows) that resemble those employed by deaf human signers. Lana has been taught the concept of naming and will spontaneously request the names of unfamiliar objects and subsequently refer to them by their correct names.

Some direct evidence for acquisition of the concept of negation has come from "conversations" between the chimpanzee Lana and her trainer (Gill, 1977). Lana is not only capable of correctly answering "yes" or "no" questions, but also can engage in a generalized and creative use of negation. Propositional use of negation was also elicited when the trainer deliberately attempted to mislead her. Rumbaugh and Gill (1977) have suggested that Lana's extended use of the word "no" indicated that cognitively the different expressions have a common meaning (i.e., negation) to her.

Furthermore, there are also anecdotal accounts of the use of prevarication by apes. Patterson (1978b) described several cases that suggest that the gorilla Koko was engaging in prevarication rather than making gestural mistakes. In each instance Koko provided false information, apparently in order to avoid being verbally reprimanded.

While this is meager evidence for prevarication through use of a symbolic code, Premack and Woodruff (1979) demonstrated that apes are capable of intentionally communicating false information to a human by nonsymbolic means. They found that the chimpanzees quickly developed behavioral strategies appropriate to coping with villainous and benevolent trainers (e.g., withholding information or giving false information in the villainous trainer condition). Premack and Woodruff hypothesized that chimpanzees, like humans, may be able to make inferences about motivations of other individuals (e.g., good versus bad) and regulate their own behavior accordingly.

Terrace and Bever (1976) have argued that an important aspect of human symbolic capacity is a well-developed concept of self. The ability to symbolize oneself is viewed as a necessary precondition for human language in that it facilitates the symbolic representation of "our relations to others and to the world" (Terrace & Bever, 1976, p. 580). At the time of their article, there was no convincing evidence that apes possessed this capacity. However, apes appear to recognize themselves in a mirror, whereas monkeys do not (Gallup, 1977). A better indicator, however, of the ape's ability to use symbols to refer to itself would be reference to its own emotional states (Terrace and Bever, 1976).

There are suggestive data that gorillas and chimpanzees are able to do this. Koko uses symbols to refer to her present and past emotional states (e.g., "happy," "mad," "sad") (Patterson, 1978b). Terrace et al. (1976) report that a 2-year-old male chimpanzee employs signs conveying emotion as a means of attenuating its physical expression. In one case, almost on the verge of biting his trainer, the ape repeatedly signed "bite," after which he showed no further evidence of anger. On many occasions signs for "bite" and "angry" were apparently made as warnings, since they occurred in the absence of any physical signs of anger or attack.

Conclusion

This brief survey suggests that chimpanzees, and perhaps gorillas, possess at least the rudiments of cognitive abilities underlying human language. The progressive elaboration of these abilities has made possible the spontaneous development of language in humans. The shared cognitive features between humans and apes (e.g., cross-modal abilities, symbol use) are not a consequence of language evolution (although they may have become enhanced by it) but were more likely prerequisites to its development.

Furthermore, these shared abilities are likely to be homologous in humans and African apes (Rumbaugh, 1977b). Complex abilities of this type are not likely to evolve independently along three separate lineages; it is more parsimonious to postulate that the common ancestor of the African apes and humans possessed these abilities. In this view, an amplification of these cognitive abilities occurred only in the human lineage and was associated with the evolutionary expansion and reorganization of the brain rather than with the addition of new areas (see Passingham, Chapter 8, and Evolution, p. 301).

Production

In this section we examine the argument that humans can produce speech because of their specialized peripheral vocal tracts and central nervous system. An intriguing implication of this perspective is that during its evolution the human linguistic communication system may have initially employed a manual sign language (Hewes, 1973a, 1977). In this view, speech evolved later and replaced gesture as the primary communicative medium (Steklis and Harnad, 1976). However, if nonhuman primates possess vocal production systems that are more capable to take on speech functions than previously thought, then it is unlikely that speech lagged behind gesture during the evolution of human language (see p. 308).

Peripheral Specializations

It is important to examine the differences in vocal and related motor abilities between humans and related primates and to analyze critically proposals concerning the vocal abilities of fossil hominids. The work of Lieberman and his colleagues (Lieberman, 1972, 1973, 1975, 1978; Lieberman and Crelin, 1971) has rekindled interest in the potential importance of peripheral modifications (e.g., alterations in the position of the hyoid bone) in the evolution of human speech. Lieberman's writings are of interest since his conclusions are often uncritically accepted, and also because his investigations raise two more general issues. One of these concerns the reconstruction of soft-tissue anatomy from osteological material, while the other pertains to inferring functional capacities from structure. Prior to discussing the contributions of Lieberman and his critics, we will comment briefly on the functional consequences of anatomical differences in the vocal tracts of humans and other primates, which have been reviewed by Negus (1949), DuBrul (1958), Keleman (1969), Wind (1970), Jordan (1971), Sutton (Chapter 3), and others.

In our view there are vocal tract differences between humans and other primates, but these probably do not represent critical limitations to the development of speech. These differences include dissimilarities in the oral cavity, lips, shape of cartilages (e.g., the cuneiform), position of the larynx, and other features (Keleman, 1969; Jordan, 1971; Lieberman and Crelin, 1971; Wind, 1976; DuBrul, 1977). Many of these alterations may represent secondary adaptations to bipedalism (Negus, 1949; DuBrul, 1976). Despite these differences, chimpanzees and (at least some) Old World monkeys can make vowel-like and vowel–consonant sounds (Jordan, 1971; Richman, 1976). Andrew (1976) has demonstrated that baboons utilize formants: They modulate glottal sounds by altering the resonance of the upper vocal tract. Sutton and his colleagues have investigated the muscle spindle morphology and innervation ratio of larnygeal muscles and have documented the muscle twitch properties of the laryngeal muscles (see Sutton, Chapter 3). He has concluded that there is a great deal of intra- and interspecific variability in the anatomical and physiological characteristics of the upper vocal tract. It is unlikely, however, that changes in one or several of these features are prerequisite for the evolution of speech. In any case, nonhuman primate vocal tracts permit the production of a complex variety of sounds, including vowels and formants. As Nottebohm (1976) has observed, in these animals vocal tract limitations do not preclude the evolution of communicative behavior relying on easily recognizable sounds that can be readily reordered to generate a large number of words. Thus, it is unlikely that peripheral limitations represented substantive barriers to the evolution of human speech.

A viewpoint at variance with this perspective has been set forth by Lieberman and his associates. They have maintained that some fossil hominids, extant nonhuman primates, and newborn humans have supralaryngeal vocal tracts that limit the efficiency of their speech. Lieberman and Crelin (1971) and Lieberman *et al.* (1972) have utilized a variety of basicranial features (e.g., angle of the styloid process) to reconstruct the soft tissues of the supralaryngeal tracts of several fossil hominids. In their reconstructions, the La Chappelle aux Saints (Neanderthal) larynx is positioned near the base of the skull, the pharynx lies behind the entrance to the larynx, and the tongue is contained almost entirely in the oral cavity. This reconstruction was subjected to a computer modeling, and it was concluded that classic Neanderthals, as represented by this fossil, could not produce the vowel sounds /i/, /u/, and /a/, which are supposedly language universals, and that Neanderthal speech probably had a nasal quality. Lieberman has asserted that these characteristics reduced the efficiency of the speech of the classic Neanderthal. Based on a similar analysis, Lieberman has also concluded that, while the Broken Hill fossil possessed a modern vocal tract, this individual's pharyngeal region was smaller than that of contemporary humans. This individual may have exhibited reduced vocal stability and therefore has been proposed as a transitional form between classic Neanderthals and modern humans in the evolution of speech (Lieberman, 1973).

There are many problems with this argument, one of which concerns the "reconstruction" of soft tissues from osteological characters. As Burr (1976a) has noted, the placement of the La Chapelle aux Saints larynx may be erroneous due to the effects of deformation, drying, and fossilization on the skull. Furthermore, data from a variety of anthropometric measures have demonstrated that, except for the shape of the palate, the peripheral vocal apparatus of the Broken Hill fossil is within the range of modern humans (Burr, 1976b,c). In Burr's view, "with respect to vocal tract anatomy, Broken Hill is modern *Homo sapiens*" (1976c, p. 763). Falk (1975) has argued that, in the reconstruction of the La Chapelle fossil, the hyoid bond is placed too close to the base of the skull, impairing the individual's ability to swallow. DuBrul (1976, 1977) has maintained that Lieberman and Crelin have disregarded the variability inherent in the features on which their reconstructions were based. For instance, even in carefully prepared specimens of modern humans, the angle of the styloid process varies greatly. According to DuBrul (1977), the reconstruction of La Chapelle is such that the vocal tract would have penetrated the vertebral column. Many investigators (e.g., Carlisle and Siegel, 1974; Falk, 1975; DuBrul, 1976; Burr, 1976a) have criticized the use of newborn rather than adult humans as models for reconstructing the vocal tract of fossil hominids. The questionable nature of this type of comparison was also underscored by LeGros Clark's (1959) observation that if the skulls of infant apes were compared with those of adult humans, it might be concluded that infant apes could speak.

A second difficulty with Lieberman's presentations relates to the attempt to infer functional capabilities from structural features. Even when large sample sizes are available, it is often tenuous to interpret structural differences in functional terms. This difficulty arises whether the functional conclusions are based on differences in osteological or dental structures. For instance, conclusions about the diet of early hominids on the basis of their teeth have provoked intense debate (Howell, 1972; Robinson, 1971). Unlike efforts to reach functional conclusions on the basis of hard morphology, Lieberman's inferences are contingent on another assumption: He must presuppose that the vocal tracts have been accurately reconstructed, which, as the preceding discussion indicates, is not likely. Because neither behavior nor soft structures fossilize, claims about the reduced efficiency of fossil hominids cannot be tested directly. Finally, since nonhuman primates can produce formants and vowel-like sounds, it is unlikely that these abilities would be absent in fossil hominids.

In sum, during human evolution there have been alterations in the peripheral mechanisms involved in the production of speech. Still, it is improbable that peripheral limitations represent major obstacles to an organism's developing speech. This is apparent in patients with malformation of vocal structures (e.g., cleft palate, morphological defects of tongue) who produce acoustically normal or near normal speech (Bosma, 1975). In addition, modern Kabardian language utilizes two vowels (Spuhler, 1977), a smaller number than Lieberman and colleagues have attributed to classic Neanderthal speech. Humanlike speech can be produced with a vocal tract not much different from that of a modern ape. Furthermore, rather than speculating on the functional consequences of peripheral dissimilarities, it may be more profitable to focus on the relationship between differences in the central nervous system and contrasts in vocal behavior.

CNS Specializations

We will consider the evolutionary implication of two perspectives on the specialization of the human brain for speech production. A traditional view derived largely from clinical observations is that specialized neocortical areas (such as Broca's area) provide a unique anatomical basis for volitional motor control of the human vocal apparatus. A second viewpoint is that speech represents an outgrowth of a more general functional specialization of the left hemisphere for fine serial motor praxis (Kimura, Chapter 7). In discussing these perspectives, we will attempt to specify the ways in which the human brain may differ from those of other primates.

Discussions of the role of specialized neocortical areas in human speech have frequently focused on Broca's area. However, the nature and importance of Broca's area in speech production remain controversial [see Whitaker and Selnes (1975) and Brown (Chapter 6) for reviews]. In part, this is due to reports that Broca's aphasia (which includes defects in articulation, prosody, and grammatical usage) results from lesions outside of the traditionally defined Broca's region, that is, the posterior portion of the third or inferior frontal convolution (Mohr, 1976). Furthermore, smaller lesions and surgical excisions restricted to Broca's area (or the underlying white matter) may produce neither lasting nor severe dysphasia (Zangwill, 1975). In combination with brain stimulation data (Penfield and Roberts, 1959; Rasmussen and Milner, 1975; Ojemann and Mateer, Chapter 5), these observations indicate that speech and language production depend not only on the integrity of Broca's area but also on other cortical areas and subcortical structures.

Nonetheless, Broca's area may play an important role in speech production. Mohr (1976) has suggested that Broca's area mediates "a more traditionally postulated role as a pre-motor association cortex region concerned with acquired skilled oral, pharyngeal, and respiratory movements, involving speaking as well as other behaviors, but not essentially language or graphic behavior per se. These latter behaviors seem to involve the entire cerebral operculum, including insula, supramarginal gyrus, and deeper white matter" (p. 222).

This view of Broca's area dovetails with observations that stimulation of Broca's area in the dominant hemisphere may interfere with or arrest speech as well as disrupt sequential oral-facial movements (Ojemann and Mateer, Chapter 5). However, identical deficits also result from stimulation of dominant peri-Sylvian parietal or temporal cortex. This indicates that posterior cortex also plays a role in the motor organization of speech and related movements. These data may account for the observation that damage limited to Broca's area can result in only mild, short-term dysphasia.

In contrast to humans, among nonhuman primates neither inferior frontal nor parietotemporal neocortical regions are involved in either spontaneous or learned call production (Chapters 2 and 9). Vocalization appears to be dependent upon the orbital, cingulate, premotor, and motor cortex in both hemispheres. In squirrel monkeys, the anterior cingulate and pre-Rolandic cortex may be the most critical to voluntary call initiation (Jürgens, Chapter 2). These broad aspects of cortical organization for call production appear to be shared by humans and other higher primates, since both lesions and electrical stimulation in either hemisphere interfere in similar ways with human phonation (Chapters 2 and 9). Thus, nonhuman primates lack neither neocortical control nor the voluntary aspects of call initiation (Chapter 9).

What appears to be uniquely human is (a) added functional specialization of inferior frontal (i.e., Broca's region) and posterior parietotemporal neocortex and (b) hemispheric lateralization of speech-related functions in these regions as well as associated thalamic nuclei. While these regions may have unique functions among humans, these areas do not appear to be morphologically unusual. Their cytoarchitecture and fiber anatomy do not differ distinctly from homologous regions in other higher primates, although, cytoarchitectonically, Broca's area may be an exception to this (Passingham, Chapter 8). This would be of some interest, since of all cortical sites participating in speech production, it is the least variable between individuals (Ojemann and Mateer, Chapter 5). It may be speculated that Broca's area was the first language production area to have evolved by functional differentiation of premotor association cortex. In this sense, it may be the most specialized (but not necessarily the most important) among cortical areas for motor programming of speech.

Asymmetric speech functions have been correlated with left–right fissural asymmetries: The Sylvian fissure is longer and its end point lower in the left hemisphere than in the right (Witelson, 1977). The end-point asymmetry may indicate a longer left planum temporale (Rubens, 1977). The latter area, because it is a part of auditory association cortex, is more likely to exhibit specializations related to speech perception than to production (see Perception, p. 297). On the other hand, left inferior parietal cortex plays an important role in the motor programming of speech (Ojemann and Mateer, Chapter 5), and this may relate to its larger size on the left.

It may be that this left parietal expansion is not solely correlated with speech but may also be tied to more general cognitive functions (such as sound processing). This perspective is supported by the asymmetries in the Sylvian sulcus of the great apes (LeMay and Geschwind, 1975) and by the Sylvian and superior temporal fissural asymmetries in several Old World monkeys (Falk, 1978a). Falk (1978b) has suggested that left parietal expansion may have occurred early in anthropoid evolution in connection with increasing complexity of vocal and visual communication systems of monkeys. Alternatively, it is possible that these cerebral asymmetries [like other bodily asymmetries (Morgan, 1977)] may not have any particular functional consequence.

Functional laterality for human speech processes has also been described for thalamic nuclei (Ojemann and Mateer, Chapter 5). Electrical stimulation data indicate that the ventrolateral thalamus and anterior pulvinar of the language-dominant hemisphere function in verbal attention, memory, and motor speech mechanisms. Homologous thalamic regions in the nondominant hemisphere are involved in attention to and storing of visuospatial material. The uniqueness of such laterality has not been adequately investigated. Stimulation of these areas in either hemisphere of gibbons or monkeys produces vocalization (Jürgens, Chapter 2). It is not known whether stimulation interferes with ongoing vocalization as it does in humans.

It is also uncertain whether the thalamic functional asymmetry in humans has a gross morphological basis. Lateral volumetric asymmetries (favoring the right) have been reported for striatum, that is, globus pallidus (Orthner and Sendler, 1975), but not for thalamus. Speech and other functional asymmetries, however, may be supported by biochemical asymmetries, such as those recently reported for human thalamus: There are higher concentrations of norepinephrine in right versus left ventral tier thalamic nuclei and higher left versus right concentrations in pulvinar (Oke *et al.*, 1978). A naturally occurring left–right biochemical asymmetry, which may be linked to behavioral turning tendencies, has been reported for rat striatal dopamine concentrations (Glick *et al.*, 1977). No investigations of such asymmetries in nonhuman primates have been reported. Thus, the question of the uniqueness of biochemical–behavioral asymmetries in humans remains open.

Kimura (Chapter 7) has provided much evidence for the view that human speech is founded in a more general left hemisphere specialization for serial motor praxis. In her view, complex articulatory postures of speech (including oral-facial-pharyngeal musculature) are guided by neural mechanisms also involved in the selection and execution of upper limb and hand movements. This model may explain the frequent association between receptive or fluent aphasia (including interference with deaf sign) and bilateral oral and limb apraxia. Kimura has suggested that the critical locus for serial motor praxis is in the left posterior parietal area. Through its interaction with the anterior Broca's area, this region guides the articulatory apparatus.

Several aspects of this model are controversial. While bilateral, complex limb movements depend on the left parietal region, it is not certain that speech and other sequential movements have a common neural substrate. The association between aphasia (either fluent or nonfluent) and apraxia is apparently not as strong as predicted by Kimura's model (Poeck and Huber, 1977). Furthermore, the observed association between aphasia and apraxia is due to the anatomical proximity of underlying structural correlates, rather than to a common disorder (Hécaen, 1975). This possibility is supported by Kolb and Milner's (1979a) observations. They compared patients with surgical excisions of either frontal lobe, temporal lobe, parietal lobe, or sensorimotor cortex and normal control subjects on their ability to copy arm or facial movements. Right frontal, left frontal, or left parietal patients were equally impaired on the arm movement task, but only patients with either right or left frontal excisions were impaired in copying facial movement sequences. Furthermore, speech disturbances and movement copying disorders were completely dissociated. The investigators concluded that the frontal cortex anterior to Broca's area in both hemispheres is specialized for sequential oral-facial and arm movements. The latter are additionally controlled by the left posterior parietal region.

The importance of the left prefrontal region to sequential oral-facial movement is supported also by stimulation data (Ojemann and Mateer, Chapter 5). However, parietal and, to a lesser degree, superior temporal sites in the dominant (left) hemisphere also contributed to oral-facial praxis, a finding at variance with Kolb and Milner's study. Furthermore, these sites in prefrontal parietal and temporal zones overlapped greatly with those from which naming disturbances could be elicited, which suggests at least a close association between the two systems. In the nondominant hemisphere of one patient, Ojemann and Mateer found no prefrontal sites that interfered with copying oral-facial movement. Thus, it is premature to reach conclusions regarding the nature of these left hemispheric motor mechanisms. It appears that both prefrontal (perhaps bilaterally) and posterior parietal areas are important in the execution of speech-related oral-facial movements. The neural mechanisms subserving these are at least closely linked to those underlying speech.

Specializations for serial motor praxis may not be unique to humans. Kolb and Milner (1979b) have speculated that the human prefrontal lobe's role in the programming of oral-facial movement may have its evolutionary roots in similar control mechanisms present in the frontal lobes of monkeys (Chapter 9). Deuel (1977) has provided experimental evidence for the importance of prefrontal cortex in the execution of sequential limb movements by monkeys (*Macaca mulatta*). She compared the effects of bilateral removals of either periarcuate, precentral, or posterior parietal cortex on ability to open a complex latch-box. Only periarcuate or precentral lesions reduced this performance significantly. Relatedly, Passingham (1978) found that dorsal prefrontal lesions impaired memory for movements in rhesus monkeys. It has not been ascertained whether unilateral damage would produce an impairment similar to that seen in humans.

Trevarthen (1974) has suggested that complex manipulative skills in baboons are more lateralized to the left hemisphere than to the right; however, since only one subject was involved in the study, no firm conclusions can be reached.

In conclusion, the human brain is specialized for the control of sequential oral-facial and arm movements. However, the regions underlying these abilities are not likely to be unique but are apparently also present in other primates. What may be unique about the human brain is the relative functional lateralization of this praxic system and the degree to which the vocal production system has become coupled to it.

Perception

In this section we will discuss some of the perceptual abilities associated with speech and their putative neural bases. After considering human perceptual abilities and cerebral asymmetries, we will examine the degree to which these are present in nonhuman primates. This section will conclude by considering whether language-related perceptual abilities are best regarded as outgrowths of more general perceptual processes. As in the section on production, our intention is to specify the similarities and contrasts between humans and other primates.

Although speech sounds are graded, adult humans perceive them in a discontinuous or categorical fashion (Liberman *et al.*, 1967). Prelinguistic infants show similar phonemic categorization, suggesting that this perceptual property is largely innate (Eimas *et al.*, 1971). This ability may be a manifestation of a speech feature detection system in the language-dominant hemisphere (Studdert-Kennedy and Shankweiler, 1970). However, if such a specialized system exists, its feature detection properties are probably not rigidly specified at birth. The system has to allow for some "perceptual tuning" to occur during development since the phonetic boundaries vary slightly between languages (Cutting and Eimas, 1975).

Discussions of the potential morphological substrates of this specialized speech perception system frequently focus on the posterior Sylvian asymmetries. The left Sylvian fissure is significantly longer in humans and great apes (LeMay and Geschwind, 1975). In neonates (Witelson and Pallie, 1973; Wada *et al.*, 1975) and most right-handed adults (Geschwind and Levitsky, 1968), the planum temporale is longer and larger in the left hemisphere [see Witelson (1977) for a review]. Histological investigations indicate that this asymmetry may result from expansion of a subdivision of the posterior temporoparietal association cortex, a region frequently affected in Wernicke's aphasia (Galaburda *et al.*, 1978). These observations suggest that these asymmetries are related to the perception of human speech and language.

As many investigators have noted, categorical speech sound perception is probably not unique to humans (Sinnott *et al.*, 1976; Miller and Morse, 1976; Passingham, Chapter 8). Pastore (1976) has further suggested that categorical perception is exclusive neither to speech sounds nor to the human auditory modality, but rather is a characteristic of the mammalian auditory and other sensory (especially visual) systems.

Experimental evidence indicates that among nonhuman primates categorical perception is not limited to human speech sounds. Categorical perception enables pygmy marmosets (*Cebuella pygmaea*) to decode meaningful variation in their vocalizations. Snowdon and Pola (1978) have demonstrated that, when the duration of synthetic trills was varied, the animals responded differently (i.e., presence or absence of antiphony). There was a sharp boundary between 248 and 257 msec in duration, a division that apparently corresponds to the functionally distinct closed and open mouth trill calls. The investigators interpreted this as evidence of categorical responding to continuous variations in trill calls, a phenomenon analogous to human speech processing (see also Snowdon, 1978).

Marler (1975) has hypothesized that in nonhuman primate species whose repertoires are primarily graded, categorical vocal perception may be associated with hemispheric laterality. Japanese macaques (*Macaca fuscata*) manifested a significant right ear (left hemisphere) advantage in the ability to discriminate between communicatively relevant acoustic dimensions (i.e., early versus late temporal position of the peak fundamental frequency) of field-recorded tonal coo calls (Petersen *et al.*, 1978). Other Old World monkeys did not exhibit a consistent hemispheric superiority. Furthermore, when the Japanese macaques were tested on a different (and presumably communicatively less significant) variant (i.e., pitch) of the coo call, no laterality was apparent. Thus, monkeys, like humans, appear to possess both lateralized mechanisms of sound processing and categorical perception of their vocalization.

It is of interest to determine whether these functional capacities correlate with structural asymmetries similar to those found in humans. The size of the planum temporale has not been investigated systematically in nonhuman primates. Witelson (1977) mentions that in one baboon (*Papio papio*) the area most likely to be homologous to the human planum was of "similar" size in both hemispheres. A less direct measure of planum temporale size is Sylvian fissure length. In contrast to humans and great apes, in monkeys the left fissure is not significantly longer (Yeni-Komshian and Benson, 1976). If temporal asymmetry is tied to auditory decoding of vocalization, then it should be present in Japanese macaques. Furthermore, if planum asymmetry is associated with more general, complex auditory functions, it should be exhibited by macaques, since Dewson (1977) has shown that left but not right auditory association cortex lesions impair auditory-delayed matching-to-sample learning in these animals. In any case, inasmuch as Sylvian length accurately reflects planum size, this asymmetry cannot be exclusively associated with speech, since it is present also in the brains of great apes. Since neither primary nor association cortex has been measured in either monkeys or apes, the significance of the Sylvian asymmetry in apes cannot be determined. Furthermore, because areas outside of the auditory cortex, such as the frontal and parietal lobes, also contribute to the decoding of vocalization (including speech) in humans and monkeys, these regions may be related to the known functional capacities and asymmetries. Newman (1978 and Chapter 4) has suggested that the determination of communicative meaning of squirrel monkey vocalization is made in areas outside of the auditory cortex (e.g., frontal lobe). Similarly, in humans, linguistic processing of speech (e.g., phonological analysis) involves frontal and parietal regions in addition to temporal areas

(Ojemann and Mateer, Chapter 5). Since many monkey species, including macaques, show enlarged left parietal regions (see Production, p. 290) and inferior frontal regions (Falk, 1978b), perhaps these areas are related to the known functional asymmetries. It is also premature to conclude that monkeys lack asymmetries in auditory or other brain areas that contribute to vocal decoding.

As with production (p. 290), it is possible that the perception of speech and language may be based in more general cognitive abilities, which are present in other primates. Humans may be characterized by the development of a considerable amount of hemispheric specialization for these abilities. Many types of nonlinguistic stimuli appear to be processed preferentially by the left hemispheric mechanisms, which are also employed in speech decoding. Apparently the left (language-dominant) hemisphere is better equipped to detect the temporal order of stimuli, whereas the right is superior in resolving the spatial components of stimuli [see Krashen (1976) for a review]. Davis and Wada (1977, 1978) have interpreted evoked potential data as suggesting that human cerebral asymmetries are related to general cognitive processes of which language is only a component. Normal adults and prelinguistic infants exhibit higher left than right hemisphere involvement in processing of temporally structured auditory (click) stimuli, whereas spatially structured visual (flash) stimuli are processed primarily in the right hemisphere. Davis and Wada have suggested that in most individuals the left hemisphere is better at processing the temporal structure of stimuli (e.g., comparison of stimuli with previous experience). The right hemisphere is superior in processing simultaneous stimuli that are spatially distinct. Since any stimulus has both temporal and spatial components, this hypothesis predicts that even purely verbal stimuli would be processed in parallel by both hemispheres. An auditory evoked potential study demonstrated that this occurs (Molfese, 1978). Some aspects of phoneme identification (i.e., categorical speech perception) were common to both hemispheres, whereas other aspects were distinguished between the hemispheres.

If language is "overlaid" on mental abilities also utilized in nonlinguistic ways (Krashen, 1976), then other species may show hemispheric specializations in relation to such more fundamental abilities (e.g., see also Witelson, 1977). Hamilton's investigations (1977, review) of learning abilities in split-brain rhesus monkeys are of interest in this regard. Stimuli were selected from materials that tend to be lateralized in humans (e.g., faces, two-dimensional drawings). The subjects were also taught visual discrimination tasks using spatial and sequential stimuli. Both hemispheres learned the discriminations equally well. However, Hamilton (1977) suggests that different tests might distinguish between the hemispheres, and some preliminary results indicate that there are hemispheric preferences for viewing visual stimuli. Furthermore, Hamilton used only visual stimuli. As shown by Dewson (1977), macaques exhibit a left hemispheric superiority in processing sequential auditory stimuli. Thus, in combination with the Japanese macaque data (Petersen *et al.*, 1978), Dewson's study suggests that lateralized mechanisms involved in the processing of auditory stimuli could have evolved independently of human language.

Evolution

In this section we will set forth an account of language evolution. Since this proposal is based on the contrasts and similarities between humans and other primates in abilities and neural mechanisms requisite to language, we will begin by briefly summarizing our previous discussion. Our subsequent evolutionary speculations will be organized into two sections, with each representing a postulated evolutionary stage. In the first section we discuss faculties and neural systems that are essential for language and that evolved either during or prior to a period of common ancestry between the African ape and hominid lineages. In the final section we propose a possible course of events by which these traits were transformed into human language. We are concerned only with the reconstruction of a probable sequence of events in the evolution of human language as it occurred in the ancestors of modern humans. Consequently we discuss neither the time at which [in years before present] nor the taxa in which these transformations occurred. In our view, there is insufficient fossil and behavioral information to permit either reasonable time estimates or correlations with particular fossil groups, and, in any case, neither information is directly pertinent to our present account.

In previous sections we formulated four major conclusions. First, currently there is no evidence of a qualitative distinction between humans and African apes in language-related cognitive abilities (e.g., cross-modal perception, propositional symbol use). African apes possess (at least) the rudiments of these abilities, which humans have significantly elaborated and facilitated by brain expansion and language development. Furthermore, because these complex abilities are manifested by two ape genera and by humans, it is likely that they are homologous among these hominoids and therefore may be inferred to have been present in the earliest Hominidae. Second, vocal tract differences between humans and other primates are not likely to represent critical limitations to speech development. Despite differences in peripheral anatomy, several non-human primates are able to produce speech sounds (i.e., formants and vowels) not predicted from their peripheral anatomy. Third, all primates appear to have some neocortical control of the vocal apparatus; however, in humans, additional cortical association areas in the frontal, temporal, and parietal lobes support speech motor mechanisms. In addition, the control mechanisms of speech production, unlike those mediating vocalization in other primates, are hemispherically lateralized at cortical and subcortical (i.e., thalamic) levels. This speech motor system has in humans become linked closely to neural mechanisms, which perhaps in all higher primates govern sequential oral-facial and arm movements. Fourth, the categorical nature of speech perception does not appear to be unique to speech sounds or to the auditory modality. Categorical decoding of speech may be founded in information processing capacities evolved by all (at least) higher primates. Furthermore, in two primate species this vocal decoding system appears to involve hemispheric specialization, which therefore occurred independently of and perhaps prior to the development of human language.

Prehominids

Immunological and DNA hybridization studies indicate that chimpanzees, gorillas, and humans shared a period of common ancestry after the separation of the lesser ape and orangutan lineages (Sarich and Cronin, 1976). This period may have been of some 5 million years in duration, during which African Hominoidea evolved from a form that may have been very similar to extant pygmy chimpanzees (*Pan paniscus*) (Zihlman et al., 1978). This model would explain most parsimoniously the language-requisite cognitive abilities shared among humans and African apes (see Cognition, p. 284). It is reasonable to postulate that the common ancestor species was cognitively as well equipped as extant chimpanzees or gorillas. Since populations directly ancestral to later hominids were among this hypothetical common African hominoid stock, they, too, would have possessed some of the necessary language-requisite cognitive abilities. Throughout this section we will refer to these ancestral populations as prehominids.

By similar inference, this ancestral species may have had a communication system similar to that of living apes. Many aspects of ape communication, including the nature of the information conveyed and the motor simplicity of the signals themselves [see Marler and Tenaza (1977) for a review], are not substantially different from similar aspects of the signals employed by Old World monkey species [see Gautier and Gautier (1977) for a review]. The essential functions of these communicative systems (i.e., conveying information about the sender's emotional–motivational state, sex, age, and status and the presence of food and danger) may be of considerable antiquity. What may be a characteristic limited to hominoids is the complexity of the cognitive decoding of these nonverbal signals. Consequently, although the (visual, auditory) signals themselves are relatively simple, the potential information extracted is rich in the social dimension. In this vein, Savage and Rumbaugh (1977) have suggested that cognitive processes prerequisite to linguistic behavior evolved in the service of cognitive decoding of nonverbal signals.

Monkey and ape communicative signals primarily convey information about affective states, while human language is well adapted for the encoding of symbolic information. However, as Marler (1977) has pointed out, primates are capable of divorcing communicative behavior from internal affective states, and some naturally occurring elements may even be well suited for communicating symbolic information (i.e., denoting particular as opposed to classes of environmental referents or actions). For example, Japanese macaques (*Macaca fuscata*) produce vocalizations ("girneys") that apparently are relatively divorced from internal states (Green, 1975). These vocalizations are also very similar to human speech in that they combine voiced and articulated components, are morphologically variable, and are uttered with great temporal lability. Since vocalizations of this type are probably widespread among nonhuman primates (Marler, 1977), we can infer that they were present also among ancestral hominoids, where they may well have served as vocal preadaptations for the development of symbolic speech in Hominidae (Green, 1975).

The neural mechanisms underlying vocalizations among living monkeys and apes are of considerable evolutionary antiquity and thus are probably similar to those employed by prehominids. Many basic neural mechanisms of vocal production are shared among monkeys, apes, and humans (see Jürgens, Chapter 2), including the anatomical and functional organization of essential oral-pharyngeal-laryngeal (brainstem) effector mechanisms, as well as the contribution of certain cortical regions (i.e., motor–premotor cortex and subcallosal and cingulate gyri) to call production. In all primates investigated, these shared mechanisms are bilaterally organized, with each hemisphere contributing equally. Bilateral organization of motor mechanisms (at subcortical and cortical levels) appears to be a primitive feature among primates and, therefore, was probably also characteristic of the vocal production system of prehominids.

The Japanese macaque data suggest that hemispheric specialization for auditory decoding of vocalization may have already evolved in prehominids. However, since this type of hemispheric specialization has been demonstrated in only one species of Old World monkey (see Perception, p. 297), it may have been acquired independently. On the other hand, auditory specialization is probably widespread among primates. Because the nature of the hemispheric specialization appears very similar in humans and macaques, it is likely to have evolved in connection with the decoding of acoustic elements common to both human speech and macaque vocalization. Marler (1975) and Petersen *et al.* (1978) have suggested that it is the need for concurrent processing of different aspects (e.g., pitch and formant transitions) of either speech or monkey calls that resulted in the evolution of hemispheric specialization. This type of processing is likely to be similar among many terrestrial Old World monkeys and African apes, since their vocal repertoires are very alike in both structure (e.g., graded) and function (Marler, 1975). Further research must establish whether hemispheric specialization has occurred in other species of Old World monkeys and apes. It is probable, nevertheless, that prehominids, whose vocal system was similar to that of living Old World terrestrial species, employed auditory decoding mechanisms similar to those used by the latter species.

Among primates, hemispheric specialization for auditory decoding preceded that for vocal production; however, the reason for this order is not apparent, because in avian evolution the converse seems to have occurred. Several bird species exhibit lateralized control of their vocal apparatus (Nottebohm, 1977), but there is no evidence that they employ lateralized mechanisms for auditory decoding. Prior specialization in the perceptual system also figured prominently in Hockett and Ascher's (1964) model of the origin of "duality of patterning" in human speech evolution. It is possible that in monkeys and apes (and by inference in the prehominids) the mechanisms for vocal production are of insufficient motor complexity to require lateralization, but other considerations indicate that this is not a likely explanation (see Hominids, p. 307).

Hominids

In this section we will present a speculative account of the evolution of a hominid "primordial" speech system. As in contemporary speech, in this primordial system propositional information was communicated by an acquired symbolic sound code. In its early phases this system probably differed from contemporary speech in several respects (e.g., lexicon size). However, it is assumed to have been qualitatively similar to contemporary speech in its utilization of, for example, a set of rules governing phonological and semantic encoding (Lieberman, 1975).

In discussing the development of such a system, we will briefly consider several morphological and functional alterations of the nervous system that occurred early in hominid evolution. Subsequently, we will relate these transformations to potential selection pressures. Finally, we will argue that this primordial speech system emerged in the earlier phases of human evolution and was unlikely to have been preceded by a gestural language system (Hewes, 1973a).

Paleontological evidence indicates that there was rapid expansion and reorganization of the brain after the separation of the hominid lineage (Tobias, 1971). Apparently, the earliest Hominidae (e.g., *Australopithecus*) possessed cranial capacities significantly higher than those predicted for apes of similar body weight (Passingham, 1975). Furthermore, in the early hominids, brain size increased at a rate that exceeded that exhibited by any other mammalian lineage (Jerison, 1973, 1975). These neural developments permitted the enhancement of the language-related cognitive abilities that underly the spontaneous emergence of propositional utterances.

While apes possess many of the cognitive processes prerequisite to linguistic behavior, they do not spontaneously develop a propositional (or linguistic) communicative system either in the wild or in the laboratory. It has been suggested that the ape lacks sufficient intelligence "to formulate and concatenate rapidly and spontaneously, unlimited arbitrary symbols that represent its percept of the world" (Savage and Rumbaugh, 1977, p. 297). Among humans, this type of intelligence was apparently made possible by the increase in brain size (Passingham, Chapter 8; Jerison, 1975; Rumbaugh, 1977b). Linguistic abilities are exhibited by deaf children who, despite the absence of specific linguistic input, spontaneously develop a structured sign system that contains the essential properties of language (Goldin-Meadow and Feldman, 1977). Conceivably, the spontaneous emergence of linguistic behavior was contingent on the evolution of a particular brain size. This notion might be tested by examining the language capacities and related cognitive abilities in microcephalics who are not dwarfs (see Passingham, Chapter 8).

Cognitive enhancement in hominid evolution does not appear to have depended on neural reorganization beyond that which is a predictable consequence of overall size increase (see Chapter 8 for a review). Alterations in brain size have produced allometric changes in the brain's relative proportions (e.g., neocortical association areas, thalamic nuclei). In turn, this reorganization may have resulted in the elaboration of already existing capacities (e.g., memory) as well as the appearance of new ones (e.g., spontaneous language use). In addition, the rapid increase in brain size may have precluded other major types of reorganization (Sacher, 1976). The comparative neuroanatomical data are also consistent with biochemical investigations that indicate that the human–chimpanzee morphological and behavioral differences resulted from alterations in regulator rather than structural genes (King and Wilson, 1975). These types of genetic changes are likely to affect developmental growth rates.

Evolutionary changes in regulator DNA also may have altered the brain's biochemical environment. For instance, functional (behavioral) changes may result from alterations in the distribution or kinetic properties of enzymes, changes that need not necessarily be associated with concomitant morphological transformations. In this regard, it is of interest that about 20% of mouse DNA is expressed in the brain, whereas in the cortex of adult humans as much as 45% of the total DNA is utilized (Omenn and Motulsky, 1972). This type of change may increase the diversity of the RNA present in human cortex, which may contribute to the enhancement of mnemonic and other cognitive functions (Omenn and Motulsky, 1972). In humans, differences in DNA expression may also provide the basis for morphological, biochemical, and functional differentiation of the hemispheres (Levy, 1977). Thus, striking functional changes could have occurred in the absence of substantial genetic change.

The evolution of the human brain probably resulted from a variety of selective forces (e.g., hunting, tool-making, communication) interacting in a complex feedback fashion (Bielicki, 1969). However, most suggestions do not explain adequately the rapidity of the evolutionary process [see Gabow (1977) for a review]. It is likely that the social organization of Pleistocene hominids may have contributed to the rapid evolution of the brain (Gabow, 1977). The selection pressures favoring increased brain size may have been intensified by the relative isolation of separate population units (demes) surviving in this time of climatic oscillation (e.g., glacials).

Early hominid social organization also may have been a substantial factor in the evolution of vocal learning abilities. Nottebohm (1975) has suggested that significant relaxation of genetic control over the vocal system was a critical prerequisite to vocal learning. A decreased genetic influence on vocal expression is evident among extant higher primates, in which vocal signals are at least partially learned, and in which vocal traditions may develop (see Chapter 9). If this type of "openness" or "productivity" (Hockett and Altmann, 1968) was also present in early hominids, it might have provided the necessary "raw material" for the evolution of more sophisticated vocal learning abilities. More speculatively, we suggest that vocal learning evolved in early hominids as a means of promoting outbreeding. An analogous situation seems to exist among some passerine birds [see Nottebohm (1972) for a review]. For example, saddlebacks (*Philesturnus carunculatus*) are highly territorial and form permanent heterosexual pair bonds. The parental dialect is culturally transmitted to the male offspring as part of a process that contributes to outbreeding (Jenkins, 1977). Discussions of early hominid evolution have frequently suggested that male cooperative hunting would have selected for territoriality (King, 1976a,b) and long-term heterosexual bonding (Campbell, 1974). Under such conditions it is possible that selection favored those breeding units that developed vocal dialects, which, in turn, contributed to systematized outbreeding practices (Hill, 1972). It is also possible that early hominids developed auditory templates that contributed to their recognition of dialects. In certain respects the ontogeny of human speech acquisition resembles that of song birds, which appear to possess specific auditory templates that guide vocal development (Marler, 1975). The observation that contemporary postpubertal humans develop accents when acquiring a second language and are able to recognize dialect variation in other individuals is compatible with these speculations (Hill, 1972).

It cannot be ascertained what the specific neural changes were that accompanied this behavioral achievement, since the neural basis of vocal learning is insufficiently understood in any extant species [see Nottebohm (1975) for a review]. In humans, increased brain size and cognitive sophistication provided for the development of a propositional speech system from these vocal learning abilities. Since overall brain expansion resulted in an increased ratio of total area of association cortex to neocortex, further functional differentiation of association cortex in inferior frontal (e.g., Broca's area) and posterior temporal and parietal lobes could occur for both speech and related cognitive abilities. Thus, the coevolution of cognition and vocal learning in hominids freed their vocal system from the constraints, evident in nonhuman primates and song birds (Nottebohm, 1975), of communicating primarily about motivational states.

It is uncertain what selective pressures led to the evolution of hemispheric laterality in humans. Experimental studies (e.g., split-brain) suggest that functional lateralization can be induced in animals when they encounter difficult cognitive or motor tasks (Neville, 1976). However, such laterality is task specific and reversible since either hemisphere is capable of assuming a "dominant" role (e.g., Stamm et al., 1977). Currently, song birds are the only nonhuman species in which biologically based (or innate) hemispheric differences in (vocal) motor functions have been documented. Complexity of vocal learning alone, however, does not account for lateralization, since parrots, which are at least as sophisticated vocally as song birds, do not manifest hemispheric dominance (Nottebohm, 1977). Rather than vocal complexity per se, it may have been specialization of the opposite (right) hemisphere for other functions [e.g., visuospatial (Webster, 1977)] that resulted in laterality. If the right hemisphere mediated other tasks, it may have had less "neural space" available for the processing of complex motor functions (see also Le Doux et al., 1977). This speculation predicts that birds exhibiting cerebral dominance for song should also possess contralateral dominance for other functions.

The lateralization of speech motor mechanisms may have been enhanced by simultaneous selection pressures for sophisticated manual skills associated with tool-making and use. There is evidence of rudimentary stone tool-making activities as early as 2.5 million years ago, indicating that this activity was probably an important behavioral adaptation of the early hominids (Isaac, 1976). Many accounts of language evolution have emphasized the importance of the cognitive capacities underlying tool-making and other skillful motor activities (e.g., Holloway, 1969, 1976; Hewes, 1973b; Lieberman, 1975, 1976; Montagu, 1976). In our view, tool fashioning and speech are significantly related since both involve highly skilled, sequential motor activity and rely on similar left hemispheric mechanisms (see CNS Specialization, p. 293), and Kimura, Chapter 7). Furthermore, tool-making, while it employs the hands asymmetrically, is a bimanually coordinated activity for which the left hemisphere also appears specialized (Wyke, 1971). Continued selection for verbal and motor skills may have favored the evolution of closely related neural systems for serial motor praxis. However, selection for skilled *asymmetric* use of the hands (e.g., in tool-making and tool and weapon use) may have resulted in the praxic system becoming relatively lateralized (to the left hemisphere).

It is intriguing that no other mammal exhibits cerebral dominance for either hand use [see Warren (1977) for a review] or vocal production. In contrast, among humans, more than 90% of the population favors the right hand in skilled activity (Hécaen and Ajuriaguerra, 1964), and handedness is an accurate predictor of speech dominance (Levy, 1976). Apparently, right-handedness has been prevalent for at least the last 5000 years, which suggests that handedness has a strong biological basis that has not been significantly affected by varying cultural practices (Coren and Porac, 1977). Initially, right-handedness may have been favored in evolution because the speech component of the praxic system would have remained in close proximity to the previously lateralized auditory decoder (see Prehominids, p. 303).

In sum, in our view, a symbolic speech communications system appeared early in human evolution; it was made possible by brain expansion and reorganization as well as by the associated acquisition of improved cognitive and vocal learning abilities. This proposal contrasts with suggestions that a gestural language preceded speech (e.g., Hewes, 1976; Steklis and Harnad, 1976). Several language-relevant preadaptations exist in the vocal systems of extant nonhuman primates (Chapter 9). Thus, proposals such as Hockett and Ascher's (1964) suggestion that an "open" speech system may have originated through a process of verbal "blending," involving sounds essentially similar to those made by extant monkeys or apes, may be appropriate. It is clearly not reasonable to counter this suggestion with the argument that nonhuman primates, in contrast to humans, lack significant voluntary or neocortical control over their vocalizations (see Reynolds, 1968; this volume, Chapter 9).

Speech could have evolved by no more substantial neural or behavioral/ cognitive changes than would, presumably, have been required in the evolution of a gestural language. Since speech was the end point reached in the evolution of human communication, our account seems to us more parsimonious. Furthermore, gestural origin theories encounter difficulties, such as the likelihood that the hands were used for noncommunicative purposes (e.g., tool-making, use, and carrying). In addition, visual obstacles and long-distance communication become problems. Clearly, these difficulties do not arise if we postulate that gestures were not the major vehicle for linguistic communication. It has been noted that contemporary deaf humans who use sign language solve most of these problems (e.g., Stokoe, 1976). However, deaf sign is mapped onto an already fully evolved cognitive and motor speech system. This imparts to contemporary sign language a measure of complexity and efficiency that was not likely to have been present in the manual sign systems of early hominids.

In our view, early hominids may have initially employed *both* manual gestures and vocalizations for linguistic communication. The problems encountered in one channel could thus have been circumvented through (redundant) use of another. The critical point is that a gestural language need not have preceded a vocal one. Later in evolution the more efficient vocal channel became the primary vehicle for language (here referred to as primordial speech), and manual as well as other gestures assumed secondary (e.g., metacommunicative) roles.

References

Andrew, R. J. (1976). *In* "Origins and Evolution of Language and Speech" (S. R. Harnad, H. D. Steklis, and J. Lancaster, eds.), Ann. N.Y. Acad. Sci., 280, pp. 673–693.

Bever, T. G. (1970). *In* "Cognition and the Development of Language" (J. R. Hayes, ed.), pp. 279–362. Wiley, New York.

Bielicki, T. (1969). *Mater. i Pr. Anthrop.* 77, 57–60.

Blank, M. and Bridger, W. H. (1964). *J. Comp. Physiol. Psychol.* 58, 277–282.

Bosma, J. F. (1975). *In* "The Role of Speech in Language" (J. F. Kavanagh and J. E. Cutting, eds.), pp. 107–108. MIT Press, Cambridge, Mass.

Bryant, P. E., Jones, P., Claxton, V., and Perkins, G. M. (1972). *Nature 240*, 303–304.

Burr, D. B. (1976a). *J. Human Evolution 5*, 285–290.

Burr, D. B. (1976b). *Man 2*, 104–110.

Burr, D. B. (1976c). *Current Anthropology 17*, 762–763.

Campbell, B. (1974). "Human Evolution." 2nd ed. Aldine, Chicago.

Carlisle, R. C. and Siegel, M. I. (1974). *American Anthropologist 76*, 319–327.

Chomsky, N. (1957). "Syntactic Structures." Mouton, The Hague, Paris.

Clark, E. V. (1977). *In* "Psycholinguistics. Developmental and Pathological" (J. Morton and J. C. Marshall, eds.), pp. 1–72. Cornell Univ. Press, Ithaca, N.Y.

Coren, S. and Porac, C. (1977). *Science 198*, 631–632.

Cowey, A. and Weiskrantz, L. (1975). *Neuropsychologyia 13*, 117–120.

Cutting, J. E. and Eimas, P. D. (1975). *In* "The Role of Speech in Language" (J. F. Kavanagh and J. E. Cutting, eds.) pp. 127–148. MIT Press, Cambridge, Mass.

Davenport, R. K. (1976). *In* "Origins and Evolution of Language and Speech" (S. R. Harnad, H. D. Steklis, and J. Lancaster, eds.), Ann. N.Y. Acad. Sci., 280, pp. 143–149.

Davenport, R. K. (1977). *In* "Language Learning by a Chimpanzee. The Lana Project" (D. M. Rumbaugh, ed.), pp. 73–86. Academic Press, New York.

Davis, A. E. and Wada, J. A. (1977). *Brain and Language 4*, 23–31.

Davis, A. E. and Wada, J. A. (1978). *Brain and Language 5*, 42–55.

Deuel, R. K. (1977). *Neuropsychologia 15*, 205–215.

Dewson, J. H. (1977). *In* "Lateralization in the Nervous System" (S. R. Harnad, R. W. Doty, L. Goldstein, J. Jaynes, and G. Krauthamer, eds.), pp. 63–74. Academic Press, New York.

DuBrul, E. L. (1958). "Evolution of the Speech Apparatus." C. C. Thomas, Springfield, Ill.

DuBrul, E. L. (1976). *In* "Origins and Evolution of Language and Speech" (S. R. Harnad, H. D. Steklis, and J. Lancaster, eds.), Ann. N.Y. Acad. Sci., 280, pp. 631–642.

DuBrul, E. L. (1977). *Brain and Language 4*, 365–381.

Eimas, P. D., Siqueland, E. R., Jusczyk, P., and Vigorito, J. (1971). *Science 171*, 303–306.

Elliott, R. C. (1977). *Neuropsychologia 15*, 183–186.

Ettlinger, G. (1967). *In* "Brain Mechanisms Underlying Speech and Language" (F. L. Darley and C. H. Millikan, eds.), pp. 53–60, Grune & Stratton, New York.

Ettlinger, G. (1973). *In* "Memory and Transfer of Information" (H. P. Zippel, ed.), pp. 43–64. Plenum, New York.

Ettlinger, G. (1977). *In* "Behavioral Primatology" (A. M. Schrier, ed.), vol. 1, pp. 71–104. Wiley, New York.

Falk, D. (1975). *Am. J. Phys. Anthropol. 43*, 123–132.

Falk, D. (1978a). "External Neuroanatomy of Old World Monkeys (Cercopithecoidea)" Contrib. primat. 15, Karger, Basel.

Falk, D. (1978b). *Acta anat. 101*, 334–339.

Fouts, R. S. (1974). *J. Human Evolution 3*, 475–482.

Fouts, R. S. (1976). *In* "Origins and Evolution of Language and Speech" (S. R. Harnad, H. D. Steklis, and J. Lancaster, eds.), Ann. N.Y. Acad. Sci., 280, pp. 589–591.

Gabow, S. L. (1977). *J. Human Evolution 6*, 643–665.

Galaburda, A. M., LeMay, M., Kemper, T. L., and Geschwind, N. (1978). *Science 199*, 852–856.

Gallup, G. G. Jr., Boren, J. L., Gagliardi, G. J., and Wallnau, L. B. (1977). *J. Human Evolution 6*, 303–313.

Gallup, G. G. Jr. (1977). *American Psychologist 32*, 329–338.

Gautier, J. P. and Gautier, A. (1977). *In* "How Animals Communicate" (T. A. Sebeok, ed.), pp. 890–964. Indiana Univ. Press, Bloomington.

Geschwind, N. (1964). *In* "Report of the 15th Annual Round Table Meeting on Linguistics and Language Studies" (C. I. J. M. Stuart, ed.), Monograph Series on Language and Linguistics No. 17, pp. 155–169.

Geschwind, N. and Levitsky, W. (1968). *Science 161*, 186–187.

Geschwind, N., Quadfasel, F. A., and Segarra, J. M. (1968). *Neuropsychologia 6*, 327–340.

Gill, T. V. (1977). *In* "Language Learning by a Chimpanzee. The Lana Project" (D. M. Rumbaugh, ed.), pp. 225–246. Academic Press, New York.

Glick, S. D., Jerussi, T. P., and Zimmerberg, B. (1977). *In* "Lateralization in the Nervous System" (S. R. Harnad, R. W. Doty, L. Goldstein, J. Jaynes, and G. Krauthamer, eds.), pp. 213–250. Academic Press, New York.

Goldin-Meadow, S. and Feldman, H. (1977). *Science 197*, 401–403.

Green, S. (1975). *Z. Tierpsychol. 38*, 304–314.

Hamilton, C. R. (1977). *In* "Evolution and Lateralization of the Brain" (S. J. Dimond and D. A. Blizard, eds.), Ann. N. Y. Acad. Sci., 299, pp. 222–232.

Hécaen, H. (1975). *In* "Foundations of Language Development" (E. H. Lenneberg and E. Lenneberg, eds.), vol. 2, pp. 117–133. Academic Press, New York.

Hécaen, H. and Ajuriaguerra, J. de. (1964). "Left-Handedness." Grune & Stratton, New York.

Hewes, G. W. (1973a). *Current Anthropology 14*, 5–24.

Hewes, G. W. (1973b). *Visible Language 7*, 101–127.

Hewes, G. W. (1976). *In* "Origins and Evolution of Language and Speech" (S. R. Harnad, H. D. Steklis, and J. Lancaster, eds.), Ann. N.Y. Acad. Sci., 280, pp. 482–504.

Hewes, G. W. (1977). *In* "Language Learning by a Chimpanzee. The Lana Project" (D. M. Rumbaugh, ed.), pp. 3–54. Academic Press, New York.

Hill, J. H. (1972). *American Anthropologist 74*, 308–317.

Hill, J. H. (1978). *Annual Review of Anthroology 7*, 89–112.

Hockett, C. F. (1959). *In* "The Evolution of Man's Capacity for Culture" (J. N. Spuhler, ed.), pp. 32–39. Wayne State Univ. Press, Detroit.

Hockett, C. F. and Ascher, R. (1964). *Current Anthropology 5*, 135–168.

Hockett, C. F. and Altmann, S. A. (1968). *In* "Animal Communication" (T. A. Sebeok, ed.), pp. 61–72. Indiana University Press, Bloomington.

Holloway, R. L. Jr. (1969). *Current Anthropology 10*, 395–412.

Holloway, R. L. Jr. (1976). *In* "Origins and Evolution of Language and Speech" (S. R. Harnad, H. D. Steklis, and J. Lancaster, eds.), Ann. N.Y. Acad. Sci., 280, pp. 330–348.

Howell, F. C. (1972). *In* "Perspectives on Human Evolution" (S. L. Washburn and P. Dolhinow, eds.), pp. 51–128. Holt, Rinehart and Winston, New York.

Isaac, G. L. (1976). *In* "Origins and Evolution of Language and Speech" (S. R. Harnad, H. D. Steklis, and J. Lancaster, eds.), Ann. N.Y. Acad. Sci., 280, pp. 275–288.

Jenkins, P. F. (1977). *Anim. Behav. 25*, 50–78.

Jerison, H. J. (1973). "Evolution of the Brain and Intelligence." Academic Press, New York.

Jerison, H. J. (1975). *Ann. Rev. Anthropology 4*, 27–58.

Jones, E. G., and Powell, T. P. S. (1970). *Brain 93*, 793–820.

Jordan, J. (1971). *Folia Morph. 30*, 323–340.

Kelemen, G. (1969). *In* "The Chimpanzee" (G. Bourne, ed.), Vol. 1, pp. 165–186. Karger, Basel, New York.

King, G. E. (1976a). *J. Anthropological Research 32*, 276–284.

King, G. E. (1976b). *J. Human Evolution 5*, 323–332.

King, M-C. and Wilson, A. C. (1975). *Science 188*, 107–116.

Kolb, B. and Milner, B. (1979a). *Neuropsychologia*, in press.

Kolb, B. and Milner, B. (1979b). *Neuropsychologia*, in press.

Krashen, S. D. (1976). *In* "Studies in Neurolinguistics" (H. Whitaker and H. A. Whitaker, eds.), vol. 2, pp. 157–191. Academic Press, New York.

Lancaster, J. B. (1968). *In* "Primates. Studies in Adaptation and Variability" (P. C. Jay, ed.), pp. 439–457. Holt, Rinehart and Winston, New York.

Langer, S. K. (1972). "Mind: An Essay on Human Feeling." Vol. 2. Johns Hopkins Univ. Press, Baltimore.

Laughlin, C. D. Jr. and D'Aquili, E. G. (1974). "Biogenetic Structuralism." Columbia Univ. Press, New York.

Le Doux, J. E., Wilson, D. H., and Gazzaniga, M. S. (1977). *Neuropsychologia 15*, 743–750.

LeGros Clark, W. E. (1959). "The Antecedents of Man." Harper and Row, New York.

Lenneberg, E. H. (1967). "Biological Foundations of Language." Wiley, New York.

Lenneberg, E. H. (1971). *J. Psycholinguistic Research, 1*, 1–29.

LeMay, M. and Geschwind, N. (1975). *Brain Behav. Evol. 11*, 48–52.

Levy, J. (1976). *Behavior Genetics 6*, 429–453.

Levy, J. (1977). *In* "Lateralization in the Nervous System" (S. R. Harnad, R. W. Doty, L. Goldstein, J. Jaynes, and G. Krauthamer, eds.), pp. 195–212. Academic Press, New York.

Liberman, A. M., Cooper, F. S., Shankweiler, D., and Studdert-Kennedy, M. (1967). *Psychological Review 74*, 431–461.

Lieberman, P. (1972). "The Speech of Primates." Mouton, The Hague, Paris.

Lieberman, P. (1973). *Cognition 2*, 59–94.

Lieberman, P. (1975). *In* "The Role of Speech in Language" (J. F. Kavanagh and J. E. Cutting, eds.), pp. 83–106. MIT Press, Cambridge, Mass.

Lieberman, P. (1976). *In* "Origins and Evolution of Language and Speech" (S. R. Harnad, H. D. Steklis, and J. Lancaster, eds.), Ann. N. Y. Acad. Sci., 280, pp. 660–672.

Lieberman, P. (1978). *American Anthropologist 80*, 676–681.

Lieberman, P. and Crelin, E. S. (1971). *Ling. Inq. 2*, 203–222.

Lieberman, P., Crelin, E. S., and Klatt, D. H. (1972). *American Anthropologist 74*, 287–307.

Luria, A. R. (1970). "Traumatic Aphasia, Its Syndromes, Psychology, and Treatment." Mouton, The Hague, Paris.

Marler, P. (1975). *In* "The Role of Speech in Language" (J. F. Kavanagh and J. E. Cutting, eds.), pp. 11–37. MIT Press, Cambridge, Mass.

Marler, P. (1977). *In* "Progress in Ape Research" (G. H. Bourne, ed.), pp. 85–96. Academic Press, New York.

Marler, P. and Tenaza, R. (1977). *In* "How Animals Communicate" (T. A. Sebeok, ed.), pp. 965–1033. Indiana Univ. Press, Bloomington.

Mason, W. A. (1976). *American Psychologist, 31*, 284–294.

McNeill, D. (1970). "The Acquisition of Language: The Study of Developmental Psycholinguistics." Harper and Row, New York.

McNeill, D. (1973). *In* "The Growth of Competence" (K. Connolly and J. Bruner, eds.), pp. 75–96. Academic Press, New York.

Menzel, E. W. Jr., Premack, D., and Woodruff, G. (1978). *Folia Primatol. 29*, 241–249.

Miller, C. L. and Morse, P. A. (1976). *J. Speech and Hearing Research 19*, 578–589.

Mohr, J. P. (1976). *In* "Studies in Neurolinguistics" (H. Whitaker and H. A. Whitaker, eds.), vol. 1, pp. 201–292. Academic Press, New York.

Molfese, D. L. (1978). *Brain and Language 5*, 25–35.

Montagu, A. (1976). *In* "Origins and Evolution of Language and Speech" (S. R. Harnad, H. D. Steklis, and J. Lancaster, eds.), Ann. N.Y. Acad. Sci., 280, pp. 266–274.

Morgan, M. (1977). *In* "Lateralization in the Nervous System" (S. R. Harnad, R. W. Doty, L. Goldstein, J. Jaynes, and G. Krauthamer, eds.), pp. 173–194. Academic Press, New York.

Morton, J. (1970). *In* "Biological and Social Factors in Psycholinguistics" (J. Morton, ed.), pp. 82–97. Logos Press, Ltd., London.

Mounin, G. (1976). *Current Anthropology 17*, 1–21.

Negus, V. E. (1949). "The Comparative Anatomy and Physiology of the Larynx." Heinemann, London.

Neville, H. J. (1976). *In* "The Neuropsychology of Language" (R. W. Rieber, ed.), pp. 193–227. Plenum, New York.

Newman, J. D. (1978). *J. Med. Primatol. 7*, 98–105.

Norris, E. and Ettlinger, G. (1978). *Neuropsychologia 16*, 99–102.

Nottebohm, F. (1972). *The American Naturalist 106*, 116–140.

Nottebohm, F. (1975). *In* "Foundations of Language Development" (E. H. Lenneberg and E. Lenneberg, eds.), pp. 61–103. Academic Press, New York.

Nottebohm, F. (1976). *In* "Origins and Evolution of Language and Speech" (S. R. Harnad, H. D. Steklis, and J. Lancaster, eds.), Ann. N.Y. Acad. Sci., 280, pp. 643–649.

Nottebohm, F. (1977). *In* "Lateralization in the Nervous System" (S. R. Harnad, R. W. Doty, L. Goldstein, J. Jaynes, and G. Krauthamer, eds.), pp. 23–44. Academic Press, New York.

Orthner, H. and Sendler, W. (1975). *Fortschr. Neurol. Pschiat. 43*, 191–209.

Oke, A., Keller, R., Mefford, I., and Adams, R. N. (1978). *Science 200*, 1411–1413.

Omenn, G. S. and Motulsky, A. G. (1972). *In* "Genetics, Environment, and Behavior" (L. Ehrman, G. S. Omenn, and E. Caspari, eds.), pp. 129–171. Academic Press, New York.

Pandya, D. and Kuypers, H. G. J. M. (1969). *Brain Res. 13*, 13–36.

Passingham, R. E. (1975). *Brain Behav. Evol. 11*, 73–90.

Passingham, R. E. (1978). *Brain Research 152*, 313–328.

Passingham, R. E. and Ettlinger, G. (1974). *Int. Rev. Neurobiol. 16*, 233–299.

Pastore, R. E. (1976). *In* "Hearing and Davis: Essays Honoring Hallowell Davis" (S. K. Hirsh, D. H. Eldredge, I. J. Hirsh, and S. R. Silverman, eds.), Washington Univ. Press, Mo.

Patterson, F. (1978a). *Brain and Language 5*, 72–97.

Patterson, F. (1978b). *National Geographic 154*, 440–465.

Penfield, W. and Roberts, L. (1959). "Speech and Brain-Mechanisms." Princeton Univ. Press, Princeton.

Petersen, M. R., Beecher, M. D., Zoloth, S. R., Moody, D. B., and Stebbins, W. C. (1978). *Science 202*, 324–327.

Petrides, M. and Iversen, S. D. (1976). *Science 192*, 1023–1024.

Poeck, K. and Huber, W. (1977). *Neuropsychologia 15*, 359–363.

Premack, D. (1976). *In* "Origins and Evolution of Language and Speech" (S. R. Harnad, H. D. Steklis, and J. Lancaster, eds.), Ann. N.Y. Acad. Sci., 280, pp. 544–561.

Premack, D. and Woodruff, G. (1978). *Science 202*, 532–535.

Premack, D. and Woodruff, G. (1978). *J. Behavioral and Brain Sciences, 1* (4), 515–526.

Rasmussen, T., and Milner, B. (1975). *In* "Cerebral Localization" (K. J. Zülch, O. Creutzfeldt, and G. C. Galbraith, eds.), pp. 238–257. Springer, New York.

Reynolds, P. C. (1968). *American Anthropologist 70*, 300–308.

Richman, B. (1976). *Current Anthropology 17*, 523–524.

Robinson, J. T. (1971). *In* "Background for Man" (P. Dolhinow and V. Sarich, eds.), pp. 122–155. Little, Brown, Boston.

Rubens, A. B. (1977). *In* "Lateralization in the Nervous System" (S. R. Harnad, R. W. Doty, L. Goldstein, J. Jaynes, and G. Krauthamer, eds.), pp. 503–516. Academic Press, New York.

Rumbaugh, D. M. (1977a). "Language Learning by a Chimpanzee. The Lana Project." Academic Press, New York.

Rumbaugh, D. M. (1977b). *In* "Behavioral Primatology" (A. M. Schrier, ed.), vol. 1, pp. 105–138. Wiley, New York.

Rumbaugh, D. M. and Gill, T. V. (1977). *In* "Language Learning by a Chimpanzee. The Lana Project" (D. M. Rumbaugh, ed.), pp. 165–192. Academic Press, New York.

Sacher, G. A. (1976). *In* "Language and Man. Anthropological Issues" (W. C. McCormach and S. A. Wurm, eds.), pp. 417–441. Mouton, The Hague, Paris.

Sahgal, A., Petrides, M., and Iversen, S. D. (1975). *Nature 257*, 672–674.

Sarich, V. M. and Cronin, J. E. (1976). *In* "Molecular Anthropology" (M. Goodman and R. E. Tashian, eds.), pp. 141–170. Plenum, New York.

Savage, E. S. and Rumbaugh, D. M. (1977). *In* "Language Learning by a Chimpanzee. The Lana Project" (D. M. Rumbaugh, ed.), pp. 287–310. Academic Press, New York.

Savage-Rumbaugh, E. S., Rumbaugh, D. M., and Boysen, S. (1978a). *Science 201*, 641–644.

Savage-Rumbaugh, E. S., Rumbaugh, D. M., and Boysen, S. (1978b). *J. Behavioral and Brain Sciences*, 1 (4), 539–554.

Sinnott, J. M., Beecher, M. D., Moody, D. B., and Stebbins, W. C. (1976). *J. Acoust. Soc. Am. 60*, 687–695.

Snowdon, C. T. (1978). *Prim. Med. 10*, 225–231.

Snowdon, C. T. and Pola, Y. V. (1978). *Anim. Behav. 26*, 192–206.

Spuhler, J. N. (1977). *Annual Review of Anthropology 6*, 509–562.

Stamm, J. S., Rosen, S. C., and Gadotti, A. (1977). *In* "Lateralization in the Nervous System" (S. R. Harnad, R. W. Doty, L. Goldstein, J. Jaynes, and G. Krauthamer, eds.), pp. 385–402. Academic Press, New York.

Steklis, H. D. and Harnad, S. R. (1976). *In* "Origins and Evolution of Language and Speech" (S. R. Harnad, H. D. Steklis, and J. Lancaster, eds.), Ann. N.Y. Acad. Sci., 280, pp. 445–455.

Stoke, W. C. (1976). *In* "Origins and Evolution of Language and Speech" (S. R. Harnad, H. D. Steklis, and J. Lancaster, eds.), Ann. N.Y. Acad. Sci., 280, p. 481.

Studdert-Kennedy, M. and Shankweiler, D. (1970). *J. Acoust. Soc. Am. 48*, 579–594.

Terrace, H. S. and Bever, T. G. (1976). *In* "Origins and Evolution of Language and Speech" (S. R. Harnad, H. D. Steklis, and J. Lancaster, eds.), Ann. N. Y. Acad. Sci., 280, pp. 579–588.

Terrace, H. C., Petitto, L., and Bever, T. G., (1976). Project NIM, Unpublished Progress Report II.

Tobias, P. V. (1971). "The Brain in Hominid Evolution." Columbia Univ. Press, New York.

Trevarthen, C. (1974). *In* "Hemispheric Disconnection and Cerebral Function" (M. Kinsbourne and W. L. Smith, eds.), pp. 187–207. C. C. Thomas, Springfield, Ill.

Wada, J. A., Clarke, R., and Hamm, A. (1975). *Arch. Neurol. 32*, 239–246.

Warren, J. M. (1977). *In* "Evolution and Lateralization of the Brain" (S. J. Dimond and D. A. Blizard, eds.), Ann. N.Y. Acad. Sci., 299, pp. 273–280.

Webster, W. G. (1977). *In* "Evolution and Lateralization of the Brain" (S. J. Dimond and D. A. Blizard, eds.), pp. 213–221, Ann. N.Y. Acad. Sci., 299.

Whitaker, H. A. (1969). "On the Representation of Language in the Human Brain." Work. Pap. Phonet. No. 12. Univ. California Press, Los Angeles.

Whitaker, H. (1976). *In* "Studies in Neurolalinguistics" (H. Whitaker and H. A. Whitaker, eds.), Vol. 2, pp. 1–58. Academic Press, New York.

Whitaker, H. A. and Selnes, O. A. (1975). *Linguistics 154/155*, 91–103.

White, L. A. (1959). *In* "The Evolution of Man's Capacity for Culture" (J. N. Spuhler, ed.), pp. 74–79. Wayne State Univ. Press, Detroit.

Wind, J. (1970). "On the Phylogeny and the Ontogeny of the Human Larynx." Wolters-Noordhoff, Groningen.

Wind, J. (1976). *In* "Origins and Evolution of Language and Speech" (S. R. Harnad, H. D. Steklis, and J. Lancaster, eds.), Ann. N.Y. Acad. Sci., 280, pp. 612–630.

Witelson, S. F. (1977). *In* "Evolution and Lateralization of the Brain" (S. J. Dimond and D. A. Blizard, eds.), Ann. N.Y. Acad. Sci., 299, pp. 328–354.

Witelson, S. F. and Pallie, W. (1973). *Brain 96*, 641–647.

Wyke, M. (1971). *Cortex 7*, 59–72.

Yeni-Komshian, G. H. and Benson, D. A. (1976). *Science 192*, 387–389.

Zangwill, O. L. (1975). *In* "Cerebral Localization" (K. J. Zülch, O. Creutzfeldt, and G. C. Galbraith, eds.), pp. 258–263. Springer, New York.

Zihlman, A. L., Cronin, J. E., Cramer, D. L., and Sarich, V. M. (1978). *Nature 275*, 744–745.

Index